BEYOND the DARK HORIZON

Book 3

J.B. Kingsley-Lauren

BEYOND THE DARK HORIZON
First published in Australia by Jean Thompson 2019

National Library of Australia Pre publication Data Service entry:

Creator: Kingsley-Lauren, J.B., author
Title: BEYOND THE DARK HORIZON

ISBN: 978-0-6481255-4-9 (pbk)

Subjects: FICTION / War & Military
FICTION / Family Saga see Sagas
HISTORY / Europe / Germany

Also available as an ebook: 978-0-6481255-5-6 (ebk)

Typesetting and design by Publicious Book Publishing
Published in collaboration with Publicious Book Publishing
www.publicious.com.au

All characters and events in this publication are fictitious, any resemblance to real persons, living or dead, or any events past or present are purely coincidental.

Dedication.

I wish to dedicate this third book of my trilogy to: My son Graham Leslie. My son believed I could master adversity and keep my interest in creating characters for my novels. This I have accomplished over seven decades of my life.

And also to Judith Killingback. Jude encouraged me to keep creating stories and never to forget those less fortunate than myself. This I have endeavoured to accomplish with grace and humility.

1

Early summer of Christmas 1972 in Australia: The instant their Cessna touched down in Canberra, the semiretired diplomat on compassionate leave was eagerly greeted by his secretary. During the drive to his home in Yass, Deidre Sanderson outlined what she knew of the situation on Abergeldie.

Her boss, Senator Peter Bucknell didn't expect grim news on his arrival home from England. Nevertheless, he immediately swung into action by contacting Abergeldie where he spoke to Doctor Ross. It came as a shock to learn from Kendall of his father's Nazi history. Hearing these facts from the English-born Australian physician, Peter knew the cable he'd received was authentic.

He understood the Ross family would need his assistance; although they spoke a mere few seconds he reflected on his discussion with Kendall. There were limits of what could be discussed on the phone. Still, he'd heard enough to advise his Canberran boss of its contents, devastating as it seemed. The news of an ex-Nazi living in their midst was utterly unbelievable to this diplomat. Bucknell knew stringent procedures couldn't be set in place until he conferred with the Australian Federal Police (AFP) and the Australian Security Intelligence Officers (ASIO). After discussing this serious agenda with his superiors at the crack of dawn, Peter set his sights on Abergeldie. Within twenty minutes his vehicle pulled in through the homestead's iron gates and proceeded down the floral garden-edged curvaceous drive which ended at the homestead's front door.

Feeling a cool breeze brush against his face on this hot summer morning, Peter zealously braced Abergeldie's steps in time to hear Father

Brady's damning declaration. Shocked at the loud voice raised in bitter debate Senator Bucknell was dumbstruck.

This heated one-sided discussion related to the years the priest had spent in war-ravaged Germany. The religious man's caustic affirmations struck him as extremely odd. Consumed with curiosity and interest, Peter paused in the hall to listen.

'No Kendall, give them photos ta me. The miniatures are mine. I must destroy them like I shoulda done last night. They're evil and guilt is driving me mind to madness. What have I done?' The priest's tone rose in anger and the strain of yelling made him hoarse. Still, it didn't prevent Father George Brady's voice from escalating in disgust as he ranted on about this tragedy being his fault because of his own wanton stupidity.

From what Bucknell could gather, the speaker seemed to be reliving his innermost turmoil. It conveyed to him that the subject related to victims of the Nazi regime whose horrific history he would soon discover. He did hear the name of von Breusch mentioned several times in the course of this argumentatively haunting conversation.

Huddled in a corner the priest refused to speak to or look at Kendall. Silently he prayed: *When will this agonising pain I've caused this lovable family cease? That damn Nazi deserves no pity because of all the innocent people he sent to their deaths in camps all over Germany.* A flush of unsaid words travelled through his brain. *How many other souls must be crucified by that evil criminal's wicked deeds?*

Turning his head, Father's brimming eyes tried to focus on the stranger who'd wandered into the lounge room. Yet through the haze of confused torment he vaguely saw the figure as a misty apparition. By then, his deranged constant ramblings were beginning to subside.

'Hush Father, it's over for the moment. Come and sit down,' Doctor Ross whispered. Then turning to address his female companion he nodded. 'Kirsten, put the kettle on please.' Kendall knelt beside the agonised priest to quietly comfort him. 'From what you've said we have a fair idea of how cruelly you suffered at that criminal's hands. Now it's time to forget all those horrible memories. Put them to rest for now, please Father George.'

Leaving the priest to become composed Kendall's nod indicated for the diplomat to follow him through to the kitchen. He needed to bring

Peter Bucknell up to date with the most recent events. Alone in the warm room and in the still of night they talked over coffee.

After being fully acquainted with the previous night's happenings, Bucknell asked which room he might use to make a private call from.

'Use Mum's office, it's by far the safest. From there nobody can hear you in the front hall. Besides, the main study or mine are easily accessible if someone walks in from the backyard. I suppose *HIS* den from here on will be out of bounds to everyone.' Kendall couldn't bring himself to say "father". Intensive and repeated arguments over his father's cruelty to his mother made this devoted physician resent and hate his male parent.

'Afraid so, at least until the federal boys scour it for any trace of …' Peter almost said "your father, Ian", but curbed his tongue in time. Feeling embarrassed he made a hasty exit towards Gigi's office to put a call through to his superiors in Canberra. Minutes later he sidled back to the kitchen.

'Well, at least now, the federal police are fully aware of the facts and what we will be up against here. The powers above will contact ASIO; they will ring me here before noon to confirm what I mentioned earlier. In the meantime, I hope the AFP has procured more conclusive information on the Nazi absconder's latest movements. Nasty business Kendall, and I can't see how your mother will cope under excessive strain, well not until the damn blighter is caught.'

'Peter, not one of us had a clue to his real identity. My mother has aged twenty years or so it seems since she found out how he'd deceived her down the years. It nearly killed her to discover she'd married a bastard, a detestable and evil war criminal.'

'Worse than that I'm afraid,' Bucknell confirmed. 'War criminals are hard to trace. By now the majority of their victims, or military associates are dead, or too feeble-minded to remember their distant past. The conniving murderers like him, usually have their getaway prepared well in advance, long before their escape takes place. Kendall, from what I can gather he would've plotted his escape right down to the finest detail. Military men are well-trained strategists, or the ones I know are.' Peter preferred not to name Ian David Ross under his Germanic identity of Rolf von Breusch.

Kirsten Svensson stood silently outside the door until Kendall had left then quietly walked to the kitchen. On the way through she retrieved a notepad from the hall table.

'If it's a help Peter, I've jotted down Father George's declaration in shorthand.' She passed the pad over. 'Every word spoken by those present is written on these pages. It should alleviate some of the anguish for all concerned here. I don't know what drove me to note it all down. The priest's horrific story of frustration, hurt and truth, I suppose.'

'Brilliant thinking Kirsten.' The diplomat scanned her notes. 'While you check on Father Brady, I'll let the others know we're going down to Ian's office.' Then he added, 'Kendall has briefly filled me in on what's happened here over the past two weeks. Now I'll relay the details through to my boss in Canberra, before interviewing you all in turn. Excuse me Kirsten, the federal boys will be waiting for me to call. Your notations, plus everyone's depositions will be beneficial for their dual teams to fully assess the situation here on Abergeldie.'

Approaching the study Peter stopped in the rear hall to confer with Doctor Ross, who'd just finished writing up his mother's medication.

'Kendall, can you think of any other particulars I should know about? I've just received some valuable information from Mrs Svensson on your discussion with Father George.'

'Well, yes there is in fact. I had to force this lock, because that Nazi swine always kept the den keyed. Now it's obvious why he did so. I've not touched a thing. The den is still as he walked out and left it.' Kendall tried to relive his previous moment of entry, to confirm if this statement was correct. He recalled having seen an oil painting that concealed the safe. Now a picture of an English pasture in summertime took its place. Thinking it over, he realised he had missed something else in this small dingy office, but couldn't bring it to mind.

'What is it, Kendall? You look a bit puzzled.'

'On leaving this room earlier I caught sight of something behind the leg of that chair. I retrieved the scrap and discovered the combination of the wall safe scribbled in his nondescript hand.' Kendall produced the scrap from his shirt pocket. 'That Nazi must've dropped it in his haste to leave here on the night of my engagement to Brianna.'

'Most likely,' Bucknell agreed. 'It's fortunate you found these figures. Having the safe combination will save endless hassle and worry. I imagine calling a Chubb mechanic to re-tumble a sequence of digits on this safe would run to a small fortune these days.'

'Shit, no way would I let a local bloke do the job.' Another quick scan of the room revealed what Kendall had forgotten. 'Ah, there's something else. That rotten briefcase with its distinctive Nazi eagle insignia is missing. That's apart from the 10 000 quid we'd put aside to pay all his outstanding debts, under Abergeldie's name. It, along with our recent wool clip money he pinched from the master safe in Mum's office. Fortunately, her exquisite gold and priceless pieces of jewellery are in mine.'

With handkerchief-wrapped fingers clasping the brass doorknob so as not to disturb any fingerprints, Peter absorbed every item in Ian's study. He also scrutinised the minutest scraps of paper in the wastepaper bin with extreme interest. One piece scribbled with abstract digits caught his trained eye. 'This looks like a nautical chart of some kind. Boy oh boy … won't the Feds have loads of fun decoding these squiggles. Eyes only, don't touch!'

'Struth, they're German naval coordinates. I missed that scrap. Peter, this could mean my old man left somewhere along our coastline by submarine. It's more than possible. It means he must've had a local German helping him to escape the net ready to snare him. Why would he leave this state by sea? Yass is two hundred kilometres inland.'

'In that case, I think it advisable to class this whole area out of bounds, at least until the authorities have finished ferreting in this office. One never knows what will come to light once the two-legged bloodhounds begin their search. Things we wouldn't dream of looking for usually show up. Can we have this and further discussions in the main office, Kendall?'

'As I've said, it will be better to interview everyone on staff in Mum's office,' reiterated Abergeldie's part-owner. 'Neither Gigi nor Brianna will be going in there, not in their distressed states. And definitely not with Mum's red setter pup Amber on duty. Her office is well away from all the bedrooms and our domestic staff and outdoor employees. I'll personally vouch for our workmen, shearers and the stable lads and muckers' trustworthiness.'

'How did the Nazi come by the wool clip funds?' Peter's curiosity jumped to the fore. And still he chose not to mention Ian or his true identity by name. 'Surely nobody here would be stupid enough to leave an important safe unlocked,' Bucknell queried, reading over Mrs Svensson's transcribed shorthand.

'Not as a rule we wouldn't, Peter. Mum and I trust every member of our staff. No one here on Abergeldie knew of his criminal past. Over time that mongrel has stolen bundles of cash, unbeknown to either Mum or myself, until I realised the master safe in her office was unlocked. Then I asked her if she'd finished in there. Mum confirmed she'd forgotten to reset the tumblers. In hindsight, I think it might've been five minutes. Gigi told me she needed to use the ladies' urgently.'

'Your absconder must've been awfully quick to snatch a huge wad of notes.'

'Yes, to snatch fifty rolls of one hundred dollar notes in a short time. I didn't discover anything was missing until after he had absconded. Peter, he knew every item in her safe. My guess is the bastard helped himself to the bundles when Gigi was speaking to you on the hall phone.' Kendall pointed to the wall contraption on their way back to the kitchen. 'From here it would be impossible to see the Nazi sneak into her office. Mum isn't aware the money is missing. If she queries it, I'll cover his theft by telling her a trivial lie.'

Mulling the problem over, and discussing it with Kendall, the Canberran diplomat queried, 'Would any of your staff have entered your mother's office without her knowing? A local ringer or visitor may've pilfered the money. A casual workman, out-of-town shearer passing through here or a roustabout could have been the culprit?'

'Peter, I trust every member of our staff on Abergeldie and also all those on Jalna. I can vouch that everyone authorised to enter this house that day was, and is, trustworthy. The point is … I know damn well the kleptomaniac pilfered our wool clip money, as well as that huge roll of notes Mum and I had put aside to pay the water and municipal rates. No one will convince me otherwise.'

Doctor Ross, accompanied by Bucknell, exited the kitchen and diverted down to a small annex that led off the main front hall. These conjoined halls formed a huge "L" with Gigi's study, not a metre away from where the men were now standing.

Kendall accepted the challenge to continue this conversation on their way through to the main office. 'We are fortunate the deeds to this property, its holdings and legal drafts are with our family solicitor in Yass.' Reefing a creased envelope from his trouser pocket he passed it to Peter. 'This appeared yesterday around teatime in the pocket of my suitcoat. I found it while clearing our RMB on the way over to Jalna for our engagement party. Bridy and I had stopped to see if more cards had arrived from Ireland for her twenty-first birthday.'

Peter cautiously handled the grungy brown envelope between handkerchief-cloaked fingers. 'Have you opened this envelope, Kendall? The seal's broken.' The item fluctuated in his fingers.

'No Peter, I haven't nor did I read the content. I only handled the envelope by that torn corner, just above the stamp.'

'This letter is addressed to a private box in Yass. If it wasn't you, then I can only assume it was wrongly delivered to some other landholder. Upon discovery this envelope must've been thrown in your roadside mail drum.' Closer scrutiny divulged the smudged, faint postal mark. 'This franked imprint is Suisse. Yes, from Zurich. Now, that is interesting. Ooh brother, won't ASIO or our federal boys have a field day researching this decisive wafer of evidence.' Instinctively Peter, with his diplomatic intuition running wild, figured there could be an intricate urgency behind this letter. The sender's name intrigued him. Perhaps it will tie in with the case unfolding here in our neighbourhood. His eyes thoroughly scanned the small ecru envelope.

'I wouldn't mind guessing this originated from a banking house in Zurich. The Feds will be keen to trace its origin. Leave this with me please, Kendall. Don't mention it to your mother. Gigi has enough things to worry about at present. I'll tell her myself, when something conclusive is proven on its contents.'

'Peter, that damn envelope was never slit open by me. I guarantee neither Mum nor Bridy were the culprits. I inadvertently shoved it in my pocket last night, and forgot that I put it there, until this morning.'

The men dawdled back to the kitchen as Bucknell spoke. 'I'll require a deep drawer in Gigi's office. One that locks preferably.' He saw Kendall nod. 'Good, I'll leave it in your capable hands to arrange. Let me know the moment it's ready.'

In the interim he bumped headlong into his colleague, Doctor Stephen Jarvis's staunch figure. His sad smile caught Kendall unawares. 'Mum's eurhythmic pulse rate fluctuated dramatically this morning and her heartbeats were erratic.' His languorous smile amused Kirsten, who's frown deepened over him flaunting his medical terminology.

'Fine, I'll attend to your request now Peter. If you need anything just ask.'

'I'm not staying Kendall,' Doctor Jarvis said as he nodded to the friendly diplomat. 'I'm on my way to an emergency somewhere in or around town. I shouldn't be long. There are a few things I need to discuss with you about your mother's present medications.'

'Excuse me gentlemen, I'll be back in a minute.' Mrs Svensson smiled, scurrying away to speak with their ailing patient, Gigi Ross, in her bedroom.

'Before you go Stephen, when you arrive back I want to show you something I think you will find quite fascinating in my office ...'

Bucknell interjected. 'This concerns you Kendall, and you also, Doctor Jarvis. What I'm about to say doesn't come from me. It's a federal ruling. No one is to enter, or leave these premises tomorrow, not without contacting either myself or their men, when they do arrive.'

About to call Kirsten back, Bucknell realised she'd hurried off to the master bedroom, so he naturally assumed she'd gone to see if Gigi, or their guest Brianna needed anything.

'Your local police must not be contacted either.' Addressing Kendall, he continued, 'Make sure nobody, and I mean *nobody,* discloses what we suspect may have evolved here over the past thirty-eight hours. Advise your immediate household staff and workmen of those rulings. Should anyone have a query of any kind refer that person and all messages to me.'

Kendall nodded then remarked as Jarvis walked past Abergeldie's front door, 'Hell, Stephen could be in town for some time. He said he's been called out to attend an accident.'

'Well, it's possible. Vehicular traffic is a never-ending snake with a double head threading its way through Yass. A busy town, it can't cope with speed hogs using the narrow roads as racetracks.' About to walk off, Peter paused until Kirsten caught up. He expected a comment. Nothing was forthcoming, so he informed her of the same demands he'd given to Kendall and Stephen Jarvis.

The men talked until Harriet Graham, Abergeldie's housekeeper-cum-cook announced in her English-cum-Aussie twang, 'The cuppa tea is ready, Doctor Ross. I'll brew a fresh pot for ya mum and the ladies. I'll carry this tray in to Gigi when the kettle re-boils.'

Kendall's mind travelled on the harrowing road of despondency. After discussing multiple worrisome problems with the diplomat his curiosity bubbled over. 'Tell me Peter, what the hell do the federal blokes expect me to tell my staff? I'm at a loss to explain our dilemma with the shearers and drifters in the bunkhouses. It'll be damn awkward trying to think up a legitimate excuse. Will the Feds turn up around noon tomorrow, do you think?'

'Don't have to think. Several officers accompanied by a female stenographer will be landing here once dawn breaks. It will take their superiors that amount of time to organise permits and arrange for extra personnel with background knowledge of the awkward situation like yours here. Kendall, their delay was in finding a reliable and trustworthy pilot who'll be willing to remain on Abergeldie for an extended period.'

'I expected that. Will their inquisitions be drawn out do you think? Peter, I don't want my mother suffering more stress than is necessary. In her condition, her health won't stand more indecisive or extensive questioning.'

Bucknell didn't respond. His attention was drawn to something this young physician had stated earlier. It concerned more than worried him. After some consideration he confirmed, 'Well, I can't speak for my colleagues, though I'm sure the federal blokes and ASIO's men will delve into every avenue possible to procure more constructive leads.'

Kendall nodded in agreement, rebuttoning his shirt. *It's going to be sheer hell for mum.*

Sensing his discomfort, the diplomat's frown dissipated. 'Regarding your mother, it might be advisable to say Gigi has sustained a severe injury while riding. Either Stephen or yourself, as physicians, will come up with something positive,' Bucknell declared passing the notepad back to Kirsten. 'Perhaps you could put forward that proposition tomorrow over breakfast. Keep only to the facts, as we know them. The Feds can't afford to let the Nazi deviant escape their net by listening to or accepting subversive innuendos from gossiping or overzealous tongues.'

Kendall glanced down at his dust-encrusted wristwatch. 'Hell, where has the morning flown? It's almost smoko. Why did you say tomorrow morning, Peter? Why not tonight?'

'I suggested tomorrow, because Gigi confided to me how she enjoyed riding in those top paddocks early most mornings. Around six, I believe she said from memory.'

Kirsten broke in to admit, 'In good weather she does. So do I, over home on Jalna.'

'Sometimes mum rides bareback to which I strongly object. Her mare's a feisty beast and could easily throw her. It's dangerous being up there, especially when swirling dust rides in with the force of a cyclone.'

Peter Bucknell, a man who never pulled punches, agreed with Kendall. 'It's extremely dangerous to ride in a dust storm over corrugated furrowed ground. In this dry spell it's absurd.' It sounded an unusually harsh remark for this polite, honest politician to make while a guest in their home.

'After you've phoned Stephen Jarvis in town, it will be the final call going out, and coming in to Abergeldie. Shit, I need to give my Canberra boss a bell about the coordinated digits in your old man's den. Okay if I speak on this phone alone?' With the potential for interruption Bucknell raised a hand. 'Let me finish please, Kendall. This morning I rang the head bloke at the Post Master General's office in Sydney. He's arranging to have double-ended plugs connected to every PABX switchboard from our local Yass Post Office down to the capital. The plugs will automatically transfer all calls directly from your line here at Abergeldie, straight through to our top brass and the federal boys in Canberra.'

'Why? I realise this is serious, but to cut Abergeldie off completely is ridiculous.'

A tad incensed, Bucknell nodded. 'Probably it'll be for only a short period. This move is vital, I can assure you, Kendall.' Peter chose to defer speaking to clarify what he meant to confirm. 'The double-ended plugs will eliminate ominous or unsolicited ears from listening to official conversations. You do understand if leakage of this delicate situation becomes public knowledge it could be detrimental in tracking down the criminal? It's the main reason why I suggested Jarvis should stay here at Abergeldie. You have agreed that Gigi will need constant monitoring by an independent physician, other than yourself.'

Peter realised he'd been discussing this topic with a qualified practitioner. Kendall Ross, as an inaugural part of this mess, was forbidden by law to prescribe his mother's drugs. 'You and your family are about to embark on the most horrendous journey. All kinds of unsavoury inquisitions and innuendos will be bartered back and forth for those involved. Remember this Kendall, I'll be ready at a moment's notice to assist you all with the future ordeals.'

Championing a crusade of this nature was always a challenge to Bucknell. He revelled in the idea of closing in on their quarry. Whether it be soon or in the near future, his intuitive brain worked in conjunction with dog's teeth. Once set, his mind clamped down on informative data to track his prey. When caught, his quarry would not be released without positive proof.

'Thanks a ton, Peter. There's one more thing. Bjorn must be away at the crack of dawn. He needs to replenish his breeding stock from the yearling sales. This buy means a lot to Jalna, and he's sweating on new bloodlines to bulk his stables. I'd appreciate it if he could still travel with his manager down to the sales in Canberra first thing tomorrow. By gee you're thorough. A good man to have around when trouble looms or things go wrong.'

'One has to be precise in my job. Even though I'm on extended leave and temporarily out of circulation, I must observe the rules and keep alert. I'll arrange it with the Feds. I'll vouch for Bjorn Svensson. I know he'll keep this matter under his battered Akubra, so they say.' Having given his promise, Bucknell's crusade to do his part in capturing the Nazi war criminal would begin in earnest within twenty-four hours. Excusing himself, Peter hurried along the main hall to where Kirsten stood, tapping her index finger on Abergeldie's rose-decaled doorknob over its brass plate.

Kendall caught up with him in the small "L-shaped" alcove leading to his mother's study. 'Peter, I hate to detain you, and I have a special favour to ask. Not now, but when it's safe, I'd appreciate you calling this number.' He withdrew a fountain pen from his shirt pocket, and scribbled a series of digits on a scrap from his wallet, which he passed to Senator Bucknell.

'That is Monsieur Bouvier's home phone number in Paris. Pierre will be frantic if he hasn't heard from my mother in weeks. It's a clandestine pact they contrived to keep the nosey paparazzi at bay. The devious press hounds would love to dig their teeth into Pierre's private business, him being a well-established French entrepreneur and a philanthropist. He adores Gigi and she idolises him. Foolish woman, she doesn't realise the depth of his love for her.'

'I'll pass this message on to my colleagues, once they arrive here tomorrow. They apologise for their delay. Reliable RAAF pilots are harder to obtain than ordinary blokes who lack security clearances in our business.' Bucknell sighed as Mrs Svensson unlocked the office door.

Kirsten embraced the subject. 'Kendall, I agree with you wholeheartedly. Phoning Pierre will solve one problem, before it becomes a genuine problem.'

Laughingly, Bucknell intervened. 'What say I call the Frenchman myself in a day or so? You realise Kendall, that it may not be my decision to make. I'll check with the Feds, but I can't go over their heads. They will let you know tomorrow. I trust you both not to disclose this, or any conversations we have in private. Don't mention what I promised to Gigi. She will discover our secret in due course. It depends on the outcome of this investigation, Kendall.' Bucknell looked at Kirsten whose smile indicated she understood his demands.

'Be more than my professional honour is worth to disclose a vow received in confidence. Your secret is safe with us Peter,' confirmed Kendall, who watched the diplomat step over the threshold of his mother's office. Reviewing Kirsten's notes, plus the statements taken from each family member would take hours to decipher. This allowed Kendall time to review his deposition. Houseguests present at the time of Ian Ross's sudden departure and their current guests would be interviewed in turn by the federal team once they settled down on Abergeldie.

His Canberran boss left it in Bucknell's capable hands to collect all possible data he could extract from staff members and workmen on both properties. Knowing most of the workmen, Peter would be discrete in confronting them, without raising their suspicions. He promised to sanction whatever excuse both Kendall and Doctor Jarvis devised to tell the men and domestic staff why they must keep their distance from the homestead. Writing his notes on Gigi's health problems while transcribing Mrs Svensson's shorthand he mentally confirmed, *What horrific times this caring family must confront. Suppression of all unfounded gossip by muckraking old fogies will be withheld from the nosey media hounds. I'll do my dandiest to keep that lot of stringers, including their cameramen, off this property and away from its outer boundaries.*

Under extreme stress, Gigi greatly feared that her London and Parisienne friends and local acquaintances would discover their dilemma at this critical time. To quell her fears Bucknell, through diplomatic channels, advised a mutual friend, Homer Ellis in London of Abergeldie's problems. Peter didn't disclose the real reason regarding the dramatic situation under investigation at Yass in New South Wales. Secretary Ellis in turn, privately raised the subject with his boss, the British charge´ d'affaires, who promised to quell gossip filtering through from foreign sources. Bucknell had mentioned this to Kendall earlier in the day, but it was of little consolation to him.

'I'd hate Mum to get wind of what Homer Ellis *thinks* he knows of our predicament. If even one local gossipmonger latches on to our plight it will be devastating. It's unthinkable of the town gossips spreading untruths about what they assume to be the truth. I know it would have a catastrophic effect on this homestead and Jalna. It needs only one landholder to leak the tiniest scrap of vital information to ragbags in the press contingents who haunt innocent folks in town. I dread the prospects as it'll undoubtedly benefit that deceitful Nazi absconder; a despicable creature, whom I will never forgive nor forget, for the beastly way he brutalised and terrorised my mother, and how he tried to drown me in my early childhood. Nor his attempted murder and maltreatment of Mum up in those top paddocks. I do know he hates both of us.'

As his colleague Doctor Jarvis walked in, Kendall met him outside her bedroom door. On Stephen's arrival he insisted on taking over Gigi's medication. This allowed Kendall time to relax. Complete bedrest for his mother was the predominant factor now. The future would unfold its own dramas soon enough, to cruelly impart unwarranted tales of anguish on unprepared countrified and hardworking landholders on both properties. They were struggling farmers and herdsmen all striving to keep their private lives and anguishes within their silent domains.

Conferring with the diplomat in Gigi's office, Kendall, Jarvis and Bjorn were trying to work out a strategy to prevent the rugged male workers from invading Abergeldie's woolsheds.

Stephen Jarvis hit on the solution. 'What I propose is this. You found your mother lying on the bathroom floor. She collapsed and her symptoms indicate a mild form of meningitis. Kendall, I'll agree with your diagnosis by admitting her temperature is dangerously high, and she needs complete bed rest. Until my diagnosis is confirmed by blood tests, all outsiders must keep their distance. I'll recommend that Gigi needs ample time alone to recuperate.'

This wasn't far from the actual truth. In Jarvis's professional opinion there was a genuine risk of Mrs Ross suffering a massive stroke. Traumatised by recent events, complete rest was vital for several weeks, and maybe months to allow her mind to settle and her heart to cope with further unwarranted stress.

2

From that instant on Abergeldie's cook, Mrs Graham and her daughter Rose, were the only domestic staff allowed in the house. They had chosen to bunk down in the back study, used as a storeroom. This uncluttered room lay within a second's walk from Ian's office, or as he'd called it, his den. It only took the women a short time to make their new bedroom comfortable. The narrow rear hallway led to the kitchen, and could be easily accessible at a moment's notice.

Stephen Jarvis left to check on Gigi, as the three remaining men took their time to mull over the proposals he'd put forward. It included the arrival of the federal team, plus Colonel Montsard's private secretary and a technical expert, also a senior agent from ASIO.

'While we're discussing this subject, I suggest that well before sunrise your manager should organise a crew to clear an area in Abergeldie's top paddocks, somewhere flat to use as a temporary airstrip.'

'You know, I imagined their blokes would come by car. How naive can a man be?' They cracked a smile as Kendall said, 'No need, Peter. Out on the far side, opposite Jalna's border fences there's a huge plateau flatter than a cheeseboard. A rough grassed area, it's stable and suitable for a serviceable airstrip. It'll be perfect. It will be a bit risky, even dusty leading up through the lower soil-eroded paddocks.'

'Sounds an ideal spot,' Bucknell replied taking notes. 'Good! Now tell me honestly Kendall, can all the upper paddocks be seen from either here or your neighbour's property? I always wanted to trek up there and never have until …'

'Well, now's your chance sport,' Kendall interjected. 'I'll arrange for a grader to be bought in early this afternoon. We'll have the work well underway before dusk. And no, not one dry blade or spike of grass can be seen from either here or Jalna, or their top paddocks.'

'When will you warn everyone about the house and area being out of bounds Kendall? It might be advisable to tell your staff and workmen now, before they hit the sack.'

'I'll stir my foreman at five tomorrow to warn him why the homestead is forbidden territory. I might give the excuse that Stephen put forward about Mum collapsing. I'll also let the drifting drovers and my blokes think I've arranged for a contractor to finish grading the top paddocks before the blistering heat of day sets in. They'll know it would be impossible for him to grade such a large area in one afternoon.' Having reconsidered this Kendall laughed.

'Hey, what's so funny sport?' Bjorn, who'd remained quiet to listen, chimed in. 'Explain what ya mean mate. Don't leave us in the dark. We're not all bloody mushrooms.'

'Did I say something to amuse you?' Bucknell seemed also at a loss to think of a reason why Kendall's smirk broadened.

'It's ironic, Peter. Yesterday, I cursed the blasted grader for packing up. Now, I'm blessing the damn thing. Your old machine is out on lease, Bjorn?' He nodded. 'Dust coming from that direction will make the men think it's the contractor I've hired working up there.'

'Brilliant! Now if you gentlemen will excuse me, I might catch my boss in Canberra before he finishes work. He advised me a while ago that the Cessna's due to land at approximately seven tomorrow morning. If we're up there early Kendall, we can wave the plane in. Thank goodness the forecast has been upgraded to fine.' A harrowing scowl inflicted his nearest companion's jaw.

'Don't whinge mate,' Bjorn said. 'I need water, a bloody lot more than you cow cockies do hereabouts. My water tanks have almost run dry.'

'I agree BJ. Yeah, our top dam's just holding. We're lucky to have the artesian bore working.' Ready to head off to his room Kendall faltered. 'I forgot to mention Peter, you can doss down in the third bedroom. It's along the front hall. I've given Stephen the smaller room further down.'

'Any chance of a swap?' Bucknell lifted the finger ready to dial. 'It'll be easier for me, near this phone. Jarvis will be closer to the ladies' bedrooms, just in case he's needed.' Kendall nodded as Bjorn sauntered down the main hall to find his wife, Kirsten.

Thinking over the diplomat's request he responded, 'I can't think of a reason why not. Probably suit Jarvis better and all the bedrooms are ready. Harriet got stuck into them when she landed back in a while ago. I'll tell Stephen about the swap before he settles. Being handy to this hall phone should please him. I forgot to tell you; Bjorn just mentioned he and Kirsty have offered to stay on here until mum's health improves. Guess that'll relieve everyone's mind, including mine.'

Bjorn Svensson had previously arranged for a shipment of his cattle to be delivered to the Canberra saleyards. He sweated on the money his heifers and calves would bring. His men intended to truck them down before daybreak. Jointly the manager of Jalna and his station hands would run his mob. Kirsten would care for Gigi's personal needs on Abergeldie while he supervised the stockmen and their part-time shearers. With nine bedrooms and three bathrooms in the homestead it wasn't a hassle. Most of the rooms were seldom used, unless for guests. Gigi had insisted the antique furniture be covered with disused, though breathable sheeting.

Before settling down for a well-earned rest, Kendall breezed in to see if Father Brady needed anything. Away with the hogs and in the process of shooing the pigs to market he was oblivious to anyone standing by his bed. Kendall smiled and left him snoring. He decided to put his feet up before someone ruined this brief chance to relax.

No sooner had his heels touched the cretonne bedspread when the phone buzzed in his mother's office. With Peter Bucknell there to answer the call, his eyelids drooped to a close.

Somewhat startled ten minutes later, he jumped when a finger tapped his shoulder. Half asleep, he almost rolled off the bed, then stood erect and looked blankly at the diplomat.

'Kendall, you're wanted on the phone in Gigi's study. An international call from Paris. I must insist on being there while you're speaking. Also make the reply brief, please.'

Curious, he approached the study, with Bucknell hot on his heels. 'Hi there, Pierre. I wanted you to know my mother is okay. She's asleep. I can't talk for long. We're having trouble with the phone lines here.' At Kendall's remark Peter's azure blue eyes enlarged. Nevertheless, he ignored his host's smug grin to listen.

'No, don't ring us. We're having difficulty at this end; you'll not get through. Oui, yes, there is trouble with the lines. I'll be in touch when possible, Pierre. Mum's heading down to Canberra tomorrow for a week or two, so don't worry.' Kendall hated lying to their oldest and dearest friend, but what else could he do? 'You've sent us an engagement gift. Thank you, I'll tell Bridy. You couldn't have heard with the racket at the party anyway. Catch you then. Bye.'

Replacing the receiver, Kendall turned to address the diplomat. 'Peter, I owe you one. It amused me when you frowned as I mentioned trouble. Pierre caught my meaning though. Now it's over to you, old sport.'

'Well, that's one less hassle you, I and the federal boys will have to worry over.'

'Yeah, I agree. Now if you'll excuse me, I'll pass Pierre's message on to my mother. Peter, will you call this number for me, before the double-ended plugs in the Yass exchange are reconnected?' Kendall slid a note across the desk. 'Then I'll speak to the contractor Bjorn and I normally phone. Bert owes me a whopping favour and he's been slack lately, so he'll probably bring the grader in after lunch. The quicker we get the job up top underway the better. I heard the announcer say wet weather is predicted over the next couple of days.'

With this matter taken care of, Kendall Ross left Senator Bucknell to contact his boss in Canberra while he hunted down his colleague. Doctor Jarvis advised him that he'd secured a second locum to commence in his surgery from midmorning. With three reliable doctors managing the practice it allowed him the freedom to constantly monitor Gigi's condition. Stephen assumed all private or urgent messages for him would go to the diplomat's home number. All long distance or local calls would be redirected to the homestead and all messages recorded.

Once the federal team moved in with loads of equipment, they would connect a temporary phone line. This freed Abergeldie's private line. All incoming and outgoing calls were to be constantly monitored by one of their phone operators.

Silent and brooding, Father Brady sat on the lounge. He enjoyed dinner, but didn't take kindly to the refusal of a second nip of whisky. The cooling coffee he sipped with disdain. The sullen scowl inflicting his weathered jowls gradually increased.

'Not a tiddly drop of the smooth nectar can ya spare fa an old clergy? Look Kendall, me tongue's so dry it's choking me. I'd not be abusin your hospitality by askin for just one wee dram more, would I?' he pleaded with a mournful frown.

'Sorry Father George, once the authorities have interviewed you, I promise to pour you two good nips. Might I suggest you go and lie down? This terrible business has come as a huge shock to you. You'll need a full night's rest to face lengthy inquisitions. There's movement afoot for the Feds to land in here first thing tomorrow. It'll be around breakfast time when they do arrive. A good night's kip will help you to cope with their tireless questioning.'

With a kindly pat on the old celibate's shoulder, Kendall assisted Father George to his feet. A little unstable, the priest shuffled towards his bedroom.

A quiet room, it overlooked Jalna's nearside paddocks. Moonbeams filtering in through the flimsy green curtains lent his cosy domain an ambience of peace. Leaves wrestling in their magical manoeuvres rustled and created a soothing lullaby in the warm night air; scented with magnolia and the aroma of apple blossoms helped to dispel his sombre mood. Soft bronzed rays of the crescent orb comforted him. A crimson moon indicated another sweltering hot day.

Ready to draw the curtains Kendall hesitated when he pleaded, 'Please let them be, me boy. The rays have a calmin effect on me nerves. Since the terrible war years, I need some light in me room. What better light than the one our Lord gives us, Kendall.' Smiling, he removed the cord of his dressing-gown. Kendall caught it as Father George let his tired body sink into the soft goose-down padding atop a quilted mattress. Snuggled underneath the warm eiderdown, it created a feeling of security. Slowly his tormented mind relinquished its fight for peace.

'I never thought of moonlight giving anyone comfort, Father. I can understand to a point how you feel.' Kendall's serene tone conveyed an element of quiescence to his elderly friend, whose drawn face expressed an appreciative hint of relief.

'One moment before ya go, me boy,' he pleaded in a drawn moan. 'Will ya mother forgive me fa bringin her family name inta disrepute? A man should've burned them mementos.'

'There's nothing to forgive, Father. And destroy them; never! Then you would've done not only yourself an injustice, but also my family. You owe it to those innocent people that damned Nazi inflicted terror upon. Don't ever mistrust your own judgement, Father George. It's important for those miniatures to be intact. Pity they hadn't sprung to life before my mother met that Nazi b …'

'Bastard, ya can say it, Kendall. I have many a time. Often in me nightmares I still hear screams of tortured prisoners near my cell below the Chancellery building. Now everything's out in the open, perhaps those terrible dreams will leave a stupid old man in peace.' He paused to reflect. 'Ya know, that criminal looked quite handsome and kindly in his Reich uniform. No wonder he captured many an innocent woman's heart in his web of cruelty. Do me a favour will ya lad, put a flame under the snaps when they're done with.' Constant recall of the Nazi's sadistic smirk chilled the Irishman's blood as an icy shiver travelled down his osteoporosis-riddled spine.

'Sorry, I can't promise you anything. Well, not at this stage Father.' Kendall sat on the bed beside the priest. 'All three miniatures are evidence. It's an indictable crime to destroy or tamper with solid evidence.' He eased the covers up around the shivering elderly man's shoulders. 'You're very special to me Father George. I can speak for Bridy, my mother and myself in saying that. You're heaven blessed. Amid all the dangerous years of suffering from Nazi beatings, you considered others. Now try to rest.' A steady hand gently embraced the ailing man's knuckles.

The aged priest turned his head away from his impenetrable gaze. Unrestrained tears welled up in both men's eyes; neither wished the other to witness his individual pain.

Quiet and reserved, this young English-born physician crept from the priest's private domain. In silence he mused: *God only knows how the recent cruel events have wrenched all our lives asunder, but none more so than his. It's been said and I agree Father Brady's wrinkled features have seasoned well, way beyond the age of seventy.*

4 am: Abergeldie's homestead closed down. Every person in the house deserved this short respite. An hour later the alarm clock disturbed an extremely tired physician. Clad in his riding jodhpurs and an earth-toned cotton shirt Kendall approached his undermanager in the main hall. 'Tom, how about lowering the top fences butting onto Jalna's lower paddocks? They all need reinforcing and pretty soon they will collapse.'

'I agree with ya boss. If not, they'll cost ya a packet of dosh to fix.'

Doctor Ross considered the bushman's proposition before answering. 'Hey, what a bonza idea. Begin lowering them today, before noon. I'll give BJ a buzz on Jalna. He'll send his crew up there as soon as they've eaten a solid breakfast.'

Kendall promised more in his crew's pay packets if they completed dropping the fences by noon. He knew their hard yakka might linger for a week at least. A worthwhile job, it would keep their noses down, their curious minds busy and gossiping tongues silent. Idle men prattled over a couple of beers much more so than women.

'Tom, in case you haven't heard, my mother collapsed this morning. Stephen and I aren't sure if she's suffering from meningitis. Keep all our men busy today. Nobody must come within a whip's curse of the house until we're sure. Mum needs sleep after her horrific fall. Sorry to off-load extra work on your shoulders. None of us have control over accidents. Well, I damn well don't!' About to angle back to the homestead Kendall halted. 'If you need me, or more blokes to cope with the excessive workload, give me a buzz on your two-way walkie-talkie. I'll leave mine tuned in back at the house.'

'Will she be okay?' the manager frowned and saw his boss nod. 'Tell ya mum, I'm sorry about her fall. Her horse is usually a stable, steady beast. I'll see to them fences now, Kendall. It sure is awful that Mrs Ro … err … ya mum's crook. I'll keep the shearers and those drifters informed, so no worries there.'

On his way out Peter Bucknell met Kendall on the back verandah. 'We'd better get moving. The Fed's Cessna is due to land shortly. A brief message landed in on your mum's phone line a while ago. The constant buzz woke me up actually.'

'My crew's only minutes ago left for Jalna's top paddock, so they won't have a clue about a plane landing up in that region,' confirmed Kendall, dusting his tattered Akubra against the threadbare knees of his mud-stained leather chaps.

'Peter, you looked so uncomfortable in the rickety wicker chair. Why the hell didn't you doss down on the sofa in Mum's room? I bet your neck aches after lying in such a crooked position.' Kendall smirked as the diplomat shrugged his stiff left shoulder; it served as a reply.

3

With a strong tail wind, the twin engine Cessna touched down fifteen minutes ahead of time at 0645 hours. Two male federal officers, plus a woman clad in flimsy clothes climbed down from their temporary icebox into an inferno, as one of the ASIO agents termed this sweltering heat.

An officer in an army uniform hung back to confer with the pilot. Then he joined the group standing in shade under one wing. With brief introductions made, Abergeldie's guests all moved out into the furnace and angled towards Kendall's idling jeep.

'Afraid it won't be a comfortable run across in this heat, Colonel. Jeeps are notorious for being rugged on bums and strides. Oops, sorry Miss! I wasn't trying to be rude or sexist.' Kendall apologised, glancing up in the rear-vision mirror at the colonel's secretary, whose brief miniskirt revealed a delectable amount of raw flesh on her thighs.

'Charlene's the name, Doctor Ross. I must say this is a huge property. I have never travelled in the outback territory before.' Flicking her eyelashes at Kendall, she winked. He smiled.

This young titter has a seductive manner; is she trying to latch onto me? Bridy won't be pleased with her kinswoman batting her peepers in my direction. Her accent sounds Irish, but her name indicates American or French origin. I wonder if she is Irish.

'This seems a well-run stud, the little I know of sheep stations,' Charlene remarked quite casually, edging her skirt down a fraction.

Little did she realise how her seductive voice had intruded on his thoughts. 'Abergeldie is an extremely large property. It doesn't compare with other homesteads further out of town. They are massive.' Kendall's next polite, though sharp retort caught her off guard as he studied her topaz eyes. 'How long do you folks anticipate staying here on Abergeldie?'

'Just long enough to delve into all facets of the Nazi absconder's life with your family. Our aim is to collect every bit of evidence and complete an intriguing jigsaw. Once we've weeded out fallacy from truth in this serious matter, you'll be left with two men, both trained in combat skills. They'll remain here for some while. It's safer talking in the jeep. One never knows what ears listen inside four walls. Right, Consular?' stated the colonel who glanced across at Peter Bucknell. Both their eyes focused on the unusual bulky baggage stacked above the jeep's spare wheel hub. 'That is why all this sensitive, expensive equipment is necessary.'

As the plain-clothed officer next to him half-turned to gawk at the rear passengers, Kendall spied a thirty-eight-calibre pistol sheathed in his shoulder holster. *Hell, what a savage looking weapon*, he shuddered, yet refrained from commenting.

'Class my team as unexpected, temporary tourists, Doctor Ross,' the colonel then declared, rebuttoning his uniform coat. 'Never can tell whether your family is on the absconded Nazi's hit list as targets. If the war criminal is who we think he might be, he's capable of murdering everyone who trespasses on his turf, and he must not be underestimated.'

'Ah!' exclaimed Kendall, 'he's capable of murder all right. I can vouch for that. I've seen firsthand what he's done to my mother more than once. The swine tried to murder me as a child.' *This pussyfooting around is obviously to spare me discomfort or distress.*

'What! Your father actually targeted you, Kendall? How and when?' Peter found this news hard to absorb. Although none of the other men seemed surprised by this revelation, it floored Bucknell. He didn't expect to hear this declaration from a man he'd known most of his life.

'That swine's no father of mine. I can't bring myself to mention his name. Neither can my mother. Hang on! We're going to hit a whopping big dustbowl. Tuck your heads down between your knees and cover your eyes, or you'll cop a lung full of fine grit.'

The jeep bucked. Its wheels bit into the raw earth. Kendall held the steering wheel fast and hard on lock to pull it out of the powdery silt. This manoeuvre worked. The jeep twisted, wrenched itself free and then steadied on all four wheels.

'That insensitive swine tried to drown me. The bastard held my head underwater. Oh, sorry Miss.' Charlene smiled. 'Hey, don't you blokes mention it to my mother. She's battling enough trying to cope, without more muck being dragged through the quagmire of time. Just thought you should know what the mongrel's capable of, by perpetrating evil against the ones he professed to love. The bastard never loved me. I've always been shit, an idiot in his eyes.' Gritting his teeth on sighting the huge rut they were about to encounter, Kendall apologised, 'Sorry I swore, Miss. It's become a habit working with men all the time.' He caught her nod in the mirror.

'That's mild,' decreed the colonel, 'compared with some of the choice words we hear and use in our field of work. We need every bit of information, even if it seems insignificant or trivial to you or your mother in the slightest way. Always be truthful with your answers, Doctor Ross.'

Colonel Montsard wiped grit from his lips and eyes, then coughed. 'Our questions will be to the point and reasonable. Once we've compiled all the required information, we'll leave and let you move on with your lives,' he uttered, biting his tongue as the jeep struck a dust-filled pothole.

When able he looked across at Senator Bucknell. 'You have given us reams of data relevant to this criminal, should it be proven that he is one.' After a good spit over his arm, the senior Australian agent spoke in a strong overtone. 'With a concise idea of how the ex-Nazi behaved and now thinks, it will help us to anticipate his next and every subsequent move.'

7.30 am: As promised on the following morning two rather staunch-looking bodyguards in plain clothes arrived by car. They immediately swung into action by interviewing Abergeldie's household and its staff which took an extended period of the day.

Heavily drugged and quite confused, Gigi Ross managed to survive the stringent grilling well. Kendall knew Father George couldn't take another bout of being constantly traumatised by repeating unpleasant memories. Accepting his advice, Stephen Jarvis begrudgingly sedated the Irish priest.

Brianna battled through the questions lobbed in her direction, and weathered the storm of discreetness with a mere shrug of her shoulders. 'I have not spoken to my fiancée's father at all, although I did see him glaring at me on our arrival. I disliked his snarl and thought he looked brutish.'

By the end of that first week an extraordinary memorandum had surfaced. Intrigued by the report Senator Bucknell received, he confided in Doctor Jarvis at dinner that evening. 'An elderly woman from Düsseldorf in Germany has passed on important information to a member of our investigative team. He's a British agent whom I've known for years. He, on occasions has worked in conjunction with the Jewish Mossad. It appears her brother Hans went missing from Berlin during the latter part of 1945. Recently she produced a letter. Enclosed in the envelope are two photographs. One is of a Nazi in a captain's uniform, which compares with the miniature in our files. The other photo is of her brother, taken sitting opposite a Reich Captain in a Hamburg café. The woman has circled Hans dressed in civilian clothes. It seems Stephen, that Rolf von Breusch and Hans Selig were identical in appearance.'

'I have seen the snaps and also read that letter, and it's now on the colonel's desk. What a horrific time it must've been for the woman living in Düsseldorf's destroyed city. Keep me informed, Peter. You know I will never reveal a confidence. This news has the potential to cause Gigi and Kendall more grief. And neither one is in a fit state to cope. He's far stronger than his mother. I don't need to tell you how worried we are about Gigi's health.'

Bucknell went on to reveal, 'She stated, after reading her brother's letter she knew by its postmark it was written in Hamburg, but posted in Berlin. Possibly by him or a friend after Hans had resigned from the munitions factory where he'd worked. The woman reported that she hadn't heard from him at all, not until the Israeli authorities contacted her a month ago. However, the Düsseldorf Security Police have refused to elaborate on the reason for Herr Selig's brutal murder.'

Interested by what he'd overheard, Bjorn Svensson chimed in to say, 'Going by what one of the Feds conveyed to Kendall, it tallies with what the colonel told him on their arrival. They've scrounged another important, unusual letter and a photograph, but I'm not sure who from. Everything you've described ties in with Father Brady's miniatures of that Nazi bastard. Why the hell his foolish mother didn't drown the bastard at birth astounds me.'

'Pity she didn't BJ. I've also seen those snaps. Father Brady showed them to me the night of Kendall's engagement party, just after this shocking news first broke here on Abergeldie.'

'It's just occurred to me, earlier in the year I tackled Ian about trying to swindle Gigi out of a lot of cash. We were in this room when I heard him swear in German after I accused him. As usual, the liar denied stealing money from her wall safe. I threatened to kill Ian, or von Breusch as we now know him, that night. Now I'm sorry I didn't shoot the bastard.'

While discussing this indelicate matter on the quiet with Jarvis, it brought to mind a photo Bjorn recalled seeing, prior to his and Kirsten's arrival in Australia. 'From memory it appeared someone had taken it twenty-odd years previously in Switzerland. The engagement photo displayed von Breusch, then also known as Ian Ross, with a goatee beard and a moustache.' Bjorn preferred not to comment or ask Gigi about the snaps in her locket, in case it reopened a string of raw wounds.

All data officialdom had received thus far confirmed what the Feds expected. The absconded fugitive was, without a doubt an ex-Nazi, Captain Rolf von Breusch. Further conclusive evidence of his distinctive writing on the signed document transferring Abergeldie to his Welsh-born Australian wife Gigi, verified their suspicions. It compared with several scraps of signed papers discovered in his den. The combination of his safe, written in Ian's distinctive style proved invaluable. The second, a small yellow note with a Reichstag imprint Kendall had retrieved from under the lining of his father's cufflink case with his initials RMB inscribed on its gold surface. If nothing more than a historical piece, this item would be classed as evidence. It proved mandatory in tying together all the evidence ASIO and the AFP had received in recent weeks. This information, plus fresh incoming data was sufficient for their teams to put the Nazi absconder on their most wanted list.

All federal agencies were independently building healthy dossiers on the fugitive. Even members of the Jewish Mossad in Australia were constantly monitoring his movements. The AFP's last report followed his trail from Yass in the Southern Highlands up to Kings Cross, a suburb of Sydney. All facets of compiled data would be reclassified, and set in files now the teams had begun working in their individual huts on Abergeldie's homestead. The only laymen aware of these facts were Kendall Ross and his esteemed colleague. He and Jarvis had initialled an official document stating they would never reveal the secret information listed on their signed depositions, unless requested by court authorities to verbally amend both their previous statements.

Traumatic in nature, the federal agent's stringent questioning continued on through the second week. A brief respite allowed the residents and guests on Abergeldie to recoup their sanity.

Two days later, at the conclusion of being questioned by the AFP surveillance team, Senator Bucknell accompanied Father Brady on a private flight from Yass down to Canberra. Before leaving Abergeldie they were advised, "At a later date, you will all be required for further interviews to clarify the incoming data, especially once our net begins to tighten".

Colonel Monstard, the officer in charge of their team on Abergeldie, refused Kendall's request for permission to leave for Paris, likewise Miss O'Shea for Ireland. Peter Bucknell, while visiting the homestead had asked Brianna to supply him with all relevant data relating to her dance company policies. He intended to pursue the matter during his visit to Canberra. When done, everything would be put on hold. Two unexpected AFP representatives, who'd just arrived at the homestead, had confiscated the couple's passports until they considered it safe for either person to travel overseas.

At first bell next morning, Bucknell rang Abergeldie and advised Brianna that her contract had been secured. The company guaranteed her option of resigning would be held in abeyance until notified otherwise. Personal trauma and illness were cited as the reasons for her delay in Australia. None of their top echelon considered this a fabrication. Lies were simply necessary half-truths to maintain the highest standard of security that ASIO and its confederate agencies demanded.

Limited time demanded further urgent investigations of all personnel on Abergeldie, thus its boundaries remained under strict quarantine. The reason given by Doctor Jarvis for Mrs Ross' sudden collapse was a fictitious illness: suspected meningitis.

This wasn't far from the truth. Around noon that day while everyone dined, a long penetrating scream echoed through the house. Kendall, standing by the kitchen table, looked at his colleague, who in turn frowned at Bjorn. His wife Kirsten, who'd just walked from the breezeway almost collapsed. Kendall hurried close on her heels.

'The scream came from Mum's room. She must have fallen off the bed. Kirsten, fetch my medical bag. It's locked in my wardrobe. Here's the keys.' They flew through the warm midday air with the speed of a fox chasing its dinner. 'Good catch.'

An hour earlier, Gigi had stirred. In her dream-imbued state she imagined five soldiers wearing nothing but Nazi armbands, pinning her against a wall. She tried running down a long dark corridor to reach a row of locked doors. As she reached the first door it swung back on rusty hinges, to reveal a naked man she recognised as Ian. His taut blood-engorged probe was ready to penetrate her. Fearing he'd rape her, she fled to the next door. There three faceless naked creatures reached out to grab her with their hawk-like talons. The same occurred at the third and fourth doors. As she touched the last knob a ghostly figure pulled her inside where a group of naked man groped her loins. Spread-eagled, she tried to fight each one.

The last offender whispered in her ear while raping her. 'You won't get away from me this time. Not like you tried to do after I piddled on your ugly face, before I left you to die up in our top paddock. Gigi, I loved you, until you brought that wretched brat into this world. I've always hated *your* son. Kendall was the blight of my miserable life. I'm sorry that I didn't drown him in the toilet. He struggled to scream, but I held his head under the flushing water. Then you intervened and pushed me, not only out of there, but also out of your life.'

In a battle to escape more cruel marauding hands, she crawled on hands and knees to a cavernous hole that swallowed her. More fiery claws reached out of the darkness to grab her. Gigi had collapsed on her way to the toilet. She now lay in a pool of her own blood. She held a wet facecloth against her mouth in fear as she writhed and whimpered in pain.

'Kendall, I can't push this damn door back,' growled Bjorn. 'There's something jammed behind it. It might be your mother. Gigi may have fallen against the blasted thing.'

'Move out of the way, Bjorn. No, better still, stay here. I'll race around and break her front bedroom window. There's a broom on the verandah. I'll soon make good use of it. Kirsten, make sure Stephen has my medical kit. We also might need your assistance, if Mum is injured.'

A ghastly situation confronted Kendall as he entered his mother's room. He couldn't believe his eyes. Bending over her body, he felt for a pulse. Trained fingers detected the faint irregular beat of her carotid artery. Edging her back away from the door it sprung open of its own accord. Bjorn, a burly bloke of solid build, lifted her to the bed and then left the room. The bed covers were pulled down by his wife.

Jarvis appeared clutching both their medical kits. Kirsten nodded and continued to disrobe her friend. Gigi moaned, partly opening one eye, yet her mind lingered in the fragile world of unconsciousness.

'Stephen, has the bleeding stopped?'

'Eased to a trickle, a little spotting now. I'll leave this surgical gauze between her thighs in case your mother haemorrhages again. I've finished dressing her head wound. You and Kirsten can change her bloodstained night attire.' Stephen's stern frown dissipated on replacing the surgical instruments in his battered and aged brown Gladstone bag.

Minutes later Jalna's boss caught Kendall leaving his mother's room. 'Hey sport, I've given the ambulance a buzz. They're on their way. Reckon they'll be here in the swift twist of a possum's tail.' Bjorn's grin outshone that of a Cheshire cat. 'I better fetch one of the federal blokes. Kendall, before I do, I bet ya didn't know their top gun was a medic in the army. Yeah, a woman's quack of some renown in London, so his mate told me yesterday.'

The officer in question appeared with a briefcase of signed documents. He spoke to Jarvis in the main hall outside Gigi's door. 'Can I beg a ride with you to the hospital, Doctor? My car's in dock. Needs new tyres and brakes tested. I'm here to relieve you and Kendall, as his mother's watchdog, not a yappy one either. Where is the pup?'

Kendall laughed. 'She'll be here in a tick. Amber is very protective of Mum. The red setter doesn't bite, unless provoked. Colonel, your officers will need a break every five hours, otherwise they could collapse in this heat. I hear you earned your surgeon's stripes in London. Which university, Greg?'

'In St Bartholomew's Hospital. I worked in A&E there. Long hours of hard slog to save accident victims. Good facility. Bonza food and staff. My colleagues kept me sane.'

'I hear you've caught Bjorn's disease. Aussie slang! He's a character. One never knows what he'll say, especially if he's riled. Taken me ages to understand his weird ocker sayings. We better be moving. There's the ambulance. Will you let their blokes in the front door? It's unlocked Greg. I'll nick in and make sure Mum's respectable. Stephen Jarvis agrees with my diagnosis of an ovarian cyst. We'll know for sure, once she comes out of theatre. Listen to me, if you're only going to the hospital to interrogate my mother, forget it. You or the colonel can speak to Mum when she's well and then in *my* presence.'

'Weren't you going overseas this weekend, Kendall?'

'I postponed my holiday. Peter Bucknell will reorganise my travel arrangements to Paris later. My mother's health is my primary concern now, Greg. Our flights can be rebooked. Brianna feels the same as I do. Please, don't ask me again.'

As instructed, the ambos ferried their patient straight to a private annex, prearranged by Doctor Jarvis ten minutes earlier. He thought it advisable, rather than having his special patient being examined by undergraduate interns, or junior nurses. Privacy, not neglect was his main concern.

Before Kendall publicised his mother's suspected condition, Jalna and Abergeldie had stocked up with extra supplies, enough to last months. This wasn't unusual because country homesteads resorted to bulking their larders in order to keep their properties viable, functioning through droughts, floods and blustery, difficult, prolonged winter periods.

Bjorn's property housed all his neighbour's workmen in the abandoned shearing quarters, behind its current buildings. All the units were self-sufficient. Together the managers took over running their sheep in Jalna's fertile lower paddocks. Wire strands between these properties were released and fences lowered to allow the sheep a generous run. Feisty rams, ewes and newborn lambs were easier to control under these conditions. It made the men's work less hazardous in this extreme summer heat. Not a solitary outsider suspected anything was wrong with these arrangements set in place. It also deterred busybodies and local gossipmongers from prying into Abergeldie's privacy and their business deals in this small community.

Over the past few days budgets of paperwork had surfaced which began to build this mysterious puzzle with bulging dossiers. Difficult though their jobs were, ASIO and their associated agencies had produced enough facts on von Breusch's past history to potentially bring him to trial.

'Good news from Young,' declared Bjorn, landing in with mail from home. 'My uncle and aunt are on their way down, Kendall. It should alleviate some of the pressure off both our men. With Sam's expertise, it'll ease the burden on all your top ringers.'

Kendall frowned at his neighbour. 'Bjorn, you haven't said anything …'

Irate, Bjorn stamped his foot. 'What the hell do you take me for, a bloody fool or an idiot? I'm neither, *mate*. The Feds have spoken to Peter and through him to both your "watch dogs". My uncle Sam's been worried why I haven't contacted him. That's all I know at this stage.'

Kendall Ross ignored his display of anger and glared at the cantankerous neighbour. 'Bjorn, one of the Feds mentioned to me over breakfast that Peter and Father George should be home soon. He's really been distressed by this lot of strict interrogations.'

'Yeah, it seems the old clergy blames himself for this entire shebang. Struth, I hope he doesn't cark it before he goes home to Ireland.' Bjorn's harsh remarks sounded unwarranted to Kendall.

Ignoring his rudeness Jarvis addressed his colleague. 'Will you be called down to Canberra soon do you think, Kendall?'

'Afraid so, Stephen. Not that I want to leave Mum. When is anyone's guess. Suppose I'll be kept in the dark like a damn mushroom. You know how tight lipped the Feds are.' A shrugged shoulder displayed his disapproval to leave his mother. Gigi needed him and would do so until she recovered from the operation and shock of hearing her husband's Nazi history.

With restrictions lifted to some extent on their freedom, everyone on both properties were warned how dangerous it would be if the locals discovered Abergeldie's secret. Small country towns were notorious for spreading gossip. And every town had at least one over-eager gossiper who took supposition as fact.

One of the AFP blokes suggested it may be possible that the absconded criminal procured help to escape their net without warning. Everyone on Abergeldie knew the traitor wasn't one of them. Kendall suspected this supposition was rubbish, fantasy borne of ignorance. He considered ignoramuses were the worst tongue waggers. They could swing the power of government to bring an innocent man to the gallows on unsubstantiated gossip.

Their gossiping tongues infuriated Kendall, who tried to think of excuses to deter them from learning the truth behind an excessive amount of innuendos flying around town.

The following morning Senator Bucknell, with Father Brady trotting close on his heels walked in unannounced. 'Hi there all!' Peter sprouted tossing his Akubra on the table. 'It's good to be home, even if it is for a short stopover this time. Much to my dismay.'

'Rough flight home, Peter?' Kendall tossed in his bid, with Kirsten dogging his footsteps to the kitchen.

'Rather,' acknowledged the diplomat whose attention diverted to Mrs Svensson. 'Kirsten, please assist Father Brady to his room. Give him anything he wants, while I speak to these men in private.'

'Okay and thanks for taking good care of him, Peter.' Excusing herself, Kirsten steered the weary traveller to his room. Supporting the elderly priest up two narrow front hall steps she commented on his drawn features. 'You look tired, Father. Here, let me carry this case to your room. You should rest, at least until lunch is ready.'

''Tis kind of ya, dear lady. T'was a doin they put a man through down in Canberra.'

Bjorn, who'd just joined the group, nodded to their guest as he entered the kitchen. 'Hey Peter, it's good to have you home. I'm off to wash. See you here, when Mrs Graham finishes causing that racket by bashing the damn luncheon gong.'

Intrigued, the exhausted diplomat frowned. 'Where's your mother, Kendall?'

'She's lying down after being sedated. Mum's not well today and I'm quite worried about her health. Surely you won't need to disturb her, Peter? She's just come home from hospital after surgery. We didn't expect you to arrive here until tomorrow at noon.'

'No, let her rest. I'm not staying. Once I collect several documents I'll be off again. That's if the Cessna's okay. Blessed thing missed a beat on the way up from Canberra.'

'What happened?' Kendall's puzzled frown deepened. 'Let's adjourn to Mum's office where we can discuss what occurred in private. The Feds have taken over mine. It'll be bonza when we have some space to relax, without someone always listening to our conversations.'

'It will, I suppose.' Bucknell excused himself and hurried off to confer with his federal colleagues. Their replacements would be down soon with the pilot, all of whom intended to stay on Abergeldie for a week. If more problems arose, it may be a lengthy sojourn.

Re-joining the men in his allotted office, Bucknell repeated the same directive prior to his departure for Canberra. 'Even though restrictions have relaxed a little, it's still imperative to continue under the cloak of silence. Everything I tell you goes no further than this room.' The consular smiled. On the cusp of adding something he hesitated when Jarvis and Bjorn walked into the office.

'Your mother's asleep, Kendall. You'll be relieved to know her breathing has improved. I'm pleased that her general health has shown a marked improvement since she arrived home from hospital this morning.'

With everyone on hand, except the two indisposed women, Gigi and Brianna, Peter Bucknell took the floor. Kirsten took notes as he bought the men up to speed regarding.

Abergeldie's situation. When finished, he expressed the desire to speak with Doctor Ross.

'Be prepared to leave for Canberra at a moment's notice. Don't scowl, it spoils your ugly features.' They laughed. 'ASIO's top brass intends running through your deposition again. They prefer to tackle their informants personally. Don't panic Kendall. It's routine. By clarifying situations as they crop up it eliminates hiccups later.'

'Okay Peter, and thanks for warning me. Will you rebook Brianna's flight as soon as possible? She can't afford to be left in limbo, because of official red tape disrupting her travel plans.'

'Leave it with me. Kendall, I hope your mother will be well again soon. If the Feds need more information from Gigi, one of their blokes will sound her out later. The data they've gleaned from her thus far still stands.' Jotting down a few ideas and making notes Bucknell ceased speaking momentarily. 'The top blokes in Canberra should know something positive about your flight tomorrow at ten. After I've spoken to my boss, I'll reconfirm those details or as much as I can with you.'

'Suppose I better keep a travel-bag packed. Brianna's worried that the company won't honour her contract and may not pay her long service leave in full.'

Bucknell assured Kendall the Irish Dance Company delegates were prepared to forego her contract. However, they'd stipulated Brianna must return to Ireland for two weeks, or until her understudy could take over. As Peter spoke, a sly smile infiltrated his pleasant, though wizen features.

Thrilled with this news, Kendall failed to notice the mischievous glint in his eyes.

'Brianna is free to leave Australia the instant she signs this declaration, as you've already done,' he confirmed passing a folder to Kendall. 'Once those papers are back in my hands, I'll rearrange her flight to Ireland. There shouldn't be any hassles.'

'Thanks, and I appreciate your efforts, Peter. While we have some privacy, I want to ask you something important.' Kendall forced back the tears.

'Hey, I'm aware of your predicament,' Bucknell said, munching on a ham sandwich and sipping the dregs of his cold tea. 'It's to do with your surname, if I'm not mistaken.'

'Well, sort of. Mum won't admit it, using that Nazi's surname is causing her untold stress and constant depression.' Trying to keep composed, Kendall suppressed the urge to splutter and make a fool of himself. It wasn't easy. But then, what was in this dreadful business.

'Is there a possibility of our surname reverting back to her maiden name? I'll need it reclassified for my profession. What patient would want a Nazi's son attending to their needs in a professional capacity?' Anguish depleted this young physician's ability to control his emotions. It failed. Lost beyond the realm of sanity, he caved in and wept.

Bucknell remained silent, which gave Kendall ample time to relax. Peter signed the documents he needed, then pushed a note across the desk. Through weeks of untold anguish, he'd observed how this quietly spoken physician's personal stress had increased tenfold. Now it bubbled to the surface. *Time for relief. He needs the time to become composed without official harassment.*

The Australian diplomat vacated his chair to allow Kendall to reflect in peace. Quiet and unassuming he closed the door behind him. Mourning the loss of his own wife's death, Bucknell understood how solitude would be this young physician's only line of defence.

He isn't going anywhere at present. There'll be time to continue our discussion later. By then the papal hierarchy in Rome, plus our government may have granted my requests.

Important questions had been shelved in lieu of more lengthy inquisitions that were about to resume. Without some form of release, the grilling weeks ahead could destroy Doctor Ross, whose only thoughts were for his mother's welfare. Dedication to his medical profession made his talents far too valuable to be sacrificed needlessly.

On her return from Ireland Brianna would be a great help to this distressed family, who through no fault of their own, must still tread a long road of misery.

After meditating for fifteen minutes in his mother's study, Kendall emerged feeling refreshed. This brief time alone had relieved the stress from harrowing weeks of self-incrimination. If nothing more, it encouraged him to face the inevitable, the unknown. Reflecting on his moments of peace, Kendall bumped into Bucknell in the main hall.

'You needed that release, Kendall.' The diplomat's neat rustic moustache twitched as he continued to address his friend in a sedate manner. 'As I said before, if I can help to alleviate your anguish give me a bell. I'm at your disposal, so please don't hesitate to phone me.'

'There is something you can do for my family in Canberra. Peter, I'll just fetch the documents you require and return here in a minute or two.'

'Go ahead. I have time before my flight is due to leave.' Peter Bucknell's ruddy cheeks beamed. 'The pilot's managed to detect the engine's trouble. Something minor, he reckons. We won't need to stay here overnight. Around four this afternoon we're scheduled to fly out. Enjoy your lunch, Kendall.'

'Uh-uh, what I have in mind won't take long.' Kendall darted to his room where Gigi's private documents were located. Collecting the folder pertaining to their recent problems, Kendall hurried back to the lounge room.

Being out of breath prohibited him from speaking for a second. 'This envelope contains my mother's birth certificate and marriage lines. They are pinned together. My papers you'll need are under that lot. Brianna's signed affidavit is also in there.'

Kendall passed the manila envelope to Bucknell. 'Don't say a thing to Mum. She'll hit the roof if she discovers I've given you her documents. And she'll buck the system, if she even suspects that I found them.'

'Father Brady mentioned to me this morning that he intends to submit his recommendation to the Vatican. Dispensation for your mother. When he's able to cope Kendall, he's bound to broach the subject with you personally. That rascally old priest has really taken a shine to Gigi.'

The rapport between these two friends couldn't be faulted. Formidable in their approach, each considered the other akin to his extended family. Welcomed in this family, Bucknell couldn't appraise their hospitality enough, even now in their time of need. Skipping through each page his lips curled ever so slightly. *Finally, the papers have disclosed Gigi's personal secret. In all the years I've known the Svenssons, they've never revealed her maiden name.*

The smile on Kendall's face broadened. 'Unlike you, everyone here and in Yass would love to know her secret. Mum will tongue lash my ears for giving you those documents.'

'Enough said! Ours to know …others to find out, if they can, Kendall.'

'Well put, my friend. Best forgotten than acknowledged.'

4

Glossing over the recent events while sitting in Gigi's office, Senator Bucknell glanced through their family documents. A glint in his eye expressed surprise on reading a trio of names. The stern glower adorning his face lessened. 'Everything here seems in perfect order. It amazes me,' he gasped having known the Svenssons and this family for almost a decade.

'Never in my life have I seen such political correctness. The summaries Gigi has written for Kendall in the event of her death can't be faulted. Well, not by me.'

Peter failed to understand why such wealthy landowners of this lucrative property would leave important documents lying unattended on a desk. *Every item in this folder and their wills, plus their valuable jewellery should be in the bank. I'll advise Kendall to be more cautious with their valuables in future. Oddly, he doesn't seem the kind of man who'd be neglectful of his duty in keeping these irreplaceable documents in a safe. I could be wrong in surmising his neglect.*

What this Canberran diplomat didn't know was all their family documents were normally housed on Jalna in the Svenssons' safe. Neither Gigi nor her son trusted the local roustabouts, or casual shearers working to accrue money to send to their families overseas.

After perusing their personal information Peter sighed. *Evidently her scoundrel of a husband wasn't aware these documents existed. With his criminal mind, that coward von Breusch would've had a field day, if he discovered even one of these papers before he absconded. Well, everyone here on Abergeldie is carrying the fort in absentia.*

Compiling her will and private data had become an obsession to Gigi, a sad, though important lesson she'd learned after being injured and abandoned in the top paddocks in the dead of night. At the time she expected to never see her son again.

After thrashing his wife with a horse whip, Ian had gloated over her dying alone in the wilderness. Cunningly he boasted as she lay unconscious at his feet, "The crows, or birds of prey are welcome to finish the job before morning. If they don't, the freezing night air and your injuries will accomplish what I've started, my *darling* wife".

In fear of being humiliated Gigi had never spoken of this event nor the threat Ian held over her. Nor had she disclosed his aim was to ruin Kendall's prospects for a promising career.

She now knew the reason why Ian hated her. By her not aborting Kendall, whom he detested, he'd taken control of the family finances. Funds were scavenged over time for his own devious purposes which Ian considered rightly his. Should this criminal's Nazi history in the Reich emerge, it would destroy not only their lives but also cause his fall from grace. In revenge the absconder had sworn to enjoy watching his family cringe.

A gentle rap on her door caught Gigi unawares. In turmoil, her mind balanced on the brink of despair. Hearing the knock, in a husky voice she called, 'Come on in, the door's unlocked.' As it swung inwards, she gasped. 'Oh, it's only you, Kendall. I thought it may've been your pig-sticking colleague. Jarvis delights in jabbing my sore bottom with blunt needles.'

'Mum, how can you criticise a man who cares for you intently?'

'Quite easily. Come sit by me.' Gigi tapped the bedspread. 'I've harboured a secret through the years, which no one here knows. Don't argue Kendall and listen. During my early pregnancy in the Cotswolds your father tried to force me into aborting you. On my refusal he threatened to kill our unborn babe if I didn't comply with his horrible demands. Later, he deserted me in my seventh month to consort with street women in Stratford-on-Avon. I believe he conducted himself in the most inappropriate manner. He spent most nights drinking with young prostitutes who exploited him solely for monetary gains and left me almost penniless and in pain.'

Gigi couldn't accept the fact that her husband still controlled the power to destroy Kendall: to strip him of all dignity as a physician and bring him down before his peers. This threat she still feared most of all.

'Mum, I have known about his exploits in the Cotswolds for some time. Why rake up his despicable past by telling me all this now? I don't consider him as being my male parent. Scum born of ignorance is all he means to me, an ugly memory that I filed in a lost memorandum. Bridy means more to me than that Nazi murderer ever could.'

Tears welling in Gigi's sensitive eyes overflowed. Their lilac hues clouded as she spoke. 'Son, your fiancée has stolen a huge piece of my heart. Brianna reminds me of an Irish rose. In the dew of morn her petals unfold and glisten in the early sunlight. Everyone who has the pleasure of knowing her, admires her gentle nature. The softness of her voice soothes the most hardened of hearts. Her personality is carefree, the essence of desire and my greatest wish is to see her in a wedding gown. Now you know how Brianna Skye keeps my whole sorry world from unravelling. Misery borne of ignorance lies in my youthful past.'

Miffed by his mother's description of the woman he loved dearly Kendall could feel tears welling in his eyes. 'Mum, the way you vented your soul and expressed your love for my Irish rose has absolutely overwhelmed me. Bridy will be delighted to hear how you have expressed your adoration for her.'

Gigi gave her son a gracious smile. 'Darling, she knows already. We talked for hours yesterday before she went to pack her little bags. I think she also dreads the idea of leaving you to travel back to her precious green Isle. Brianna is the sweetest young lady I have ever had the pleasure of accepting into my home. No, it's yours and her home with me here on Abergeldie.'

Saved from disgrace by the promises secured from the diplomat and federal police, Gigi knew her fear of living alone in a country she'd grown to love would never be realised. This ill and incapacitated woman desired peace, especially for her son. Free of a belligerent and over-burdensome father, Kendall could now begin practicing in his dedicated profession without the fear of reprisals from the unknown and unseen enemy.

Around three-thirty that afternoon Kendall drove Senator Bucknell and his pilot up through the top paddocks. Standing by the Cessna, with its sole passenger, the engines purred to life, ready to roll off Abergeldie's soil.

On mounting the steps Bucknell had paused. 'Kendall, I'll keep you posted about the situation you've both placed in my care to correct. You do realise that you and Gigi should make new wills? Once everything's correctly documented, I think you should keep all your business and personal documents in a safety deposit box at the bank. Believe me, it can and will save you both a lot of unnecessary hassles …' The deafening roar of two engines cutting in caused him to hesitate. Once the throbbing knocked back Peter finished speaking, '… especially in your case.'

'We shall accept your advice,' Kendall yelled. As the roar dulled and throttle eased off a little, he then took the chance to shout, 'I can hardly wait until Mum finds out why you have all her private documents. Gigi can be stubborn at times.'

'Ha, can't we all!' Peter bellowed in return as the Cessna began taxiing along the improvised runway. 'Oh, I meant to mention. Shortly I may have an unexpected surprise for you all here on Abergeldie.'

'Not a repetitious one I hope,' Kendall called, trying to top the noise of the swirling propellers. *We've had our share of nasty, or should I say Nazi surprises. It's not before time something good came our way.*

Driving back down to the homestead his mind did a jubilant back-flip. Before forging a dry creek bed, Kendall stopped the jeep to read Bucknell's note, he'd left on its front seat.

"Hey matie, I have special news lined up for you and Gigi. It'll be an unusual, though pleasant shock. Knowing your family, it will be a welcome one."

Hell, it's useless dwelling on an unexpected promise that may never eventuate. With our rotten luck I can't imagine Bucknell making a statement which he couldn't substantiate. Still, it sounds an intriguing possibility.

Weaving around potholes and through pockets of dust, Kendall let his mind drift into overdrive. *What has Peter heard? What's he planning now? Maybe the Feds have proof of a positive sighting of that Nazi bastard? He's always been a mongrel, why should I worry if some bastard shoots him. Good riddance to putrid rubbish it will be.*

Whatever information the AFP had procured from illegal and other sources, Kendall decided his mother wouldn't be informed of Bucknell's secret, told to him in confidence. *Why build up her hopes over some unproven gossip? No, it'd be a cruel, repetitious exercise.*

As the jeep slowed near Abergeldie's homestead, swirling dust channelled towards the house. Doused in a plume of fine grit this unexpected drift made Bjorn cough. In desperation to be free of a cloud of red dust, he waved his battered slouch hat furiously to clear the air. His free hand clutching the back door architrave, he repeated the procedure.

'Hi there, sport. It's not before time you showed your ugly mooch around here. I hear your distinguished guest left a while ago. Did that blasted old freight carrier fly out okay? Or did its wheels get bogged down in furrows of powdered silt?'

Far in the distance a thunderhead growled. Splashes of vermillion to tangerine zigzagged across the impenetrable gloom. The ascending rumble escalated well from beyond earth's corona. Heavy purple clusters of edified formations danced across an obsidian sky in tumult, blotting out the daylight. A clump of ugly storm clouds bulked with rain rolled on towards Yass with Abergeldie right in its path.

Relishing this precious sight, Bjorn pointed skywards. 'Have a gander at that. Best sight ever, aye Kendall. Going by the latest news forecast we're in for good steady rain. Wow! Worthwhile waiting for the deluge, I reckon.'

'Yep, but only if this dust doesn't turn into one gigantic puddle of mud,' he agreed. 'Otherwise, we'll lose the topsoil. If we're in for excessive rain, it'll create a moonscape and craters across this unstable landscape. Looks as though Peter flew out just in time! I wondered why he insisted on leaving earlier than planned.'

'Nah, I didn't sport. That bloke's a bushy from way back. Smells the wet coming in, does Bucknell.' Bjorn yanked at his neck-chief to release it and frowning, he laughed.

Everything had grown deathly quiet. Birds began to settle in willows and wattlebark trees, so had Abergeldie and Jalna. A ghostly mist of cobalt green engulfed the ionosphere and descended down through layers of time to inflict its wrath on the distant mountains, shaded in sombre grey on their purple peaks.

'Let's hope the rain doesn't pelt down in bucketloads, or it'll spoil ya mum's roses in the front garden.'

'Where is my mother?' Kendall asked with a solemn smirk. 'The look on your grinning dial tells me that something's brewing. Tell me the truth BJ. What's she up to now?' Not game to enter the house until his companion replied, Kendall hung back. 'We've endured enough nasty shocks to last a lifetime. Things must improve soon.'

'Kirsten says your mum had a terrible premonition. Gigi reckons something horrible will happen soon. She's unsettled. Listen mate, speak to Stephen; he's in with her now.'

Without a word, Kendall threw Bjorn the jeep's keys to shelter it before the storm rode in with fury. Confused, he bounded up four wooden steps and landed on the verandah with a deafening thud. Removing the thick mud from his blucher boots on the metal-pronged raker, he left their soles upturned. The screen door slammed as he reached her bedroom.

Meeting him in the main hall, Jarvis warned Kendall to be silent. 'Shush, your mother's dozing. Her mind is wandering and she's incoherent. Talk to her later, if it's not important.'

With an intense atmosphere building on Abergeldie, everyone within its sandstone walls felt trapped. Gigi had grown morose, edgy and ill. Daunted by the prospects of her son having to leave for Canberra within twenty-four hours made her feel quite insecure.

Begrudgingly she accepted the fact that soon she must relinquish her power over Kendall. Desperate and lonely, she considered herself an unfit, selfish mother by forcing him to remain shackled to the family yoke, burdened with a heavy work schedule.

'Kendall and Brianna must lead their own lives. How can I ruin their happiness, or destroy their love for one another? Happiness to me is a scant memory, crushed by the cruelty of time.'

One afternoon recently she'd confided in Brianna about the sale price of wool falling and how an uneasy calm had befallen her. Although she didn't mention the missing wool clip money, Gigi wondered who could've stolen such a thick wad of notes from her safe. *Perhaps Kendall has decided to keep the theft from me. I've never queried his motives in the past. So, should I begin now?* Actually, his motive revolved around not revealing the pilferage, solely to alleviate his mother's worry.

Rifling through the contents of her personal safe, Gigi couldn't find her valuable jewellery. She then recalled Kendall insisted on keeping all their valuables in his office safe. *Ah of course, I remember my son clearing that safe long before his father left home.*

Gigi sighed with relief. She knew of Ian's acquaintance with the assayed purchase price plus the value of her emerald jewellery. 'Thank goodness he didn't take it. I wouldn't put anything past that man. Not with his Nazi history … or what we now know to be the truth.'

After mentioning the stolen money to Bjorn, he covered its loss. 'Kendall, lucky you lent me 10 000 quid to purchase my new yearling. It came in handy and just in time,' he casually confirmed within Gigi's hearing, not to make it sound so obvious. 'Mate, I'll pay you back next week. Wouldn't you know it, my damn stayer ran last yesterday, blasted nag!'

Not conversant with Bjorn's betting habits nor his horses, Gigi naturally accepted his word as gospel. By strengthening the bond between her son and this Swedish-born, naturalised Aussie it had saved her from tumbling over the brink of insanity. She considered their family's friendship invaluable.

'After all he and Kirsten have done for us over recent months during our ordeals, they deserve our loyalty,' she confirmed at lunch. 'Son, the money owed to us by Bjorn is now paid in full. Write it off our slate please. We still owe them a huge debt of gratitude. One day soon we will return the favour.'

Kendall had temporarily deferred his flight to Paris. *How can I leave for foreign shores when I'm worried about mum's health?* He haggled over having to leave Brianna for a few days before she returned to Ireland. Kendall considered it unnecessary to leave until his jaunt to be interrogated in Canberra. Even the idea of another separation was causing him stress.

In the privacy of her room Brianna challenged him, while he caressed her in a loving embrace. 'Once we're cleared to leave here, you are going on your holiday. If not, I'll go back to Ireland and stay there permanently. Kendall, it's your choice. Make up your mind, and then I'll abide by your decision.'

'No, it's not up to you to clarify this issue, Bridy. For me to leave now is absurd. My friends will be going again next year in the skiing season. I have no qualms in deferring my plans until then.'

'Don't fob me off with ridiculous answers. While you're away, Kirsten will be staying here on Abergeldie to care for your mother. You need this break Kendall, or you'll have a nervous breakdown. Doctors have been known to suffer them, when under excessive pressure. You should know better than me, Kendall.'

'I do not intend to keep arguing with you Brianna. I have said all I need to on this absurd topic. Let's walk up to the cedar tree. Amber could do with the exercise. She's fretting because of Mum's inability to move without pain. Stephen's going to remove her stomach sutures today. Then she'll be able to walk and stand with ease.'

The high humidity made it unbearable to breathe indoors. Feeling uncomfortable in the oppressive noonday heat, Brianna suggested they move to the front verandah. Sitting together on the swing in a cool breeze, they held hands while discussing their future.

'Darling, you need this break. You're quite crabby most of the time, so try to see the logic of this holiday,' Brianna pleaded. 'You'll have a well-deserved rest. And I'll have one without being tempted to ask you for your special love.'

'Bridy, let's not go down that path again. You know I was born out of wedlock. That beastly man ruined my mother's life. And I refuse to ruin yours. Try to understand my side of this argument.'

'I can't even talk to you. You're stressed and are angry with me over little things. We shouldn't be arguing. It's ridiculous and so are you, Kendall. If you don't think I mean to go back to Ireland and stay there, you're wrong.' Cradled in his arms and with her head resting on his shoulder, tears began running down her cheeks. He wiped them on his shirt.

'Gigi made me promise I'd talk some sense into you. She wants you to go; don't let her down sweetheart.' Blackmail under any other name didn't matter, if it worked. With teary eyes focused on his, he gently kissed her forehead. 'Well, what's your answer, Kendall?'

After listening to their little spat, his colleague who had just joined the fray, in no uncertain terms put forward his stalwart notion. 'Will you two rethink your ideas before declaring this another war zone? I think this break will do you both good. If you're sensible and look at things in the proper perspective and with a positive attitude, your time apart will fly.'

'Really Stephen …'

'Don't Stephen me!' Jarvis snapped, then lowered his voice to a whisper so they wouldn't disturb Gigi. 'If you could hear yourself bickering over the damn holiday you need Kendall, you'd feel ashamed. You're tired, crabby and exhausted. If your mother can accept her dilemmas, so should you. I wouldn't consider having you on my staff in this belligerent mood, Doctor. Accept your fiancée's suggestion with grace and go on this damn holiday.'

Fed up with their nonsense, Stephen Jarvis didn't think his intrusion an interference. 'Lower your voice, and have some respect for your mother. She's trying to sleep. The injection I gave her will be useless, otherwise.' Having laid down the law, Jarvis stormed back indoors to finish updating her medical history.

'With my colleague and you two stubborn women ripping into me, how can I resist? You all know I can't leave yet. Not until Bucknell rings to verify my trip down to Canberra. Anyway, this futile argument is premature. I'll go when I can and that's my final decision.'

'Promise me!'

'Yes, Brianna,' Kendall snapped. Exasperated, his hushed sigh sounded like thunder. 'Tell Mum I promise to leave by the end of this month. That's if the Feds have finished querying everything I say. Dorian and my other friends will be in Austria for another six weeks. Please honey, let it rest.'

Satisfied by his vow, Brianna respected his decision as final. She knew it would make his mother feel more content. Unable to suppress her anguish, she sniffed.

'You're a conniving colleen, insisting I must go to Paris then Austria. Therefore, you owe me ...' Kendall didn't finish speaking. Instead, he kissed her long and lovingly and danced Brianna Skye around the front verandah until her knees almost buckled from exhaustion.

Gigi had heard their argument from her window seat. Smiling and still unstable on her feet she quietly, yet groggily edged her way back to bed. Feeling like a voyager lost on a tumultuous blackish-green sea, her aching head slumped on the pillow of dreams.

A wilting sun on the cusp of sleep shaded her rose garden as her son escorted his fiancée up to collect the afternoon mail. At their gates Kendall saluted the local mailman who pointed to their RMB. 'There's a pile of stuff in your drum, Doctor Ross. Somebody must a forgotten to collect it yesterdey. There's also a pot of me wife's strawberry jam. She made it fresh this morning, especially for Gigi. Give ya mum our love; tell her Amy will pop in sometime next week.'

'Thanks, Max. I won't forget to pass your messages on. Gigi's still recuperating after her sudden illness. Tell your wife to make it longer than two weeks, say another month. We'll know by then if mum needs a longer stint in hospital.'

Brianna smiled and frowned at Kendall who pulled an ugly face. On their walk down to the house she giggled. 'By golly Ken, you know how to tell huge whoppers. I found it hard to keep my face still. Did he believe the fibs? His pushbike wobbled as he rode down the gutter.'

Kendall squeezed her hand. 'No Bridy, you've witnessed how sincere country people, landholders and townsfolk are when trouble strikes. Most city folk are interested in their own worries and don't care a damn about cow cockies doing it hard in tough times.'

'Cow cockies? I've seen the pink and grey galahs bathing in your birdbath in this rose garden. What did you mean by cow cockies? I heard Bjorn say Cockie's Joy yesterday. You bush people have a weird way of expressing things.'

He couldn't stop laughing, and squeezed her hand tighter. 'Bridy, a cow cocky is a farmer. We're all graziers, *our* cattle and sheep graze on the top paddocks. All landholders find it hard handfeeding animals in a drought, as we are at present. Cockie's Joy is golden syrup, like the treacle you smothered on your porridge this morning. It's what we lather on our toast, or dribble over date puddings at dinner.'

She beamed back at him, admiration glowing in her eyes and the flushed cheeks reddened. 'I love you dearly, Kendall. I will miss you more than you'll ever know.'

Tired of trying to convince her, he couldn't resist feeling sorry for Brianna. He felt guilty over dragging her down to their depth of degradation. With her ready to leave for Ireland within days he vowed, 'I promise we'll be married the moment I return home. You, my darling mean everything to me. Don't go fluttering those amber eyelashes at strange men while I'm away.' No answer crept through her pursed lips.

Friction caused by the devastating news of his father being an ex-Nazi was tearing this family apart in a time when solidarity was of the uttermost importance. The future looked bleak, uncertain for all those under Abergeldie's roof, and their neighbours on Jalna, the Svenssons.

In absentia von Breusch still controlled and detested his wretched family. Crushed under the powerful spell of his evil by conflicting with one another in torment, Kendall and his mother were unconsciously kowtowing to this conniving Nazi criminal's dominance.

Four days later Kendall stood on the front verandah. Laden with a sorrowful heart, he waved to Brianna as tears blotted his sun-bronzed cheeks. Brianna blew him a kiss through the wound-down window. Miserable on seeing Kirsten's car disappear up the drive he returned indoors.

About to enter his mother's office he heard a loud buzz behind him. 'Damn the phone. I'm not answering it. The blasted thing can ring until someone does. I've more on my mind than worrying over some inconsiderate bastard who might be sticking his nose in business which doesn't concern him.' Kendall continued to grizzle on entering the office.

Trying to find Doctors Ross and Jarvis, the federal officer on duty consorted with Bjorn. 'Sir, these messages have come in from Canberra. This is yours, Mr Svensson.' Politely he passed over the note. 'My orders are to give this one to Doctor Ross, or his colleague. Could you please inform me where they might be?'

'You'll find Kendall on the front porch or working in the main office. I'll give him the message. Jarvis is attending to an emergency in town.'

Clutching the note, Officer Brentnall shook his head. 'No sir, my orders are to give this note to Doctor Ross. If not him … to his colleague. I'll find Doctor Ross; I do know my way.' Courteous and in a refined manner, the plain-clothed federal agent hurried towards his office.

Looking up, Kendall addressed the agent by name. 'Come in Tom. Something has finally come through, by the looks of that note. Hope it's not more horrible news.'

'I've been instructed to await your reply, Doctor.'

Scanning the scrawl, he stated unequivocally, 'Please inform Senator Bucknell everything will be ready. My manager Tom Marden or my neighbour Bjorn Svensson will be in the top paddock to meet the Cessna at five pm sharp. Until then I'll be working here in case Jarvis needs me to assist him lift my mother. Ask the colonel if we can arrange for a live-in nurse to stay here while I'm away overseas. My mother will need nursing around the clock.'

Kendall watched the federal bloke retreat along the hall to his post, then swatted on trying to decipher the context of an unpaid bill. 'Why do businesses employ incompetents to type the wrong information on their accounts? Nothing here tallies.' With his mother incapacitated this extra load had fallen on his shoulders. And he hated being saddled to a desk. 'Shit, I'd rather straddle my stallion Potchkin, than be a damn pen-pusher.'

Bjorn, on reading his note, went to the kitchen to see their cook-cum-housekeeper. 'Mrs G, set three more places please. It'll be a late dinner tonight. Father George has just rolled in from Goulbourn. He's lying down. It looks as if there'll be two extra guests tonight. Oh, by the way, don't mention what I've just told you to either Kendall or his mum.'

'Certainly not, Mr Svensson. I'm not a tattler. This afternoon Gigi has a touch of pink in those peaky cheeks and a sparkle in her violet eyes. Poor dear's just gone to sleep. I've taken her in a fresh jug of ice-cold orange juice.'

'Yeah, each day Gigi's getting stronger. Tell Doctor Jarvis ...' The front door quietly clicked. 'Never mind Harriet ... Stephen's just landed in, so I better tell Kendall.'

Informed of his colleague's return, he met Jarvis in the main hall. 'You look stressed Stephen. Why? What happened in town?'

'I am rather.' A silent curse accompanied his drawn out sigh. 'That sharp corner near Bond's Hotel has claimed another life, a little fellow of four this time. Sadly, I arrived too late and couldn't save the lad.' The lump in his throat restricted Jarvis from speaking.

Kendall, who saw the tears of anguish building in his colleague's eyes remained silent.

'It's high time,' Stephen's harsh frown deepened, 'the council accepted some responsibility for causing all these unnecessary deaths. They're slack. They leave it up to blokes like us to complain. It appears none of them give a damn about the youngsters who ride their bikes on these streets. Kendall, when will motorists realise that speed kills? Some idiots use the straight stretch of our roads as racetracks. After this last horrific accident some nincompoop might listen. With a prominent citizen's son now a statistic, something might be done to upgrade that winding and steep gradience of highway.'

'Tragic accident,' Kendall agreed, observing the pain on his colleague's face. 'Stephen, I'm blowed why those in authority refuse to take notice of medical advice. Perhaps with a little persuasion from the top, they might grant our community a zebra crossing near that treacherous corner.'

He cringed to think of the deceased child lying in the morgue. Having attended to accident cases in casualty, Kendall imagined the deceased child as one of his city patients. He knew it would be useless to put forward his opinion on the matter. *In all probability and with a stroke of good fortune, I'll be in Austria or Switzerland by then. I'll post our damn council a signed letter confirming Stephen's statement and his well-deserved complaints.*

'I'll be talking to Bucknell later in the day, Stephen. A prominent and well-respected citizen, he may be able to swing some weight in the council's direction. He'll soon stir our local cops into action. They have the authority to persuade the State Traffic Department to alter the speed on all road signs to make them precise and more readable.'

'Tell him by all means. Apart from that, *no* Kendall. You've enough to worry about with an ill mother. Think over what I suggested to you earlier.'

Right on five he sat meditating in his office when a sudden blast of the phone made Kendall jump. 'Speaking,' he said nonchalantly, but quickly changed his tune. Quiet and in a transient state he took in all Bucknell said over the line. 'Terrific, things are on the upper then? Gee, the federal blokes move once positive news drifts their way.'

After a brief respite, Kendall listened to the senator. 'Yes, I did tell Jarvis. Of course, I will back him regarding that damn crossing. It's deadly, an accident in the making. You should know that better than anyone here. You prefer me not to mention it again. Your reason sounds valid, providing they begin work on that project immediately. Give me a buzz tomorrow around this time, please Peter.'

Slamming the phone down on its cradle Kendall locked the office door and ambled back to the kitchen where he caught Bjorn happily imbibing on beer. 'Good news for a change,' he bubbled sprightly. 'I'd better tell you this before mum awakens and walks in here. There's positive news filtering through from Sydney. The Feds know for certain where the mongrel is …' Kendall froze unable to say his father's name. 'Well, you know to whom I was referring.'

'Don't say his name. Refer to the monstrous beast as "he". We'll all follow your meaning, matey.' The expression in Bjorn's eyes showed great compassion for his young neighbour.

'I can't imagine that swine ever belonging to your family,' Jarvis stated, on re-joining the men. 'If you let the idiot continue to plague you

Kendall, he's won. The man thrives on making your lives hell. In time, if you do, he'll be your downfall. Besides, you know damn well if your mother hears his name mentioned, it could kill her.'

'Stephen's right mate,' concurred Bjorn. 'Forget he ever existed, otherwise it will tear your heart in two. Each time just pass on the positive news. Forget any garbage that won't be beneficial to the Feds, us and ya mum.'

How easy it sounded for him to say, but difficult for Kendall to acknowledge.

'Well gentlemen, it appears the wretched absconder has for many years travelled overseas under an assumed name and with forged passports. The hideous briefcase I detest has been recognised in transit. I can't think why he would keep it, especially if he thinks the Feds are ready to entrap him. Probably he'll toss it in a rubbish bin on his way to eat in a restaurant. I can't imagine his eating in a dosshouse or an unclean piggery or a café in Sydney's Chinatown.'

What else did Bucknell say? Overtired, Kendall tried to think. His brain refused to function while stressed. Minutes later he remembered. 'Peter told me that an ASIO officer has recently sighted the Nazi absconder in a Kings Cross hotel bar. The mongrel's been seen walking around Sydney with a goatee beard and a small, dark brown moustache.'

'Hey, that description sounds familiar,' Bjorn chimed in. 'I've seen a photo in your mother's locket. A friend took the snap of the swine Ian at their engagement party in Switzerland.'

Jarvis sniggered. 'That means nothing. Don't build your hopes up, Kendall. A man with his scholiastic talents would be a master of disguises. I'm positive of that. He fooled everyone here until he absconded. I disliked his sneaky manner and mistrusted him,' confirmed Stephen whose sense for determining undesirable traits in humans now flourished.

Satisfied with his colleague's opinion, Kendall heartily agreed.

'What's in the pipeline do you think, aye mate?' queried Bjorn, sporting a distinguished frown while parking his bottom on the office desk.

'Constant surveillance of the Kings Cross area, plus extensive searches throughout his known haunts in and around Sydney's foreshores. Everywhere he frequents Peter reckoned. The Feds expect loads of documentation to constantly filter through from their overseas and local sources, particularly now their agents are tapping in the known countries he's visited over time.'

'Shocking mess,' Jarvis sighed, leaning one elbow on the filing cabinet. 'I hope this rabbit hunt's not a waste of our precious resources. From what I can gather, it sounds as though he's gone to ground. His trail may've run cold. It's feasible, not impossible. The criminal could still be hiding in Sydney …' Stephen hesitated, preferring not to name Ian Ross for his son's sake.

'Peter assured me no, it's most unlikely. Apparently, they have an excessive amount of information on his movements all throughout the Rocks in Sydney. I have great faith in his assertions. I also believe everything possible is being undertaken to unearth the Nazi murderer.'

Swilling a gulp of cold beer Bjorn interjected. 'Shush matie, here's the Feds with ya mum in tow. We can continue this later.' In keeping with his smart witticisms, Svensson then suggested, 'Kendall, give ya mum a big hug. It'll make ya both feel better. Now if you'll excuse me, I'm going home. There's a lot of work I've neglected ta do in my stables. Are you coming, sport?' As an afterthought, Bjorn added, 'Gigi might enjoy the short run with us over to Jalna.'

'Okay, that sounds terrific. Mum, you'll be able to have a long chat with Helen.'

'Son, I'm not deaf. I heard Bjorn. Cooped up in a hot, stuffy bedroom wasn't my idea of pleasure. It'll be marvellous to have a break for an hour.' Gigi sighed. 'This place stifles me and I'm sick of being confined to four walls for weeks. This brief jaunt to Jalna might brush a few cobwebs from my mind. Fetch my coat and forget all this fiddle-faddling around.' She nodded to Kendall.

Bjorn tossed Gigi his wife's cardigan, which she donned as Kendall spoke. 'Brianna crept in and planted a kiss on your forehead Mum, before Kirsten drove her to the airport in Canberra. She only left here three hours ago and I'm already missing her. It's incredible to think of Brianna having to live under a cloud of misery since her twenty-first birthday and our engagement party. Mum, I hope in time she will forgive me for dragging her into our mess.'

'She will son. I'll also miss Brianna. This place without her spontaneous laughter and keen wit will be miserable. I suppose she told you, I have promised to behave while she's in Ireland.'

'Good, then it looks like Stephen and I will have a well-earned rest. You know *Mother dear*; you are a handful and do test our patience at times.'

She enjoyed his joke. 'Now I want a hot cupper. I'm thirsty and rather hungry. Helen will put the kettle on and make a pot of tea on our arrival at Jalna. Have your rellies arrived yet, Bjorn?'

He nodded as a scowl inflicted both the men's faces. *How did Gigi know my aunt and uncle are on the way down from Young to Yass?*

A diminutive grin bloomed on her pale cheeks. 'You shouldn't talk with a lady in earshot. Going by that smug look on both your faces, you'll be tucking into something stronger than tea with Sam.'

'I'm a bit hungry myself. A belated lunch will go down well.' Kendall laughed climbing into Bjorn's battered jeep. 'There's roast beef with all the trimmings on here tonight, and it'll be just what this doctor ordered,' he smiled, while stroking his sun-bronzed chin.

'You're a card, Doc. Or should I call you a quack?' Tempted to nudge him, Bjorn laughed. 'Ya need a good nourishing meal in that skinny gut of yours, sport. Since that shocking news broke you've hardly scratched a plate. No wonder ya look like a battle-scarred skeleton.' His tone heightened. 'Now mate, about those horses?'

Ignoring the men prattling on about their stallions, Gigi proudly sat on the back seat of the battered jeep as it wended up Abergeldie's drive to the gates, then swung right to Jalna.

5

'What a hectic day it's been crutching and drenching a thousand sheep,' Kendall replied on the phone to his neighbour as he dozed in his office at four. A sharp rap on the door rendered him awake. 'I'm indisposed, come back later if it's not urgent.'

When the door creaked back a familiar face appeared around it. 'Sorry to disturb you Kendall. I've just been advised the Cessna is due in at six-thirty or thereabouts. It depends on whether they've struck a headwind up from Canberra. I may be wrong there. Anyway, we better not dawdle. Your mother is anxious to know who its passengers might be.' Jarvis sniggered. 'I know and so do the federal blokes who've arranged their flight. It's a surprise.'

Tugging on a pair of strides he snatched a warm coat off the chair beside him. 'Here, catch the keys. I need a shave or I'll frighten a ghost looking this dishevelled. Meet you down by the stables in five minutes Stephen. Why didn't the colonel warn us at lunchtime?'

Ready to leave, Jarvis shrugged his shoulders as the phone rang. 'Don't ask me. I'm only their message boy. They lob orders in all directions and I comply. See you in five, Kendall.'

A hand gesture warned him to stay as he picked up the receiver. 'Okay, if you insist, I'll be here until what time? Six-thirty? Give me a buzz if things alter.' He waved Jarvis off. 'Sorry, it looks as though Lanky will be your driver up to meet the Cessna. I'm grounded!'

Jarvis nodded; disappointed he strode from the office. The jeep bucked as it ploughed through furrowed dust on its way to the top paddocks. With summer on the wane, twilight was the ideal time to welcome Abergeldie's new influx of guests. The aircraft with three extra-tired passengers on board touched down on Abergeldie's soil, right on the dot of five-thirty.

Dawn the previous morning: Her alarm's continual buzz coincided with the hall's antique clock striking at five-twenty. Gigi stretched, and bounded out of bed brighter than the brass buttons on her father's army coat, mothballed in the corner wardrobe. She caught a glimpse of one embossed button through its plastic protector. Feeling revitalised by a hot shower she donned her well-worn riding breeches, a faded blouse of blue cotton, grabbed a pair of old plimsolls and socks. Clad in a warm cardigan she skipped in a flippant way down the front hall. The light in her son's office sent a challenging message to her revitalised brain. 'What is he doing up so early? The cocks haven't crowed, nor have our hens stirred. The dogs are barking at something trivial. If my son's asleep I shan't disturb him. I'm damned if I know why Kendall's tolerated me and my stubbornness during these past difficult weeks.' She sighed as tears blotted her unadorned cheeks. Pushing the door open with her foot, she saw him dozing with a fist full of unpaid bills scrunched in one hand. Creeping into the office she blew a kiss on his sweat-drenched forehead.

Waking with a start Kendall glowered at the shadowy figure standing near him. Natural instinct made him reach for the metal-bladed paperknife with a mother of pearl handle.

Until the dark figure moved towards him, Kendall was prepared to use the knife. 'Mum, why are you prowling around the house at this ungodly hour? You frightened the hell out of me. Are you ill or just plain stupid? Go back to bed. I'll bring in your breakfast as soon as I've finished these blasted accounts.'

She smiled at the grimness of his unshaven face. 'No son, I'm not ill nor have I lost my marbles, as you and Bjorn often say. How about coming for an early morning ride on Potchkin with me? I'm going down to saddle Wallaby Downs now. So much for your inquisitiveness! I have never felt better. No time for mourning what doesn't exist. I have a new life etched on my calendar. I intend to make it work well, by not slumming around here. Put on your boots and leather chaps. You are riding with me. If not, I'll throw a doozy of a tantrum until you agree. You know how determined I can be. I've beaten you many a time, in a race on our horses.'

Astounded by her crass effrontery, he couldn't stop laughing. 'Mum, sit on that chair and listen to me. Have you forgotten the house rulings are yours, not mine? Those top paddocks are out of bounds. You know damn well Mum we've all been forewarned about leaving the confines of this house and property. The colonel, your close confederate in war, has forbidden us to trespass those top paddocks. If you go back to bed and promise to rest, I'll sound him out after breakfast. Until then leaving the house is forbidden for us and our staff.'

'Forget the orders, Kendall. You're not my boss. Well, I suppose you might know best for once. On such a glorious morning let's not waste time. I'll be out on the front verandah reading while you finish dressing in your riding clothes. Breakfast sounds a magical idea. Fetch mine now. Hot bacon, an egg turned easy, two slices of hot toast lathered with heaps of butter is my order. And don't forget a pot of freshly brewed green tea. I'll also have a hot crumpet oozing honey from Bjorn's beehives.'

'Any further orders, Mum? Harriet is cooking all the breakfasts now. I might pass your order on, if you promise not to throw another doozy?' Their spontaneous laughter rang through the entire homestead and echoed down to the kitchen as his sluggish footfalls followed in its wake.

Having dropped Brianna and her luggage at Canberra's airport at ten the previous day, Kirsten decided to keep her dental appointment and then do some shopping in town. After an overnight stay in the Heritage Hotel she arrived exhausted and pulled in through Abergeldie's front gates at five pm.

Walking into the kitchen she tapped on its door. 'What's on the menu, Harriet? The dinner smells delicious. I could tuck into a solid meal. I didn't stop to eat all day or on the way home.'

'Your favourite home Beef Wellington, served with freshly gathered mushroom and roast vegies, Mrs Svensson. There's apple and apricot tarts to follow, but they won't be ready for ten minutes. Why not have a warm shower? Then you'll feel refreshed to enjoy your meal.'

'I might duck under a cool shower. How's Gigi been today? She looked peaky when I left yesterday.' Kirsten frowned at Abergeldie's cook. 'What secrets are you and Kendall plotting?'

'What makes you think he's my accomplice in mischief? His mother's complaining of hunger, so she must be feeling better. Gigi's walking in her herb garden, I think. There's a flush of pink in those drawn cheeks this afternoon.'

Intrigued she looked puzzled. 'Harriet, why are the extra places set on her best damask cloth in the dining room? I noticed the silver and cut glass goblets on my way through to here.'

'Can't rightly say, Mrs Svensson.' Harriet sighed wiping her fingers on a handtowel. 'Your husband asked me to lay two more settings, apart from Father Brady's and our six federal guests. Why, I don't know!'

They must be expecting Rotary guests as well. Kirsten rubbed her left temple. *Bjorn didn't say anything to me, when I phoned Jalna from Canberra, regarding visitors. He must've forgotten and the private party and their guests have been arranged in my absence.* Shuffling her stiff legs to increase the circulation, she hurried out of doors to find Gigi.

Kirsten had just left Abergeldie's kitchen when she heard her husband's car burling down its drive. Instead of entering the house, he assisted Gigi down to the stables. Settling her on an upturned milk pail, he pointed to the row of stalls and whispered something to Kendall.

'Damn and bother, Bjorn thinks my horse is lame. How Potchkin managed to injure his fetlock's a mystery to me. He wasn't limping after I rode him up to collect our mail this afternoon. Perhaps he's thrown a shoe, or loosened it when Corey walked him to his stall. These stables need a good mucking. I'll ask him to do the job. It'll mean ten extra quid in his pay packet and keep the young boxer busy until he goes home. I hope he wins tonight's bout against his Irish opponent in town. The lad is talented and has a prosperous future if he keeps boxing.'

'While you're checking Potchkin's hind fetlock I'll give my gelding these vegies. You're a good boy, Wallaby Downs. You always love munching on the carrots from my garden.'

Without saying a word, Bjorn left Kendall to examine his horse. Gigi turned to feed her horse and missed seeing Bjorn leave and walk back up to the house. In Abergeldie's main hall he paused to speak with Doctor Jarvis who wanted to discuss the influx of their unannounced guests.

Bjorn then strode down to the kitchen. 'Mrs G, could your daughter spare a minute to run a small errand for me?'

'Sorry Mr Svensson, Rose is polishing the bedroom doorknobs, their back plates and bathroom taps. A good rub each day keeps germs at bay. The brass doesn't tarnish and a polish gives them a shine.'

'That's okay. Ta muchly Harriet. You're a good sport. A real treasure. I'm going out to the main cool room; she'll find me in there.'

Stephen Jarvis greeted the new guests in the lounge room. 'Come on in Kirsten and join the throng. Your husband should be back in a minute. He's doing me a favour.'

On entering the room, she couldn't believe the vision confronting her. 'What are you two doing here?' Shaking her head in disbelief, Kirsten glared at her husband as he walked in. 'I'm absolutely amazed. No, I must be dreaming.'

Before she could say another word, Bjorn held a finger to her lips. 'I'll explain everything later. This is an awkward and complicated situation. I can't go into how or why it occurred.'

Floundering for an explanation, she accepted her husband's rebuttal. 'Why have you come here to annoy me? But it's terrific to see you again. My husband should've let me know…'

Bjorn interjected again. 'Come old girl, let these folks rest until dinner. Bucknell organised the gentleman's visa to enter Australia a week ago. I promised Kendall I wouldn't spoil his secret by telling you. This little venture has shocked you.' Giving their male visitor a firm hand greeting, Bjorn took his case. 'Hope you enjoyed your flight to Aussieland. An old sky driver, I'm aware how severe turbulence can spoil a traveller's journey.'

'*Oui* Bjorn, where is Gigi? My precious chérie, is she not well?' The Frenchman bursting with admiration for his host smiled, as his mind focused only on the woman whom he adored.

'She's fine now, Pierre. Unfortunately, she must take things easy. No rushing, no late nights and no swigging a gut full of imported wine. Oops sorry, I apologise for my crude blunder.'

Bjorn's wife frowned at him. This faux pas sounded undignified and she disapproved of him using crude remarks. 'Gigi's in the kitchen at present. I heard her come in a moment ago. Pierre, you'll know when it's okay to sneak in there. Listen for my low whistle as your cue. We led her to believe you were home in Montmartre. Kendall's and your secret has taken a lot to keep safe. Gigi hasn't a clue that we organised this dinner on Abergeldie. I'm warning you, there will be a room full of guests here this evening.'

'Mon Amie, my chérie … she is not working?' Pierre's perturbed look changed to a frown. Aware his accent sounded unusual his creased forehead narrowed. 'My English is not perfect.'

'Heaven's no,' chimed in Kirsten. 'Gigi's reading tomorrow's menu with Mrs Graham. I asked Harriet to keep her busy arranging the flowers I bought this morning in Canberra. Bjorn and Kendall have managed to keep your dual secrets. How, I don't know.' Bewildered by the puzzled look on Pierre Bouvier's face she consoled him and the other guest with a benevolent smile. 'Stay here in the lounge room until you hear my husband whistle.'

Satisfied they understood, she prodded Bjorn's arm. 'I'm returning to the kitchen, otherwise Gigi might suspect something's afoot. We can't afford to spoil Kendall's surprise.' With a brief wave, she abandoned their guests. Pausing a second, he kicked the door closed with his boot.

'You're a sly dog, Bjorn Svensson.' She shot daggers at him, a look to kill. 'You and Kendall have organised Pierre's flight to Australia without letting Gigi and me know. I can't imagine how you or Peter Bucknell could arrange their other guest to be here again so quickly.'

'You're wrong there, Kirsten,' he retorted with a scowl. 'You know I never break a promise whatever the reason. Kendall would have threatened to kill me if he thought I'd let slip about Pierre being here tonight. Now, I'm anxious to get a squiz at the surprised look on his face when he comes in shortly. Darls, keep his mother busy in her room. He'll be furious with me otherwise. We've both known of Pierre's visit for a week. Neither of us knew about the other guest's arrival, I swear. The only message we received from Bucknell was to be in Abergeldie's top paddock at six tonight.'

'Tell me later, here's Gigi now. I'll walk her to the shower. You keep Pierre occupied in his room. We can't afford for her to discover that she has an unrequited guest. Not until everyone is seated at the dining room table.'

Bjorn gestured to his wife, as if to say, "I'll keep Kendall occupied with a couple of cold beers and savouries." On the cool back verandah they enjoyed imbibing. 'Hey matie, Kirsten's promised to give us a hoy just before the tuckers ready.' Suddenly he pointed to a huge ball of dust swirling on the reddish-purple, cloudless horizon. 'Looks like another hot day ready to ride in tomorra.'

'Yep, I agree with you BJ. We need good soaking rain, as long as it's not a torrential downpour.'

'It'll be bonza. Steady rain will break the drought. Fill the lower and top dams, also settle this dust. The cow cockies and desperate landholders hope our sheep and cattle won't die. The poddy calves suffer dreadfully in this stifling heat. More so as their mum's milk dries up. Ya know what I miss most, Kendall? The parched paddocks sprouting rustic buds of sorghum, and sage with its silvery green leaves. And fields of ripening wheat, rows of bright yellow to warm orange sunflowers in full bud. They all form a spectacular sight on these blistering hot summer days.'

Kendall sighed. 'Yeah, now they've all turned a dismal brown, naked stalks with dried husks, and a crazed earth dying for a drink of nature's gift. Rain! God only knows when we'll get a solid downpour.' Kendall pointed to the lower paddock. 'Now fill me in about Sam and Edna. Have they settled in yet on Jalna?' he asked downing the last drop of light ale. 'How are they managing over home? The old homestead won't be the same with its former boss lauding his weight around, giving you and all your staff orders.'

'Struth, I forgot to ask me uncle. Alright I reckon. They might come over later, mate.'

Bjorn's bush humour amused Kendall who frowned with a quizzical smile. 'Good, a visit from your rellies might give mum something worthwhile to rattle off her caustic tongue. Her harsh temper increases and she gets savage at times, especially after she's groomed Wallaby Downs in the stables. Mum really got stuck into me this afternoon, because I'd forgotten to post a stack of letters to *his* creditors in Yass.' Bjorn knew Kendall meant his father.

Exhausted after his long three-hour journey home by bus from a Catholic seminary in the Southern Highlands, Father Brady rested in his room until six-thirty. Hearing strange voices, he chose to mingle with the guests in the dining room, where those present eagerly awaited Harriet and Rose to serve dinner.

A lemon-iced chocolate cake with two candles took pride of place on the mahogany table. Its cut glass stand sat in front of three vacant chairs. Two tall, cut glass candelabras with a delicate gilt-edged candle in each of its silver holders, would create a festive ambience when lit. The sight tended to lift Gigi's sombre mood.

'Whose birthday is it, Harriet?' Entering the dining room Kendall noticed the oval table set with his mother's crystal goblets and best china. 'Are you and Lanky joining us for dinner?' His curiosity reached its climax as he addressed their cook, whose contented smile highlighted her demure, elegant features.

Harriet Graham moved to block his and Gigi's view of the main hall. 'Your mother insists on Rose and me dining with you and the house guests this evening, Kendall. Arthur Hancock says he's busy reshoeing your horse and hasn't the time to spare to enjoy a baked meal,' she uttered churlishly, easing a chair out for Father Brady to be seated. 'Don't be inquisitive. Just you be patient. You might be surprised.'

Harriet lit the candles, three in each cut glass holder on the two candelabras, with a taper. As the wall lights dimmed, firm fingers covered both Kendall's and his mother's eyes. 'Guess who we've invited to dinner? No cheating, and no feeling hands either,' she laughed gleefully.

Kendall could feel the excitement building and his spine tingled. 'Ah, ha, you haven't fooled me, Harriet. I know whose dainty fingers are fondling my neck.'

Setting the gramophone needle on a 78 record, she played a bush ballad. 'A penny to the Englishman with a radiant smile, who can't conceive how this event has occurred.'

In an effort to fool their host, Kirsten edged the offending fingers away and whispered in his ear. 'Who are you kidding, Doctor? You're wrong, guess again!'

'Don't have to Kirsten. I know who wears that seductive perfume, by its delicate aroma. My darling. Come and sit here beside me, Brianna.' Clasping her fingers, he looked up into her tear-filled eyes with a grin which accentuated his high cheekbones.

Gigi, who'd remained silent, sniffed. 'I know who my guest is by his Parisienne aftershave. Pierre Jean Paul, my darling, please remove your fingertips from my eyes.'

As his lips brushed her forehead, the Frenchman's sigh gave him away. '*Oui* chérie, I cannot lie to you. I am here to be by your side in all this turmoil. Now you can flutter your violet eyes at me.'

In the tall ruby Venetian glass vase in front of her were an array of delicate green ferns, amid five long spikes of white phalaenopsis orchids, their throats tinged with pastel pink. As Pierre bent over to kiss her temple, Gigi spied a single orchid in the buttonhole of his silver-grey suitcoat. A neatly tied, tone-on-tone paisley cravat highlighted the pallid flesh of his clean-shaven, unwrinkled chin.

Gently easing her up from the chair Pierre's eyes gleamed. 'Your son promised not to say a word to you, until I'd booked my flight in Paris, sweet lady.' Their eyes met in love. Focused only on each other, they expressed their deep adoration in sighs. A spoken word could never convey true love, in this, their stolen moment of ecstasy.

Hand-in-hand, Kendall and Brianna silently sneaked unnoticed from the room. In the seclusion of her room he almost burst into song. 'Heaven forbid Bridy, I don't know how this miracle occurred. Pinch me darling, I must be dreaming.'

'Ken, it *really* is me! Mr Bucknell arranged time off to meet me at the airport. Somehow customs or whoever rescinded my pass to fly out. I suspected it might happen, I didn't want to build up your hopes,' Brianna confirmed excitedly. 'I'll be here to stay with you for a while. I'm all yours, my darling. My Irish dance bosses have allowed me a month's leave, with the option of re-joining their troupe within six months.'

'Ah ah, I don't think so. Until we're married you are staying here on Abergeldie with my mother and me. Honey, I can't begin to imagine …' Speechless, he forced back the tears.

'Don't go weak at the knees. One hug and a kiss; then we should join the others.'

'Not on your life, my girl.' He kissed her as if she'd been gone months instead of less than twenty-four hours. 'I want to hold you and look at those beautiful green-flecked eyes, Bridy. This is incredible. I still can't believe you've returned to me.' Kendall cradled Brianna in his arms as his lips settled on her bronze-tipped upswept auburn hair and wished their precious moments together would never cease.

'Can we go for a walk after dinner and I'll fill you in with the details? Now I'm anxious to see the surprised look on your mother's drawn features. She didn't know of Pierre's visit. Did you forget to confide in me, before I left here yesterday morning, Kendall?'

'Well, I must confess I did forget! Peter accepted my request to persuade Pierre to book his flight from Orly to Canberra and now you know why. Although, why he or Mum neglected to forewarn me about your surprise arrival back here tonight, is a mystery to me. I love them all, but no more than I do you, Bridy.' Tears of ecstasy in his closed eyes never flowed.

Abergeldie's joyous festivities lasted all evening. Exhausted by over-indulging on a scrumptious meal, Father Brady dozed in his bedroom chair until eight. Quite unexpectedly and looking glum, he appeared at the lounge room door.

'Kendall, will ya pour a drop of me favourite whisky fa a wearisome traveller? We were given a nip with our meals in Goulburn. Ya malt whisky in Australia is like the delicious poteen we brew at home in County Clare. Good vintages ya get from the brewers' market here in Yass.'

Smiles radiated all around. This elderly man of priestly persuasion would never survive without his daily swig of the delectable amber liquid. He'd relish it more often, if possible! His facial pallor showed a marked improvement and the distress in his voice sounded less evident. A week's break at the retreat in Goulburn had worked to stabilise his heart and general health. Time ravaged, his mind often meandered down the horrific paths of a forgotten war, now non-existent.

Satisfied with his drink, he tackled the subject of their wedding. 'Brianna me darlin, when do ya and Kendall plan ta marry? Why I'd be askin, I should still be here on ya wedding day. Perhaps someone could arrange fa me, an old celebrant, ta conduct your nuptials?'

'Father, we wish our special day could be tomorrow. Once we set the date, you'll be the first to know. I'd love you to officiate at our wedding, and I'm sure Kendall will too. I'll speak to him tonight before we retire. Then he can tell you in the morning.' Brianna knew this elderly priest, who originally taught her the sacramental vows in a convent now depended on Kendall's family for support.

'Suppose this nasty problem his father has created must be clarified, before ya both can decide,' he commented, his Irish brogue more pronounced than ever. 'Has Kendall told ya we'll both be off ta Canberra soon?'

'Well, he did mention it, yes. When is still to be confirmed. Don't worry Father George, it could be a week or longer.' Brianna's despondency reflected in her eyes as she sighed. *After his stint in the refractory, Father must be trying to come to terms with his horrendous past. It may still take him an age to forget the harrowing hours spent in the gruelling interrogations. Even to think of having to revive some of the atrocities he suffered and endured in the Berlin Chancellery scares me.*

'Don't bank ya little heart on it, me dear girl. I'd reckon we'll be off ta Canberra shortly. T'will be less than a week, I'd be thinkin, Bridy.'

That soon, she pondered, *I misinterpreted the time and day that his flight leaves for France and Austria. I'll ask Kendall if he or his mother has heard the exact day.* A diplomat's daughter, Brianna chose her words carefully. 'Please don't say a thing to Mrs Ross yet Father George. She can't cope with more worry at present. Kendell can't confirm his arrangements, not until the federal officers have advised him of the correct date, day and time.'

'Brianna, I doona think Kendall wants to leave home until his mother is feeling better. Time is a necessity when illness strikes and it has the potential to ruin all our lives in this damnable world, where people do not respect their friends or neighbours. God plans what we do in our lives and how we mortals cope in difficult and unusual situations. I had to rely on me friends at St Xavier's Church in Germany. The nuns nurtured me broken body back ta health in the catacombs far below their bombed-out nunnery in Berlin. They took turns in bathing me with warm salty water and they revived me spirit ta survive. Captain von Breusch ordered his men ta whip me legs until I couldn't stand unaided. I value what this wonderful family have done to restore me feeble-mindedness. I owe Kendall and his mum a huge debt fa their patience, care and consideration.'

'They consider it their duty to have cared for you here on Abergeldie. They are caring people who spare no expense in sharing their food with the guests, including you and me in our time of need.'

6

6am the following morning: Pierre Bouvier found it inadvisable to discuss a delicate topic with Gigi at the breakfast table. His brief call on the phone, before leaving Montmartre two days earlier was indecisive. By sounding Kendall out, he could determine if his mother might oppose his offer. Until then this French aristocrat couldn't put forward his proposal.

Finding himself with time to spare on such a glorious summer morning, Pierre chose to speak with Kendall in his room. 'There's something I need to ask you in private, Kendall. Would it be possible to go somewhere quiet for an hour?'

'Hop in my car Pierre and we'll take a run down by the river. It's peaceful there and we won't be disturbed.' The concerned look on Pierre's face Kendall twigged might be because of his mother's rejection of this polite man, whom she idolised.

Pierre understood why she refused to leave Abergeldie. Disappointed, he refused to accept her excuse of still being a married woman where the church was concerned. He hoped, in time Gigi would reverse her decision and return to Paris with him or seek an annulment. He regarded this refusal of hers as foolish. *With a little gentle persuasion from her son, she may reveal her true feelings for me.*

Since his wife's death a year ago, Pierre considered they were wasting precious time, which at their age they didn't have an abundance of left. He felt guilty for not being honest with his late wife and Kendall's mother whom he'd flown across the world to visit. In Gigi's stress, he also suffered.

'What is it Pierre that's torn your world asunder, apart from our problem?' Kendall paused, to allow him time to think. 'I have some idea how you must miss Marion. I do know that you loved her deeply.'

'*Oui,* I prefer not to discuss my first wife.' Pierre hesitated. 'One day I will tell you the truth of my marriage. Your mother's ill-health is what concerns me now.' The Frenchman hesitated while trying to assess what his reply might be. 'Do you think she loves me enough to return to France with me, Kendall? I wish to marry her. I'm not sure how she feels. In the distant past we discussed marriage. Lately, she refuses to accept every suggestion I put forward and ignores to respond to me on the phone.'

Meandering under scribbly gums and native frangipani with their spreading boughs entwined, Kendall plucked a blossom. Sniffing the flower, he passed it to Pierre.

'Need I answer you Pierre?' he replied watching him tuck the sweet-scented blossom in his buttonhole. 'You must be aware that she adores you. Mum always has! Put your proposal forward then allow her time to consider it. Marriage is a lifelong proposal, one I'm not sure she's ready to commit to now. Every day she thinks how that beastly Nazi treated her here in Yass. His indescribably evil past and his despicable treatment of Jews in Germany haunt her. It could be an eon before she can come to terms with being whipped senseless with his leather belt until she could neither stand nor lie on a bed without it causing her immense pain.'

Kendall supported the Frenchman's arm to steady him until they reached a cool patch of dry earth. Trying to climb over fallen branches that lay across their path Pierre found daunting. Breathless he pleaded to sit on a log and rest.

'The horrific knowledge of discovering she married a Nazi war criminal has destroyed her faith in most men. His wickedness and controlling power that he still has over her defies description.' In a comforting gesture Kendall touched his companion's arm. 'Logic flew out the window when she learned of his criminal record in Berlin's Chancellery. He lauded his power over me as a child of four, by whipping my legs with his belt. In recent years he tried to ruin my career with fraudulent allegations. In absentia, he'll continue to disrupt our lives.'

Now wasn't the time to reveal the numerous murder attempts by his father on both their lives. *What good will it accomplish? Time is the master of our universe,* Kendall thought. *And in time the Nazi deviate will regret trying to murder Mum in the top paddock, on that bleak winter's night. If or when the Feds find the bastard, I hope they give me five minutes alone with von Breusch.*

Pierre Bouvier clicked his fingers near Kendall's ear. 'I spoke to you a moment ago and you didn't answer me. Why, my boy? Have I offended you in some way?'

'Sorry Pierre, my mind wandered off on a different tangent. We were talking about how and why the Nazi absconder had threatened to kill my mother … I think.'

'*Oui*, we were Kendall. I agree with you how this terrible business has affected her health. I know why stress is causing her illness. Now I think my chérie, she needs me. Why I didn't warn her not to marry your father in England was wrong of me. Long ago I could have prevented her marriage to that brutal man. I resented him. Such cruel eyes, they breed hate. Foolishly, I reserved my judgement and didn't advise Gigi to abandon her wedding plans.'

'Listen to me, Pierre. Neither mum, nor any of us here suspected him of being a Captain in the Wehrmacht Army in Germany. The elderly priest who's staying with my family will verify, first hand, how cruel and evil he could be. Father Brady has photos of the absconder in his Nazi uniform. A member of St Xavier's Church in Berlin managed to snap two miniatures of von Breusch in forty-four.' Kendall cringed over the idea of his father being a Nazi. 'Only his own men could understand his treacherous thinking. He fooled his superiors by escaping from Germany that year. Under oath, I'm not at liberty to explain why. Suffice to say he covered his tracks at the peril of someone who trusted and loved him… a woman in Hamburg. This is now a proven fact, Peter informed me yesterday.'

Moisture pooled in his eyes. Invisible, the pain forbade him to speak. Flopping down on a damp log close to the river bank, Kendall rested his dizzy head in between both knees.

Pierre left him to regain his composure and walked down to the water's edge. Swatting a mosquito, he recalled, *By what he's just implied this ugly business has dealt him a crushing blow. I noticed how the recent traumas are tearing this dedicated man apart. My heart aches, and I remember his innocent little face appearing from under a tablecloth amid a thunderstorm. Poor little sprite, he clung to his mother's skirts, frightened of the lightning.*

Ten minutes later Monsieur Bouvier ambled back to where Kendall sat meditating. 'Feeling better, my friend?' This quiet remark came as Pierre settled down alongside him. 'You must forget all those horrible memories. Dwelling on the past won't bring you peace or satisfaction, only misery. I know because terrible things I've done in my youth still haunt me. You will never find the peace you crave if you don't forget that horrible man.'

'I feel a lot better now, thanks Pierre,' he replied with tears rebuilding in his eyes. 'It should've been you whom my mother married, not him. I know we must all learn to forget the past. By putting all his evil deeds out of our minds, only then can we help my mother retain her sanity. Her health is my major worry now.'

'No, you are wrong, Kendall. She would never have given birth to you, a wonderful son who worships the ground she walks on. Nor would I have my two precious children. Mercedes and Raoul are my inspiration to keep motivated since their mother's death. Their love is all I have left in this dreary life of mine, apart from your mother and you of course. We're all one treasured family. Your past words, not mine.'

'You haven't fooled me for a moment, Pierre. You've known all along I was born out of wedlock … of illegitimacy. It's never been proven and never will, because my mother keeps insisting she destroyed my birth certificate years ago. She is unaware I found it hidden in her bureau drawer. The swine always called her a harlot! *Oui! La cocotte* to be precise!'

Pierre Bouvier assured Kendall that what his father had said was untrue. 'Your mother is and always will be an honourable lady. He … he, that German is evil. I want to hear of his capture. Then he will confront a court hearing for the cruelty he committed here and in Berlin. My chérie, she is a saintly *Madame*. Not a harlot …' Emotional pain prevented the Frenchman from expressing his innermost feelings.

'We should be going home. I promise you, once these proceedings have drawn to a close Mum will come to you without coercion from me and our friends. I'll bring her over to you in Paris myself, if I must.' Kendall assisted him to stand. 'Tomorrow I'll find the time to discuss my wedding plans with you. Pierre, I have a favour to ask of you.'

In the quiet lounge room, he challenged Gigi about not wanting to return to Paris for a brief time and nothing more. She refused to listen to Kendall and ignored his offer to pay her airfare.

'It's futile to even think of going home with you, Pierre. I do love you, but circumstances have changed dramatically since your last visit; it forbids me to accept …'

'Why chérie, when you know how deeply I love you?'

Regrettably, and with tear-filled eyes, she couldn't respond to this man whom she had known and loved all her adult life, now sitting next to her on the lounge. Silence reigned. Coldness in the room increased until Gigi rescinded her disapproval of his marriage proposal. Feeling guilty she cried until he apologised.

'Sorry darling, I know what you and my son have planned for me. I can't and never will go back to Paris. How can I destroy your life, like mine's been ruined by an evil man?' Keeping a stiff mode of defiance, Gigi dried her eyes on his handkerchief. 'Pierre, it would be cruel and selfish of me to even consider leaving here to be with you in Montmartre. I don't deserve your love. Not with a Nazi husband who is still alive somewhere in Sydney. Oh yes, I know he does consort with street prostitutes. Women with low morals, who depend on innocent men for greed, or to keep themselves from sleeping in its gutters.'

Disgusted with her petty attitude, Kendall fumed. 'Let Pierre speak Mother,' he pleaded. 'By being stubborn, you're not being fair to this man who idolises you, especially after he's flown more than halfway round the world to be with you.'

'Kendall, butt out of my business,' she screamed and turned to address Pierre. 'Your family doesn't deserve to have the ex-wife of a Nazi to love and respect. Forget I exist. Darling, go home to France. In time you might find the right woman to love.'

Heart-wrenched and whimpering, Gigi returned to her room and slammed the door in their faces. Turning the key, she ignored her son's plea to come out. All their cajoling fell on deaf ears.

'Leave her to me, Kendall,' whispered his colleague. 'She'll unlock the door if she thinks Father Brady is ill. You know where to find him. Then I'm going to sedate your mother and admit her to hospital. Stubborn woman.' Exasperated, Stephen Jarvis sighed. Infuriated by her belligerent behaviour, he vowed not to kowtow down to her whims.

'I think she has suffered greatly. It is distressing to see my chérie in this sorry state. Would it be fitting if I spoke to her, alone?' Pierre's concern showed in the deep crevasses on his forehead and in the kind way he'd spoken of Gigi.

'No, in this sullen mood she will refuse to listen to anyone. Monsieur Bouvier, you might be offended by what I'm going to say. Phone the airline company you travel with and book your flight to Paris. Leave Australia tomorrow, instead of in six days.'

Doctor Jarvis then confirmed this with Kendall on his return 'I suggested to Monsieur Bouvier that he should travel back to France, tomorrow if possible. Don't oppose me, Kendall. Your mother is really ill and under my care. I have the right to interfere with whatever I deem fit to make her well and her life easier. This is my choice, not yours.'

Kendall nodded in agreement. 'It's fine with me. Father Brady will be here shortly. He's indisposed.'

'Until I'm satisfied her health is improving, I will phone you in Paris. Monsieur, this is a very delicate situation. We can't force Mrs Ross to do something that later she, you or I may regret. Return to Australia when she's stable. I've dreaded her relapsing. So have you, Kendall.'

He just grunted.

'*Oui,* you are right, Doctor. My chérie, she has suffered terribly since that evil man left home weeks ago. It is agonising see her this distressed. I can't stay here until she is well again.' One hand wiped a teardrop from his cheek. 'May I see Gigi before I leave? I must say *au revoir* to her.' Fighting back the tears, Pierre's sorrowful sighs lingered on the breeze. Kendall closed the front door.

Sensing her wretchedness his heartfelt agony had increased tenfold during this tantrum. Pierre had previously confronted her anger in Appenzel. Their troupe of dancers were preparing to leave the airport until Gigi objected. Her stage make-up case and brushes were missing in transit.

'Stressed to this extent my dear lady indeed must be ill,' he gasped.

'It depends on Mrs Ross. You can ask her, Pierre. Although I doubt if she will speak to you. Personally, I think she should be in hospital.' Frankness was one of this physician's finest attributes this Frenchman admired. Feeling desolate, Pierre returned to his room and packed a small satchel.

Watching him walk down the hall Kendall fumed. *Mum's a fool if she ruins this, her last chance of happiness with Pierre. I don't know what makes her tick sometimes. A stubborn woman, she seldom accepts my advice. She might listen to what Father George has to say, especially if she believes he is ill and needs her to comfort him.*

2 pm: Conceding defeat, Gigi unlocked the door. Kicking it open with one foot she looked to see if anyone was in the front hall. As she began to

walk towards the kitchen Kendall and Bjorn restrained her. Defenceless and unable to fight she submitted to them. She didn't feel the slight prick of the syringe prick in her upper arm. The dosage of Diazepam administered by Doctor Jarvis rendered her incapable of moving one foot.

'Gigi, that drug will take a few minutes to work. Once it takes effect on your brain and nervous system you may feel faint or dizzy.'

Kendall carried his mother into her room. Kirsten hurried on ahead to prepare the bed. Helping her to undress, she folded back the top sheet.

Quietly pulling the door closed, Kendall sighed, secure in the knowledge that Kirsten would be staying with his mother overnight. Deeply ingrained grooves on his suntanned brow displayed an element of worry as he joined Jarvis and Bjorn in the lounge room. They discussed hiring a nurse, capable of attending to a patient suffering from a mental and physical collapse.

Neither physician condoned Pierre Bouvier's idea of staying on Abergeldie until Mrs Ross was well. Doctor Jarvis spoke to the colonel who, through his federal boss, sanctioned the trustworthy nursing sister he suggested to attend her in a professional capacity. Jarvis vouched the nurse could be trusted. With this problem solved, it eased all their minds.

8 am: The following day two officers were flown in to relieve their federal counterparts. On their arrival the officers informed Kendall that Monsieur Bouvier would be flown out later in the day. Senator Bucknell had arranged Pierre's flight to Canberra, then to Sydney and also his return flight Orly in Paris.

3 pm: Kendall dropped by his guest's room to say farewell. 'I'll keep in touch with you on a weekly basis Pierre, to let you know how Mum is progressing. Don't take to heart her harsh words last night. In her mental state she doesn't mean anything she says. Underlying her anger, she wishes you a prosperous future with your family. Mum intends to write a letter of apology to you. I can't promise when or if you will receive it.' Giving his godfather a manly hug, Kendall touched his shoulder in a farewell gesture. Wiping a sniffily nose on a handkerchief, he embraced the Parisienne magnate's hand.

Pierre stood back and shook his head. 'Stupid things, colds,' he spoke softly. 'I have no idea where mine originated. Mon amie, I caught it sitting on the damp log with you, yesterday.'

Kendall turned towards the door and smiled. He knew these symptoms well.

4 pm: Monsieur Bouvier, accompanied by two federal officers departed in the Cessna. With a sorrowful heart, laden with regret, Pierre strapped himself in as the aircraft headed down wind in a southerly direction on its flightpath to Canberra.

10 pm that evening: Brianna awoke with a fright and as she turned over in bed, warm fingers began fondling her naked breasts. In the darkened room she assumed it was Kendall's hand. 'Darling, you've relented to love to me. I knew my persuasive charisma would win you over …'

The hand gripped her bare shoulder. As she screamed, its mate stuffed a stocking in her mouth. A second later she felt an erected penis probe her pubic triangle of auburn hair. 'If you make a sound, I'll kill you. Your darling won't come to your aid. Kendall's asleep or he was ten minutes ago. I hated him all through our school days. His mongrel of an old man coerced mine into helping him escape from Yass. Now Dad's scared stiff for his life. Raping you will be the first stage of my revenge. I'll then kill your lover. His mother will be the next one on my list.'

Disturbed by a muffled scream at eleven-fifteen, Kendall struggled out of bed to investigate. In the main hall he noticed a shaft of moonlight streaming in through their askew front door. *Shit, what fool has left it like that?* Hearing a scuffle, Jarvis joined him outside Brianna's room.

'Don't stand there, Stephen. Grab a walking stick from the hall barrel and wait until I kick the door open. Then fetch the colonel and his crew. There's a shocking reason why Bridy screamed. The intruder who wrecked the front wire door is probably still in her room. Pass me the brolly stand. I'll soon make good use of it, by clobbering the idiot over the head.'

Kicking the door hard, all Kendall could see was a bare backside ready to jump on his defenceless fiancée. Swinging the iron stand it struck the would-be rapist's head. Dieter Hoffenberg's blurry eyes bulged and he tumbled off the bed. His head collided with the offside wall. In the most undignified way, he slithered down against the washbasin, clonking his noggin on its metal base. Embarrassed, Brianna pulled up the sheet to cover her naked breasts.

Colonel Jefferson had witnessed the latter part of this incident. Two of his men hauled the offender to his feet and tossed a towel at him. 'Wrap that around you, then get dressed. You're under arrest. There's a constable on the front verandah ready to clap you in handcuffs. Take this wretch out of my sight NOW. Or I might be forced to deal with him myself.'

Alone and frightened, Brianna pleaded with Kendall to remain with her overnight. She feared her brutal assailant might return to finish the job he intended.

'You okay, Bridy? Did that brute hurt you? You're suffering from shock and the strap of your nightdress is torn.' She sniffed. He looked at the dressing table. 'At night you usually put your engagement ring and the opal ring Mum gave you on that dressing table. They're not there.'

'They're safe in the top drawer, with the gold watch my parents gave me for my twenty-first. I nearly left them all there, Kendall. I thought better of it, before I showered.'

Kneeling on the bed beside his fiancée, he cuddled her. 'Don't be upset, darling. I'll sleep in here tonight. He can't return. By now the paddy wagon will be in Yass. First thing tomorrow, the colonel's going to sign an official complaint and a warrant for his arrest. Without revealing our predicament, he'll make sure Hoffenberg is taken, under guard, down to Canberra.'

Dozing in his arms she could feel his heart pulsating. Kendall tried to control his emotions to dampen his sex drive. 'Don't try to entice me, Bridy.'

'Let nature take its course. I need your physical love. Please don't deny me that pleasure.' Tempted to relent, he pulled away from her. Sighing, he dismissed this idea and considered hopping off the bed. 'No darling, I refuse to be the instrument of your sexual desires. My love for you will not weaken nor wither. We won't recap on the reason now. You know why.'

In her gentle, though forceful manner she tried to persuade him to love her. Tenderly, he held her close until she dozed. Brianna lay in his arms until the hall clock chimed midnight. With their arms entwined, she slept until a fine steam of sunlight warmed her face. Resting on her elbows, she looked across at the billowing chiffon curtains and frowned.

Neither his mother nor her aged priest had awoken during the rumpus.

Kendall accepted Colonel Jefferson's advice not to make a fuss over Brianna's close encounter with the intended rapist. He smiled. *Mum would be mortified if she thought I'd even considered breaking my vow of chastity.*

Still with his virginity intact, he crept from Brianna's room, bumping into one of the colonel's aides.

'Have you heard the latest news, Kendall?'

'No, I haven't Bob. Why?'

'Last night, as the paddy wagon pulled out of your drive it collided with a wild brumby. Hoffenberg's head struck the van's door. When one of the coppers unlocked it, he escaped and headed to town only to be collected by a newspaper truck. The driver's okay. Hoffenberg was deceased before the ambos arrived. I spoke to the cop in charge and he wrote it in his Day Book. A clean sheet will save an international incident from evolving.'

'The Day Book is fictitious. It doesn't exist. It's their ploy to help a friend in trouble after committing a misdemeanour, usually a minor traffic infringement. I know that for certain. Book closed!' Sighing, Kendall raised both hands in a gesture of hopefulness.

The colonel's aide smiled as though in agreement.

'Before I see how my fiancée is now, tell me if you heard Amber barking last night, before that rogue intruded on our privacy?'

'Archer and I had finished surveillance duty at midnight, and we heard him wrecking your front wire door. In this hall I heard Amber yelping before she re-entered the house after a piddle and I watched her holding up her left paw as she approached me. Suddenly she stopped whimpering and snarled at something, we assumed was a possum on the roof. I carried Amber to our office and settled her in her basket. Archer, a medic, bathed her swollen, blood-splattered paw and bound her bruised ribs. The blood wasn't hers. Probably from the rapist. Amber must've attacked him from behind on the front verandah and her teeth latched onto his lower left calf.'

'Hoffenberg kicked Amber. She's not long returned from the vets. Bob x-rayed her bruised ribs and the paw, and gave her an injection to ward off an infection. She should sleep for hours. That explains the blood on Brianna's sheets. I knew it wasn't hers. Amber's fretting for my mother. We can't let the pup go near Mum's room. She needs to rest.'

Kendall acknowledged the colonel, who walked from his mother's office. He left the federal agents to organise today's schedule. Kendall collected a newspaper off the hallstand's seat on his way through to see how his mother, and Brianna were this fine summer morning.

7

Queensland, Australia, November 1972: In Yeppoon, a small town on the Capricornia Coast, the wall phone receiver dangled aimlessly in Kurt's hand as he gasped in utter confusion. The shock he'd received caused his knees to buckle. Giddy, he grabbed the bedroom architrave to save himself from falling. Until the faintness and pain in his chest subsided, he could neither move nor relax. He staggered into the hall. A brief rest allowed him to reach the bathroom.

The ghostly apparition confronting him looked quite unlike his own image. Kurt leaned over the basin to observe, through the fog of despair, the distressed look on his features that resembled those of an aged man. The figure in the mirror stared back at him with little knowledge of his pain. Mesmerised by it and feeling ill, he considered the reflection belied his age of sixty-two. Gaunt and sallow faced, it frightened him. *How can I tell my adopted son and daughter this shocking news? They think the man whom we've feared for longer than a quarter of a century is dead. The federal officer assured me that von Breusch is alive. No, it's impossible. I must be mistaken. I could hardly hear him, due to the static on this line. I'll ask our local exchange to send one of their technicians to recheck the cable connection.*

Greatly disturbed by his conversation with the federal officer it held a devastating connotation. Fear of repercussions tore at Kurt's heartstrings. 'That demon once possessed the power to tear the fragile fabric of life apart for those I loved and treasured most. This disclosure will ruin my family's lives, as it did to Anna and mine in those dark days of war in Hamburg, Germany. Under constant rifle fire and in extreme difficulty we escaped, hidden under hessian sacks of grain in a gypsy caravan. My sister's charges, babes of three and four, lay between our feet. We took care of and nurtured those two precious infants until we reached safety. Since my wife's death two decades ago I have cared for our adopted children.' Tears of remembrance flowed. 'I do and always will miss my darling wife.'

His adult children meant everything to this naturalised Australian, all survivors of the German-Allied war. Their love created within him the sense of belonging. They would rely on Kurt's ability to help them over this forthcoming hurdle, as they had done since his beloved Anna had passed on. These two beautiful human beings had idolised Anna from a tender age. Now they remained devoted to Kurt, their adoptive father.

How can I explain to Arneka and Christian that their father is alive? All through the war years they've believed their real parents died in Hamburg, Germany. Kurt recalled how he and his family had narrowly escaped with their lives, with the help of the Urban Underground, until they reached a safe haven. In their terrible struggle to flee from under Reich domination, he and his wife had risked death a hundred times over.

Now on reflection he wondered how he and his family had managed to flee the hated Nazi regime on that fateful day late in forty-two. The only sane explanation was his family were in God's hands. Otherwise it would've been impossible for four people to have escaped from under Nazi noses. Kurt and his family finally reached Sweden. From there, on a frosty winter's night they were taken by submarine along with British pilots and other escapees to England. After a lengthy stay in London, he applied for and was granted visas, then flown to Australia, where they settled in Brisbane. Now a decade later they were located in Yeppoon, a small town on Queensland's Capricornia Coast.

Sweating and in agony he heard Christian arrive home from work, who found him in an extremely agitated state. Incapable of speaking, Kurt's skin looked waxy with dark shadows under deep-brown eyes he stared into a vacuum of emptiness.

Drawn hollowed cheeks, distorted features and eyes twisted in torment peered blankly back from the mirror at Christian. Hauntingly, they appraised him as though he didn't exist.

This eerie spectre of his father's sorrowful plight frightened the young teacher. In all the years of handling youthful addicts of all kinds, Christian Baumer had never witnessed anything like this gaunt spectacle confronting him. Now approaching Christmas and the new year of seventy-three, it didn't resemble the terror they were forced to suffer in the mid-forties.

'Come on, sit down here Dad,' Chris tenderly requested, placing a hand on Kurt's trembling shoulders. 'I don't want to know what might have occurred, not until you're feeling stronger.' Worried, he helped his father to a recliner on the front verandah. 'Where are your heart tablets?'

Kurt's eyes deflected in zombie-mode towards the bathroom, his head constantly moving in a pendulous swing with disbelief. Both his lips parted and a grunt indicated that he'd placed an Anginine tablet under his tongue. In anguish, his left hand angled towards the kitchen.

'You need a drink of water?' Christian saw a crease impinge around his father's eyes. 'Okay, I'll fetch a glass of spring water in that china jug I filtered from the rain overflow in our cement tank.'

Christian knew his father could understand everything, yet for some unknown reason he couldn't speak. Kurt's actions and weird facial expressions caused him to believe something dreadful had occurred during his absence this afternoon. 'When you're stable Dad, take your time and tell me what has distressed you to this extent.'

Settled in the cane chair with its back support upright he relaxed. This position helped him to breathe easier. It took all of five minutes before his awareness returned. In a thick voice Kurt made an effort to speak. 'Chris, I received a phone call from a federal officer in Canberra twenty minutes ago. I still can't believe what he told me.' His tone faded to a whisper.

'Speak slower, Dad. I have time to sit and listen.' While his father took a deep breath, Chris opened the louvre windows. This allowed the fresh sea breeze to waft in through their narrow glass-panes in this small room. The pungent aroma of salt air embraced his distressed lungs.

'I spoke to a diplomat,' Kurt began and a finger urged Christian to put his ear down closer to his mouth 'He's going to call tonight at seven.' As the ill man's fingers groped the air for something stable to hold onto, his eyes again glazed with shock.

'You're quite safe. You're not going to fall. You can't, Dad. Lift your feet up so I can slip this pillow under your heels.' Chris put the cushion under his father's heels, as his fingers clung to the wooden arms of his wicker armchair. 'Seven this evening, you mean? It must be urgent.'

Still holding his chest Kurt, gasped in pain. 'Yes, tonight son.'

'Take small sips of the water then you'll feel better. Don't try to talk.' Christian held the glass to his father's lips. Kurt took two sips then pushed the tumbler away. 'Your breathing isn't laboured now and the colour's returning to your pallid cheeks.'

'I'll be okay in a minute. Just sit here by me. Let me rest.'

'Time isn't important. I have an hour before my next student is due to arrive. Take it easy.'

Five minutes later Kurt regained the power to speak coherently. 'The news came as a shock. What I'm about to tell you Chris,' short of breath he paused, 'please keep to yourself. Arneka must not find out …' With less stress and no exertion, the pain in his chest eased and his breathing returned to normal. 'Sit closer. It'll be easier for both of us then. My boy, what I have to relate is horrific. It will shock you to the core, as it has done to me.'

Christian pulled the footstool closer to his father's knees. Although his voice remained in a dulcet tone he asked, 'Oh, come now Dad. Surely, the news can't be that terrible?'

'Please Chris let me speak while I can.' Kurt's head throbbed and he seemed confused. Sandwiching his temples between cupped hands he tried to prevent the constant thump from growing worse. 'This is the truth son. I only found it out an hour ago. Don't interrupt me until I finish speaking. Remember me telling you both about Erika and her husband …' tears clouded the elder man's eyes which he wiped on his shirtsleeve, '… it appears that von Breusch *is* still alive.'

'Dad, you've told us many times that our parents were killed in Germany late in the last war.' Bewildered, he challenged what Kurt had said. Mystified, Chris queried that notion. 'How could he still be alive? It's more than twenty years since you and Anna escaped with us from their terrible regime.'

Fighting to catch a breath, Kurt continued holding his throbbing temples. 'A Wehrmacht officer, he worked as an interrogator in Berlin Chancellery. A hated criminal.'

'What type of criminal? Please elaborate Dad, when you can. Having studied Reich history, I'm aware of how Hitler treated his men in the field of war, although there's a huge gap missing of those tragic years in Germany. It's something my befuddled mind cannot compute.' Cradling the man whom he and his sister had idolised most of their life, Christian continued speaking. 'Listen to me, Kurt. You're the only father we've ever loved. That Nazi officer is a nondescript memory to us. If you're hurting, remember Arneka and I also feel your pain.'

'A war criminal.' A hand gesture replaced the nod of his throbbing head.

'What! A Nazi war criminal! Ah, of course. Now it makes sense. I vaguely recall him wearing a greyish-green uniform with a swastika armband, which we called the black spider floating in a sea of red. My most vivid memory was being whipped with the buckle of his leather belt. The beatings we received were numerous and cruel. Yes, he did have the tendency of a brutal criminal. Many a time Arneka and I hid in the wardrobe or under our beds, so he wouldn't whip our bottoms red-raw with his leather cross-strap. Ranting and raving, von Breusch enjoyed lashing out without provocation. He never cared which end he used. The buckle left terrible welts on our little legs. We hated more than feared him. If Erika interfered to protect us by blocking his swing, he would strike her across the mouth with the back of his hand,' declared Christian, pouring a brandy to steady his hands and fortify his nerves.

For a long time both father and son gazed out over the Pacific Ocean. Neither man wished to intrude in the other's privacy. Squawking of seagulls, dulled by a raging surf, replicated their pounding hearts. Cocooned in this peacefully cool, narrow sunroom it helped their anguished minds to settle.

As his stress lessened, Kurt's confused and worried state of mind began to calm down. Feeling less pain, the topic under discussion resurfaced. 'Snooping in his Hamburg office, my sister unearthed a dossier, which described how von Breusch had plotted to kill subversive rabblerousers and innocent Jews. Heide referred to him as an evil man,' Kurt solemnly declared. 'In my car, she often referred to him as von Dragon, so you kids didn't know we were discussing him, although in public, I never heard her speak of him at all. Some nights in Hamburg we'd discuss how your mother, Erika loved her "brood", you two babes.'

'We called your sister Heide, Button. Erika referred to us as her mischievous imps. You know Dad, I can still see Heide's pretty face. Her love for us kids outshone the sunshine on a dismal day. I think Arneka feels the same. We joked about how our beautiful-natured nanny cared and protected us from that Nazi beast. Infants and young children seldom see the dangerous side of adults who handle them every day. Fear and love them, yes.'

'How very true, Christian.' With another insufferable pause, Kurt tried to stand. Instead, he fell back on the recliner. 'The diplomat who spoke to me on the phone asked a priest to speak to me. I remember he mentioned having met von Breusch while his thugs flayed his shoulders raw with a whip in his cell within the Berlin Chancellery.'

'Dad, we never classed that evil creature as our father. We were neglected and mistreated by von Breusch. I've known his name for some years. Delving into Germanic war history to assist my pupils, I uncovered why he pushed our real mother Erika down those stairs.'

'Where did you hear that, son?' Kurt charged in anger. 'Tell me exactly how you found out about that incident Christian. Not even my wife Anna knew of it, nor his surname.'

'One stormy night Arneka heard you and Anna discussing him. We were hiding under the kitchen table and you two were angry, but not with us.' Christian's clewed fingers covered his mouth and a thumb rested on his sweaty, sunburned cheek. 'From memory, I think it may have been around the time of Heide's birthday. One of you accused von Breusch of killing our mother, Erika. Recently I read a legal report about how and why he murdered our mother.'

Button, an endearing name Christian and Arneka often called his sister Heide, restirred poignant memories for Kurt, a distressed rehabilitee. Stricken with guilt over her death, he clasped his aching temples in anguish. 'Why didn't you tell me earlier?' he demanded, perplexed at this shocking revelation from his adoptive son's lips.

Acknowledging the family would have to travel regarding this matter, Christian got ready to pack his father's case. He withdrew two boxes from the top drawer of his lowboy. 'Suppose as kids, we thought we might've got a hiding. That beast stripped the belt from his service breeches and walloped our bottoms and legs until thick welts of burning flesh caused us to scream in agony.' This discussion now rekindled his despicable deeds. 'Is it any wonder we hated von Breusch? I curse him every time I think of Erika being brutalised by that egomaniac. The Feds will put a trace on him and may soon capture then jail that degenerate.'

'Dad, what shirts will I put in this satchel? Long sleeves or casual ones? Your brown and black shoes are in here, plus your fawn kid slippers. I'll put your shaving kit and medicine bag in last thing, with your summer robe.'

'Pack four shirts, two cream, one white and one blue. They all need cufflinks. You know where I keep my box. I'll need my memento and tie boxes as well, thanks son. Don't forget to pack my cotton pyjamas. Two pairs should be enough. Put in three sets of clean underwear and four pairs of summer socks, my navy and my dark grey suits. They won't crease in the suit-folder.'

'Everything is packed in this small case, along with your memento and tie box. I'll finish packing your suits and casual clothes in the morning, then they won't be creased.'

Acknowledging his father's distress, Christian rendered an exasperated sigh.

Kurt exploded with the king of all sighs as he recalled Heide having mentioned her Nazi boss's cruelness. Gasping in short breaths, the pains in his chest accelerated until he began to wheeze.

'Chris, neither Anna nor I ever raised our hands to either of you. Heide most certainly wouldn't have. My sister adored you two imps and she would've done anything to protect you from that beastly man. She, like Erika and my wife doted on you children. Son, I don't need to tell you how special you have always been to me.' His explanation accepted, Kurt began to feel relieved and increasingly better. With thirsty lips and a dry and parched tongue, Kurt asked for something warm to quench his thirst. 'Weak, black tea and no sugar, please, son.'

This accomplished the men continued their quiet reminiscences of the past while enjoying a summer breeze. Christian's curiosity continued to build; intrigued he frowned at Kurt. 'Hey Dad, the miniature in Arneka's locket … is it a real likeness of Erika, as you remember her?'

'Your mother was a breathtakingly beautiful woman. I couldn't understand what she saw in von Breusch. Arneka, in a passive mood looks the image of your mother.'

'No, she's too wild at times to resemble Erika. She seldom, if ever, raised her voice in anger, and with the passiveness of a dignified woman, she idolised us both. Often, I visualise her beautiful features as she kissed us goodnight. She knelt beside our beds as we said our prayers. Then she'd tuck the warm blankets around us.'

Christian's harsh account of his sister's personality, Kurt found disturbing. So much so he commented tersely, 'Oh come now son, surely you don't think your sister's wayward. Arneka is a fun-loving young woman. High spirited, I admit. She would never do anyone an injustice.'

'Maybe not! Still, I don't think there's a man on earth who will ever tame my sister, although I love her. Arneka's bright disposition keeps everyone on their toes. Occasionally she lets her tongue run away, if she's displeased. You realise she'll have to know soon?'

'Once we know all the facts, it'll be time to tell your sister. Not until then, son.'

The side door slammed and quick footsteps scurried towards the front verandah. 'Tell Arneka what? You two look as if you've won a dollar and lost a million. No, don't tell me …'

'Shut up, Arneka. Don't you know how to knock? We were having a private conversation.' Christian's livid tone should've warned his sister. It didn't. Still he got in first. 'You usually come in the back door, why the change? Go and put your bike away, it could rain.' A disdainful glare, a grunt and nod were all he received from his sister.

The men hadn't heard Arneka come indoors. Spritely in mood, she would most likely upset her father, by flouting some light-hearted or sarcastic witticism.

Worried, Chris shook his head. *If she doesn't calm down, she'll aggravate Dad's heart condition. I fear he may have another severe turn, especially if she worries him to a great degree. Whatever happens, I need to divert our conversation until Arneka leaves theverandah.*

Only, this woman whose intense blue eyes studied her menfolk, didn't wish to miss a thing. A toss of her long honey-bronzed hair over one shoulder, gave the impression she had no intention of leaving. Arneka sensed the electrified tension in the room.

'Come on give, Chris. Ha, you've got a girl preggers? You're not the wimp I thought you were, aye.' Swaying to a mystical tune, Sister Arneka Baumer all but sang the words.

'Enough Arneka. I warned you to shut up once. Must I do it again?'

A savage glare plus this caution from her brother conveyed a powerful message, one not to be ignored. Her mouth clamped shut just long enough for Kurt to speak.

'Darling, please go and change into your uniform. You're going to be late on duty again. And I don't need matron ringing to discipline you … twice in one week.'

Pulling her aside, Christian cautioned his stubborn sister not to argue. 'Dad's not well, so don't keep on antagonising him.'

'Angina again?' Arneka shuddered, knowing this was the second attack in as many days. She saw her brother nod. 'Come on Chris, help me to walk Dad down to his room.'

Pushing her aside, he sighed. 'No sis, I'll manage, while you change for work.'

Astute in her observations, Arneka came out with her favourite saying. 'Come off it and quit nagging me. Why all the mystery? I refuse to budge until you tell me the truth.'

Kneeling beside her father, she felt his pulse. Rapid and unstable in rhythm it proved his heart couldn't cope with more worry. *Something must have caused his state of anxiety.*

Kurt glanced up at his son, now standing behind him, with a pleading look. 'Don't argue. Let it rest now, please Arneka.' Ignoring her rudeness, he refused to leave the verandah.

'No Daddy,' grumbled Arneka, who'd rudely interrupted. 'I overheard enough to know something is wrong. Your face is flushed and you're going to bed now.' Her hand reached for the phone. 'You either agree, or I'll ring Doctor Phillipe. Suit yourself.' Flinging her hair back in defiance, she meant every word.

'Dad, she won't go until she knows …'

Hesitant at first, Kurt reconsidered the idea of bed. 'Sorry son, I forbid you to say a thing.' Kurt's eyes followed her pen in the process of dialling. 'Arneka please see reason. Once I've taken my Anginine tablets, I'll be fine. The bottle is on the bathroom basin.'

Having ignored his plea, Arneka flicked the next digit then paused to listen.

Christian fumed, and snatched the receiver from her hand. 'I'll tell you. Dad's received terrible news.' He hesitated for a second. Disregarding his father's protests, he continued, 'A man from the Diplomatic Corp in Canberra advised Dad that von Breusch is not dead. Don't pout at me like that. It's true. If that beastly Nazi *is* alive you know what it means, Arneka?'

'A Nazi, did you say Chris?'

'Yes, I called him a Nazi. What else would I call the swine, after the way he treated our mother. You remember the beltings he gave us over nothing, when we were little kids?'

'Well, yes I do, often. I'm sorry. What can I say?' Two fingers embraced his sister's lips. 'No wonder you feel ill, Daddy. I'll fetch your tablets plus a glass of water.'

Interjecting Kurt pleaded with his daughter. 'You're going to be late, Arneka. Let Christian fetch my tablets and the bottle.'

As usual Arneka had chosen the wrong time to be assertive. 'I have an hour before I'm due on duty. If you insist on staying here, say what you intended. Get it off your chest, Dad.'

Kurt narrated to his son and daughter the whole sorry tale, beginning from the day of Heide and Erika's horrendous deaths. Painful though the memories were, he omitted to declare anything that might be detrimental to Arneka or Christian's welfare. The truth had to be told now without any misgivings to ease the burden of being cross-questioned by the federal police and ASIO agents in Canberra.

Taking his time, they listened intently. Neither could believe their adoptive father's story. As it dismally unfolded, they both sat with their mouths agape.

'What a horrific story. It sounds unreal. Why didn't someone in authority shoot the Nazi maniac?' Arneka sensed Kurt's anguish. 'Surely, something could've been done to prevent that idiot's tyrannic cruelty.'

'None of our officers had the authority, nor the power to prevent or suppress his subversive acts of murder. If a member of the Urban Underground, myself included, had disclosed what we knew it would've placed more innocent lives in danger. Without visible proof to substantiate our beliefs, we'd have all become his targets.' Kurt paused to sip the water then put the empty glass on a small table beside his knee. 'Heide and I both knew von Breusch was guilty of his wife's murder, a revenge killing. He indirectly murdered Heide for the same reason. A clever tactician, he killed a friend, a clone of himself then escaped from Berlin to Switzerland late in forty-four.'

Kurt's throat seized with pain, which travelled down his right arm. Quietly, he slipped a tablet under his tongue and waited until it dissolved, accepting a massive headache as part of the cure. He owed it to his children to answer their questions truthfully, knowing they would face endless inquisitions in the near future. This could only be done honestly by bringing them up to date with factual truths, not with superstitions or ensconced fantasies.

Observing her father's laboured breathing, Arneka interjected while taking his pulse. 'The authorities do have undeniable proof that the Nazi *is* alive. How could this be, Dad? You've always insisted he died in Germany, late in the last war.'

'This is no fabrication; believe me it's not, Arneka. On the contrary, we'd never have been approached by the federal police without conclusive proof of his deception to murder his wife and their son.'

Rechecking her father's carotid artery's pulse, she took his temperature. 'You'd think the truth would've surfaced long before this. The war's been over for yonks. Spell it out for us, Dad.' This demand from Arneka again threw his tortured mind into a spinning vortex.

On recovery he confirmed, 'No member of our underground units would disobey orders. We were warned to keep our mouths shut. Otherwise his cronies would've targeted us and all our families. Captain von Breusch was a satirist, an evil individual. He enjoyed torturing those who stood in *his* way to progress through the ranks. He apparently thought …' Uncertain how to continue Kurt hesitated, knowing what he was about to reveal could have tragic consequences for himself and these two young adults.

'Dad, do you think he intended for us to die with Heide? I've often wondered why she died and we didn't. You showed us the photos a couple of times that she took of him releasing Erika's braided hair. Consequently, she fell down the stairs in our Hamburg home.'

Chris pleaded with his sister. 'Let Dad finish, or the stress could cause him to collapse.'

Finally, Arneka saw logic in her brother's reasoning and let Kurt finish speaking. He explained about their trek over rough terrain in dismal weather and how they had escaped with the help of partisans to England, then to Australia. Kurt mentioned the reason why Anna had drugged the babes in a gypsy caravan. This prevented them from being detected by German-Swiss border guards.

'Why can't you admit that beast meant Chris and me to be incinerated with Heide? If you can Dad, it will help you accept the dreadful atrocities you and Anna went through and suffered during the war. Then we may understand the persecution of innocent people and the shocking injustices those Nazis caused all through Germany.'

Kurt silently acknowledged his daughter was right. 'What you must remember is neither Heide nor any of my family were Jews. The Nazis didn't only segregate the Jewish people, or other races to murder. They murdered everyone who crossed them. Rolf von Breusch supported the brutal killings, to boost his authority and power of control in the Wehrmacht just to satisfy his own egotistical nature.'

Kurt accepted another glass of water from Arneka. 'Hundreds died who weren't Jews or even Germans. People never had a chance to defend their families or their principles under Reich dominance. The Nazis enjoyed living in glorious surroundings and they ruled with a fist of iron.'

Shocked over her father's revelation, Arneka cringed to think how they were forced to live under the Nazi and Gestapo's cruel regimes, before escaping from Hamburg.

'Dad, the power they wielded over innocent people sounds awful.'

'Yes, Arneka. In every town the Nazis invaded they captured, murdered, looted and raped women and young girls. They plundered the homes of innocent people. There were some good Nazis, soldiers who detested Hitler's tyranny. A lot of dedicated Reich officers hated their Nazi indoctrination. Back in forty-four I knew a few dedicated and honest Wehrmacht officers who mistrusted Hitler and loathed his cronies. My sister worked for Major Kassell as his children's nursemaid. Viktor was an honourable man whom we both trusted. Heide then entered service in the von Breusch home to be your nursemaid. She was fully qualified and he trusted her. She and your mother Erika became firm friends before that maniac killed them both in Hamburg.'

Christian, who'd remained silent looked at his father in amazement. 'Dad, I know that monster rigged his own alibi to make everyone think Erika had tripped over my toy fire-truck. Consequently, she fell down our hall stairs. We've all seen the snaps Heide took on that terrible day and others she took on our front lawn. Arneka's holding a basket full of edelweiss and Christmas flowers with Erika standing behind her. The photo of our mother isn't clear. Our solicitor here in Yeppoon has all those negatives in his office safe.'

'Yes son. And that's where they'll remain, unless the federal police need to see them.'

'After studying Germanic history, I observed how dominate and impressionable Himmler and his cronies were with Nazi policies. Before being inducted into the military academy every officer and his spouse's lineages were traced back a century and then recorded. I might be misinformed or wrong there. Did Erika have Jewish heritage? If so, that might be why we all became his targets.'

'No, von Breusch hated Heide because she refused to let him seduce her. From what I recall seeing with my own eyes, I know he instigated her murder. A group of Nazi soldiers with bayonets drawn shot Heide. Some of the younger men threw grenades and set her home ablaze. I arrived late and couldn't save my sister, who suffered from epilepsy. From that day onwards, I vowed to find von Breusch, not out of revenge though.' Tears building in his eyes caused Kurt to swallow. His throat became restricted and unable to continue, he cupped his face in shaky hands.

'We were never registered as Jews. Of that I'm positive. So why would the Nazis target or want to kill us kids, Daddy?'

Meditating over a period of time Kurt couldn't give his daughter a truthful reply. Her question came as a shock, one he'd dreaded. *How can I tell her of their Jewish ancestry? I'm not denying or hiding anything, far from it. The lineage of their mother's oma was untraceable history, written only on the pages of time. To be truthful about their great-grandmother's heritage would bring back a ghastly past, one I prefer to forget. Besides, it would make them feel guilty of knowing they were von Breusch's original targets. Not my sister, Heide.*

In silent appraisal of his father's anguish, Christian chose not to speak. Patiently he awaited Kurt's response. Astounded by his silence, he assumed he could be mulling over what he'd previously declared of Hitler and the Nationalist Socialist Party's rulings on Jewish bloodlines.

Reminiscing about the horrific life his family was forced to endure under Germanic rule and its infamous period of evil, Kurt stammered in a mournful droll, 'Sorry son, I didn't hear what you just said. My mind can't focus on problematic situations, not under this pressure and worry. What did you say about our Nazi heritage?'

'Nothing much, Dad! As you know, I have studied the Nazi regime and traced some of their ancestral history back longer than three centuries. Now I realise I was right.'

'Chris, that devious regime traced all their officers' lineage back for over a century. They delved into their spouse's heritage. Believe me, the Wehrmacht were thorough.'

'I'm in a quandary as to why von Breusch only murdered Heide and not us. It doesn't seem feasible somehow. Most Nazi officers had a barbaric streak deeply ingrained in their nature. After capturing innocent people, they forced the junior officers to torture their victims with cruel and crushing implements, like pliers and hot pokers to the spine and feet.'

'I can't imagine what von Breusch had in mind. He may've assumed you babes were with Heide when she died. He targeted her, not you. He murdered my sister because she'd spurned his advances, a revenge killing.'

Kurt felt extremely uncomfortable telling his children, now young adults, a blatant lie. A necessary lie. *There's nothing wrong in fabricating the truth if it saves them anguish and pain. I've seen and known enough suffering and horror to last me more than two lifetimes.*

'It's unbelievable to consider why the Gestapo derived pleasure from torturing their own people, or individuals whom they detested.'

'Yes Arneka, in war time families only survived by stealth. In the Underground, we cared for unfortunates who the SS and Gestapo had persecuted. Children turned against their parents, friends plotted against friends and neighbours. Hitler hated Catholics, yet the records show his supposed lineage came from a poor family with Catholic heritage. It's still a debatable subject put forward by British and Germanic historians. Their differences and debating could continue for a millennium, and it may not be solved for another hundred years.'

Inquisitively, Arneka couldn't resist the temptation to probe her father for the truth. 'So, von Breusch resented Heide to the point of hating her. Now I understand why he hated us. How horrible. To think foreign families are still going through hell, suffering after the horrific war years. I'm sure all Nazis must've been demonic by nature, or most of them.'

Kurt disagreed, adamantly. 'Far from it, Arneka. The footsloggers, or average Wehrmacht soldiers, weren't stamped with the cruelty of those SS or SD beasts. They were ordinary men caught in the hype of a German propaganda machine. Goebbels's baby! Many poor peasants and German citizens believed their propaganda. Everyone, irrespective of their nationality, fell foul to the Nazi brainwashing manoeuvres. The Gestapo men were a breed of monsters.'

Christian shook his head. He found it impossible to understand the Germanic practices and wicked ideals of yesteryear. They lacked finesse. Nor did it fit in with modern day ideology. 'A tragedy of that magnitude must never be allowed to re-occur. As for the Hitler Youth, I hear it's starting anew, Dad. It must be crushed by people of all countries in its infancy.'

Hearing this from his well-educated son astounded Kurt. He felt compelled to ask, 'Are you sure of those facts, Christian?' His closed fist arose as he solemnly declared, 'I hope to never have to witness the terrorising strategies they used against people to gain information, true or false. Nor do I wish to hear of that criminal regime being revived in this or any other country. All dictators are evil. They suppress illiterate and starving people in the cruellest of ways.'

After a lengthy hiatus to resettle the nervous tension, Christian put his hand on Kurt's shoulder. 'Dad, reaching back to forty-two,' his alert mind still ticking over at a phenomenal speed, 'did you ever meet von Breusch in person? It may have occurred after Heide began working in our home as our nanny. You do remember why Arneka and I christened your sister Button?'

'Yes, I remember. It sounds rather cute now, even after all these years. I recall seeing, not meeting him when he'd approached his mailbox one evening. I'd crouched down in Heide's car so that he couldn't see me. Attired in his Dress Greys, he drove out the gates in his Daimler. Reich officers only wore that uniform to official functions. I knew its design well. Heide and I had both seen Major Kassell in his Dress Greys when she first began in service at his home.'

In a moment of reflection, Kurt hesitated. 'Viktor Kassell was a kind and well-respected officer. Don't for one minute class all Reich officers in the same category as von Breusch. Why did you ask about him, Christian?'

'Oh, no particular reason, Dad. If you could recognise him, even after nearly a quarter of a century, why would the federal police want to drag you through hell again?'

'Yes son, I'd recognise him anywhere. I saw him another time. I'd called to give my sister a message from Anna. He drove past me on his way to work in Hamburg. I continued past his home and waited for Heide down the road. I can still see his sadistic smirk every time my eyes close. Women classed him as handsome. Believe me son, those cruel eyes of his were capable of piercing a solid piece of marble. If enraged, his downturned mouth would frighten Beelzebub and haunt the dead.'

A painful sigh emitted from Kurt's lips. *No one knows how I miss my dear wife. Looking back, I think Christian and Arneka also miss her love, attentiveness and her endearing ways. Every night in my dreams she holds out her hands, beckoning me to cross the river of no return.*

Christian touched his father's hand again then waited for a response.

Snapping out of his dream-imbued haze Kurt jumped. 'I must've dozed, sorry son. Yes, the third time I recall seeing von Breusch was in Hamburg. I'd taken Heide her epilepsy tablets to his home on the day he murdered his wife. I saw him very clearly, though he didn't recognise me. Dressed in his regular Chancellery uniform he looked quite dashing. Heide told me once that he didn't know I existed, so I preferred to keep it that way.'

An epoch gone, Kurt Baumer vowed never to forget the man he feared, more than hated. Reliving tormented memories, he shuddered violently. Every hour of every day some vision of his tumultuous past in Germany burned deep within his brain, taunting a subconscious mind. Now every feature, every mannerism came flooding back of the Reich officer whom he had greatly mistrusted.

'Come on Dad, I think it time you rested …' Christian didn't get a chance to finish. An unexpected burl of the phone startled those sitting on the breezy front verandah.

'I'll take it, son. You two keep quiet while I'm talking. The earlier call I couldn't hear properly. A crackly sound echoed down the line.'

The double-ended plugs were working overtime from Canberra all the way through to Yeppoon in Queensland. Listening for some time, Kurt hung the receiver on its wall mount then turned to address his family.

'What I have to tell you must not be repeated beyond this room.' Kurt hesitated until they absorbed what he'd put forward then continued. 'I can't believe it. How could this be?'

'What can't you reason with, Dad? Not more shocking news …' Christian ceased speaking when his father gave a dispassionate sigh.

'Rolf von Breusch has a second wife in Australia with their son. He's a doctor.' Kurt's tears no longer could be restrained. In a torrent they trickled down his drawn cheeks. 'It seems Doctor Ross and his mother are living on a sheep stud in New South Wales; a small town called Yass.' Standing by himself, he managed to grab the chair as his knees buckled.

'Arneka, turn down his bed while I walk Dad to his room.' The burl of the phone again disrupted their peace.

'Pass me the phone, Arneka,' Kurt requested in a quivering voice. Leaning back in the chair, his fingers twitched for her to place the receiver in his unsteady palm.

'Good evening, Matron. Mr Baumer speaking.' Kurt went on to advise her they were leaving for Sydney instantly. He cited illness in their family as the cause. 'I intended to call you this evening. Quite unexpected Matron, I have just received the news.'

Matron Denton, an amicable woman, stated Sister Baumer had six weeks accrued leave. Seeing this was on compassionate grounds her leave would be granted. Matron even placed herself at the family's disposal. Should they require anything Mr Baumer only had to phone.

Relaying the message to his family, he cautioned Arneka to be quiet. Kurt expected her to object over something he'd said to her superior.

Modern Miss must have her say. 'The bloody old bat. Daddy, she's the worst matron I've worked under.' She ignored both his and Christian's offensive frowns. 'I've tried for months to get time off work. Every time she's denied my request by emphatically stating, NO. The stupid cow, she couldn't run a horse to market. She buggers everything up. Yet one word from you and wow!' Her fist pounded the warm summer air, scented with a variety of citrus blossoms and the delicate aroma of frangipanis and wild boronias.

'Dad's not in the mood to argue, Arneka. Why must you always be assertive? Forget that nonsense and help me get Dad to his room,' Christian implored, wearing a savage scowl.

Kurt, in his quiet unassuming way stated, 'You're trying my patience, Miss. I'll have none of it. No respect for your superior lacks decorum. Don't make me angry. I'm stressed and your constant bickering is worrying me. Your nasty attitude towards Matron Denton is disrespectful, uncouth and discourteous.'

'Sorry Daddy. Oh! What's the use,' she grizzled and stormed off to turn down his bed. She seldom, if ever apologised. Her sharp retort had come in the form of an apology.

Stable in bed, Kurt gestured for his son to stay. 'Christian, we must be ready by seven in the morning. A commonwealth car will take us down to the Brisbane Domestic Terminal. There we'll connect with a Cessna I think the diplomat said. That's all I can tell you.'

Annoyed, Arneka growled over the idea of red tape disrupting her holiday. Knowing this journey was inevitable, she accepted Kurt's chastisement. Their hasty departure, it appeared, was vital to federal investigations. At present they possessed the power to bring this Nazi criminal to justice by supplying positive information of their previous life in Germany. She doubted whether she or Christian could remember most of their infant years in that disruptive household.

Warm and relaxing on his heated mattress, Kurt dozed. Discarding the empty syringe their local doctor had used to sedate her father, Arneka crept from his room. 'Dad might sleep until the dawn sun surfaces over those swaying palms. It's a magical night Chris. Eventually, we may be able drift off with this cool breeze on the rise.'

An hour later, after tossing sleep to the wind, she crept down the hall to her brother's room. 'Chris, I can't possibly rest with a lot of garbage rattling through my brain. Come with me to the verandah. We can talk out there, without disturbing Kurt.'

'Actually, I was nearly going to suggest the same thing to you. It's hard trying to sleep in this oppressive heat. Fetch some cold drinks, while I pull those blinds back and open the louvre windows. We'll appreciate this gentle sea breeze with moonlight streaming down over the Pacific Ocean. The palms are peaceful tonight, undisturbed by nature's wildness.'

Together they quietly discussed the day's events. The idea of having a half-brother increased Arneka's curiosity. 'His name and character have fired my overzealous imagination. Who will this young quack resemble … you or me, Chris? I can't ask Dad. He wouldn't have a clue. He warned us not to discuss anything here in this sleepy town of our flight south.' Pensively dreaming of their disruptive years in Hamburg, the missing link to the life he wished they had never endured, Christian let his tired eyes close. The memories of yesteryear continued to pervade a peaceful mind, with waves buffeting the distant foreshore. Suddenly a familiar voice disrupted his serenity.

'Hey, fancy us travelling in a fed's car and a Cessna. Boy, will we be doing it in style this time. Meant to ask you, what time did Kurt say we must all be ready tomorrow?'

'Early. Seven I think, Arneka.'

'The news about that Nazi beast has rocked him. I'm really worried about his heart.'

'Did you mean von Breusch's heart? I doubt if he has any benevolent feelings for us.'

'No stupid! I was referring to Dad's heart condition,' she commented, taking another swig of ice-cold ginger beer. 'I'm worried how Dad's taken this shocking news. It makes me cringe to even think of von Breusch still being still alive …'

Christian interjected, 'You understand the ramifications of this discovery and how the truth of it will affect Dad's health, Arneka?'

'Yes of course I do,' she snarled. 'I hate the idea of that Nazi prig coming back to disrupt our lives. He'll cause no end of trouble.' She stretched. 'Why did we get onto this topic? It's stirred up the hornet's nest in my skull. Uhr, I might as well enjoy this magical evening. The breeze is stimulating.' She almost sang the words. Breathing deeply, she watched silver dots embroider the distant quilt of sky. A horizontal shaft of intense violet, interspersed with vivid splashes of red zigzagging against the deep-indigo clouds warned of a wild storm at sea. Burnt orange moonbeams drifted down as the golden orb painted ribbons of bronze across the now passive Pacific Ocean. Gilded ripples reflecting on the purple sea enticed 'her' waters to be still. Amazed at the vibrant hues radiating from their glow Arneka exuded a contented sigh. This magical scene of tranquillity bathed her in serenity. Palm fronds swaying in the breeze seemed to sense her inner cry for peace.

'It's such a perfect night, I could stand here and dream forever. This is paradise, a place of enchantment, one in which to dream, not of nightmares that haunt the soul, just peace.'

Solemnly entrenched in thought, he didn't reply. Christian watched breakers crashing against the rock-encrusted foreshore, their foam tips rippling in the wind.

The deafening roar of a swift squall disrupted their brief platitudes. Doors slammed, window panes rattled, roof tiles creaked, glass louvres groaned. Birds hushed, seas stirred and moaned under the strain of this wind's fury.

Fleeing their comfort zone, they both rushed in different directions. Arneka headed to Kurt's room, whereas her brother rushed around to make sure everything was bolted. Windows locked, curtains drawn, things brought in from backyard clothes lines, pushbikes were anchored in safe spots under the aged Queenslander. Fortunately, their weatherboard house had withstood the battering of a vicious whirlwind that ripped everything to shreds.

'Gosh that doozy could've blown us all to hell and back.' Arneka frowned. 'I'm bushed, after running around like a dismembered cow. I'm off to bed. Dad's fine, I've just returned from his room. Call me if he needs anything, Chris. Otherwise wake me in the morning around five. I'll need to check and restock his drug satchel before we leave here at seven.'

'Okay sis. I feel a little tired myself, after battling that cyclonic bluster. I'll set the alarm for four-thirty. It should give me ample time to help Dad to the shower. I have packed enough clothes to last him a week or more. My gear can be done in the morning.'

8

9am: After a brief stopover at Yass to collect two more passengers, the Cessna 402's engines refired, its twin props revving into action. The crescendo built to a deafening roar. With their luggage stashed in the cargo bay, Kendall Ross and Father Brady greeted those already on board. Brief introductions were exchanged, belts locked in place as the aircraft slowly manoeuvred into position. When a windsock indicated a favourable wind, the small eight-seater taxied down Abergeldie's earthen runway.

The take-off was comparatively smooth, considering this rain-soaked paddock had deep pockets gouged in its mud-guttered surface. A small bump here and there seemed of little consequence to its seven passengers, none of whom felt the slightest discomfort of bucking wheels. Suddenly, their light aircraft soared skywards with the grace of an eagle in flight.

Doctor Ross buckled in alongside his half-sister and found her a challenge. For some unknown reason Arneka was amused by his solemn expression. She caught Kendall's attention by her smart witticisms and curious stares that never seemed to diminish. An occasional frown came from her brother sitting in the opposite aisle. Christian disapproved of his sister's caustic glowers and speared a harsh look in her direction. At one stage Miss Prim's pout accompanied by a pink tongue also disgusted him. Angered by her brazen antics, Christian looked through the oval window down to the desolate sundrenched country far below.

The sensual aroma of her half-brother's aftershave tantalised her senses. 'May I?' Arneka enjoyed his manly persona, as she clasped his hand. 'It's my first flight in a small plane and I'm a bit nervous.' Her deep blue eyes surveyed his dimpled chin. 'Kendall, I reckon for a qualified quack you're not really ugly.'

'Thanks for the backhanded compliment, my dear! It sounds as though you don't believe doctors can be handsome? You're a nurse aren't you, Arneka? I gather you're speaking from experience.' Slightly bending his knees, Kendall gave her a whimsical smile. 'You're not so ugly yourself.'

Even though she didn't comment, Arneka was astounded that she'd finally met her match in this young physician. She admired his scintillating manner. It appealed to her good graces and she considered that they may become more than just good friends.

Halfway through their flight dark threatening clouds had burned off. A brilliant burst of sunshine made the pilots don sunglasses to protect their eyes from this strong glare. The midsummer sun posed a threat and had a strong bite for an early December day.

Minutes before touchdown, both federal agents relayed to everyone the instructions that they'd received from their superiors. It regarded the protocol all onboard must observe and follow while in Canberra.

Father Brady slept for most of the flight. He knew the procedures having previously travelled on this route with Senator Bucknell. Yawning, he looked around the cabin, then out the window at a pair of ospreys as they winged their way towards an aeroplane travelling in the opposite direction to Young, or another town called Bathurst. He noticed a huge red kangaroo insignia on its tailfin.

'Kendall me boy, what does the logo on that aircraft mean? On me flight here ta Australia I travelled on a similar plane. I never noticed if it had the same markings.'

'Don't worry Father, the Qantas jet won't collide with us. It's set on a higher course than our plane. Heading north, I think it's probably en route to Queensland or further west.'

Strong tailwinds enabled their aircraft to arrive at the capital ahead of schedule. Once cleared to land, the Cessna glided down to a perfect three-point landing on the Canberran tarmac.

This suited Arneka who'd heaved twice. Feeling squeamish she dreaded parting company with the delicious cake she enjoyed eating half an hour previously. However, all the other passengers relaxed with comparative ease on landing.

The Baumer family's limousine was accompanied by a second vehicle with their federal "watchdogs" onboard, which followed closely. It overtook and cruised past their chauffeur-driven vehicle. Designated as their driver at all times, Max kept the car rolling until a third Plymouth slotted into view. Peter Bucknell's driver pulled in behind the two official vehicles and kept at a safe distance. The cortege then battled its way through suburbia and headed west towards their final destination.

Travelling in luxurious comfort, the five passengers noticed how their car braved heavy traffic until it reached the city limits. From there it sped past rural properties beyond Canberra. Kendall observed how the landscape had dramatically changed, which caused him to feel slightly homesick in this countrified atmosphere.

In due course the cavalcade drew up in front of heavy gates, where two sentries stood ready to challenge each driver's credentials and their right of entry. Immaculately attired in conservative grey suits and peaked caps the guards let the cortege pass. After relocking both gates they resumed their watch in the guardhouse beside these towering iron monsters.

As their entourage slowly snaked its way up the long drive, Arneka could see, through a rolling mist, the spacious double storey home set amid forest gums. This grove provided the house plus its occupants and incoming guests complete privacy.

Observing a smoky hue rising between the eucalyptus trees she whispered, 'Kendall look, doesn't the mist swirling among those tall timbers stir a mystic feeling in you? It sure does me. I can imagine within that bluish-mauve atmosphere little folk dancing under that canopy.'

'You and your fairy sprites.' He lauded a hearty laugh which echoed through the otherwise sombre interior of their car. All eyes appraised his smirk. 'Oops, sorry!' he exclaimed. Lowering his tone somewhat, Kendall smiled. 'Never looked at the misty sunlight in that way before. I suppose the ghostly shadows stir the surrealist within me. On our return home to Abergeldie, Bridy and I'll take you for a run through our ghost gum forest. Delicate hues of a eucalypt mist, similar to this one rising from the forest floor, will astound you Arneka.'

'Bridy! You spoke of her before, Kendall. Is she your girlfriend?' Arneka's curiosity grew to an overbearing pitch and she'd wanted to ask this question since leaving Yass. Envious, the green-eyed dragon's fire inflamed her jealousy as she appraised his angular features.

'More than you'll ever know. Brianna O'Shea is my fiancée. As I've said, you'll meet her later. You two will become good friends. Her nature is similar to my mother's and they both have endearing features. Mum's a tease like you, Arneka. You, my precious sister,' he squeezed her hand, 'will feel at home with my mother. I know you'll love Gigi.'

Curious about his mother being called an unusual Christian name, Arneka wondered if their family had a French connection by birth or some heritage-listed affiliation.

Kendall pointed out to her different breeds of cattle resting in the shade of peppercorn and weeping willow trees. Rivulets of spring water broke through a crack in the parched earth, as sprigs of tufted grass and dried bulrushes bent their brown heads towards a young noonday sun.

'What a pity Mum and Bridy couldn't have come up to meet your family, before the Cessna left Abergeldie. Never mind. They'll be anxious to meet you all on our arrival home.'

'Probably I'll make a fool of myself down here, with the Feds sticking their noses in our private business. I'm not used to highfaluting ways of snobbish people or smart mouthed cops.'

'Shush! The chauffeur will hear you. He's one of the highfaluting Feds', Kendall whispered. 'Try to be diplomatic in company. Otherwise you'll embarrass your father.' Another tender squeeze of her hand by Kendall comforted Arneka. 'I'll be in there with you during their inquisitions and the long hours of endless questioning, sis. This case will climax with a positive outcome and soon these stressful times will be forgotten.'

Looking directly into her misty eyes he saw terror lurking, a demon ready to pounce. His heart ached for the youthful woman beside him. Gently lifting her fingers to his lips, Kendall endowed their tips with a kiss. She gave him a languorous smile, laden with fear.

'Remember, you must be strong for your father. Kurt needs your compassion and strength more than ever now, Arneka. Remain positive for his sake. His features are drawn and Kurt looks exhausted after the rough flight down from Yeppoon, due to stress probably.'

Arneka pouted, yet she acknowledged his advice as sensible. Flipping the long strands of hair over her slim shoulders, a couple touched the drowsy priest's forehead. Running a couple of fingers through her honey-bronzed fringe, the fine filaments settled back in place.

Slightly disturbed, Father Brady grunted and like everyone onboard gathered his things ready to exit the government car. Doctor Ross supported him down its steps and over to solid ground.

Mrs O'Malley the housekeeper, a woman in her mid-fifties, greeted the five weary travellers on their arrival. She and two of the security guards made everyone feel welcome. Clad in grey uniforms, the men's coat-tails flapped free in the breeze as they greeted Kendall. Sedately returning the salute, he noticed weapons in their shoulder holsters of brown leather.

A country girl at heart, with everything Arneka had witnessed so far, she imagined scenes ready to jump out of a spy thriller. Affable, though bearing haughtiness she, in her short, soft grey Irish linen suit, considered herself the leading lady in this drama. Predominate in colour her gentle moss-green cotton blouse made her feel superior to the guards whom she encountered in their sedate attire.

Head held high and in stately grace, Sister Baumer brazened past her troops and counterspies dressed in their conservative suits. Floating down to earth mentally, Arneka hesitated before entering the private domain she now considered her citadel.

Two of her male confederates, Kendall and Christian stood aside to allow "Mata Hari", their "Lady MacLeod", the right of way. They both nodded as she paraded past them and continued down the wide rose-carpeted hall. They smiled and looked at each other while awaiting the chauffeur to collect their heavy luggage and incidentals from the limousine.

Meanwhile two security guards from the second diplomatic vehicle assisted Father Brady and Mr Baumer to alight. This group of four then followed the three officers inside the lavishly furnished home. Its crystal chandeliers caught Arneka's eye as she passed the concierge's desk.

'I find your sister enchanting, Chris. She's a woman who likes to have her own way,' remarked Kendall; carrying their coats and small suitcases, he observed her haughty stance.

'Your judgement's spot on, Doctor. Never get into an argument or a confrontation with her. Arneka is inclined to be stubborn and she wins hands down every time.'

'For a moment there, I thought we were on an official name basis. It sounded odd coming from a relative. Maybe you haven't realised that we are akin to brothers.'

'Well, sort of, I agree. We're stepbrothers, a bit different in age, if I'm not mistaken?'

'Chris, we have one parent the same, so we're actually half-brothers and you're over a decade my senior.' Kendall didn't say which parent. He imagined Christian had also accepted their birth-father as a stranger with an international warrant out for his arrest. *Sometime soon, with God on our side, that Nazi bastard will be captured and hopefully indicted for his criminal past and war crimes.*

Once the formalities concluded the guests were shown to their respective quarters. Kendall's luxurious suite was situated on the first floor, two doors along from Christian's room. Arneka observed her small private apartment with pride. Located on the ground floor, it overlooked spacious grounds and hedged gardens. Entrenched in her castle, she showed her delight by dancing around its spacious lounge room. She adored her new surroundings, and wished she could be closer to her father's bedroom. She expected her brother and Kurt's rooms might be conjoined. *What does it matter, as long as I can attend to his nightly medication?*

Father Brady, also billeted in this hallway, found it hard to relax in his tiny room, little more than an alcove. Sitting on the bed alleviated his elderly legs of stress by creating excessive exercise. He noticed a flight of stairs leading to the mezzanine between this and the first floor on which a small annex was located.

Three of their guards were stationed at base camp. Housed in rooms adjacent to Kurt's, they could survey everyone who might intrude on their privacy. Situated between his and the priest's rooms, Peter Bucknell began to settle in his suite of four rooms, one his private office, when a thud echoed from a room upstairs. *Damn, I need some peace to study these files.*

Shannon-lea, the house where they were staying, had a housekeeper and her husband who were the caretakers. They resided in a clean flatette which butted onto the staff quarters for easy access to dual kitchens, plus front and rear entrances. The remaining two guards were permanently stationed in the attic. From there they could patrol and protect the roof area from unwelcome intruders.

Nothing was left to chance regarding their guests' safety. Everything was organised to a strict timetable and rules must be adhered to while under this roof. When time and conditions permitted, all these severe restrictions might be wavered, slightly.

For the remainder of their day, this group consisting of the Baumer family, Father Brady and Kendall Ross settled in. The layover for their entire time in Canberra on this secured property, twelve kilometres from the city outskirts, would be limited. This rural retreat located midway between the capital and Cooper's Crossing could be accessed from all the vintaged towns.

Several of the guards who'd accompanied the passengers down from Abergeldie, hovered around their charges like a mother hen watching over her brood.

With familiar faces on site, it lulled each guest into a passive mode of security. Unknown to outsiders, this move had been instigated through ASIO, at Bucknell's insistence. Peter knew everyone would face intensive interrogations by ASIO. A similar organisation, the AFP plus other security agencies were involved in tracking down von Breusch. All interrogations were due to commence the following morning at ASIO headquarters, located in their Canberra building.

Tight security was in place everywhere the group of five stayed, went or travelled in this capital city. With this stratagem in operation, strict security must be maintained until all the guests were ready to fly home to Yass. The small township lay approximately a hundred kilometres south-west of their present location in Cooper's Crossing, along the Bungendore Road leading to Lake George. This countrified town with three hundred inhabitants recruited young people of all nationalities to work on their rural properties, or in Batlow's industrial area manufacturing canned fruit and fresh vegetables which were supported by local businesses.

After an early dinner, Peter Bucknell casually strode towards his friends, sipping cocktails and relaxing in the downstairs lounge. 'Kendall, I suggest you all turn in early. Tomorrow will be hectic. You'll need to be fresh minded, to comprehend questions lobbed in your direction.'

'Care to join us, Peter? You ought to try and relax a little more.'

'Unfortunately, I can't Kendall.' Bucknell thanked him for the offer. 'I still have a pile of statements to rummage through before dawn. I'll turn in once they're finished. Please see the elderly gentlemen retire at a reasonable hour. It'll be hard on them from tomorrow on. Sleep will be their only respite, once the federal blokes commence their gruelling inquisitions.'

'Will do Peter. Yeah, I reckon it will be hard on all of us. Facing an horrendous past won't be easy. I dread hearing what cruel escapades that criminal accomplished during the war years in Germany. Kurt might not be capable of listening to more tragic incidences related to his family's struggle to escape that terrible regime in forty-four. His nerves are shot and he's on the border of total exhaustion. It won't take much to push him over the precipice of insanity.'

'I understand his plight, Kendall. It'll be interesting and from what I surmise will be a heart-wrenching and gruelling time for all those concerned. The same stringent conditions still apply here, as they did on Abergeldie. Good night and hang in there, my friend. You need a good rest, so please don't ignore my warning.'

'Peter,' Kendall's fingers rested on the diplomat's grey coat sleeve, 'what time will we be awoken in the morning? I must give Kurt and Father Brady their insulin jabs before breakfast.'

'Seven for us,' replied the senator. 'Then once I've gone over the fine details with you Kendall, you can attend to their needs. Perhaps you might give the young Baumers a wake-up call at six. Be friendlier coming from you rather than me or a security officer. Not so official.'

'Christian can assist his father to shower, while I attend to Kurt's rheumatic and asthma medications as prescribed by his own practitioner. After which I'll administer an injection to Father Brady to keep his asthma and arrhythmic heartbeats stable, as per Jarvis's instructions. I mentioned this in case something unusual occurs in this lonely quadrant of Aussieland. Then no one can blame me. An insurance policy of sorts, it will secure both my medical diploma and my private life from being bought into disrepute later, by slander or malicious lies. It has been known to happen in my profession.'

Bucknell nodded. He knew the damage done by gossiping tongues. 'First thing, I need to confer with my driver before we leave at nine. I'll see him early to discuss our altered route through the city. Oh, from tomorrow on the vehicle you'll use daily has tinted windows. A mere precaution to keep you all safe from nosey reporters, who delight in preying on the vulnerable.'

With the exception of Kendall and Peter Bucknell, who briefly covered new procedures they must observe, the remainder of their group had bunked down by eleven. The elderly priest and Kurt downed their medications before settling in coke-panned warm beds.

As per security instructions only the family members on Abergeldie were informed of their progress. The federal officers staying at the homestead, under threat of instant dismissal refrained from divulging this group's whereabouts. All outside contact to and from Cooper's Crossing must be sanctioned by Canberra. These strict rulings had emulated from the Federal Minister of Internal Affairs, through his secretary.

The powers above were determined that no leaks would filter from their Canberran source, not while the absconded war criminal remained free. Surveillance of his family, friends and known associates; the latter scant in reality, would at all times be imperative.

Constantly filtering in from various sources, the data just received was enough to warrant further investigation into von Breusch's current movements in and around Sydney and later Switzerland. Without bugs lurking within the system all federal agencies were confident that within days von Breusch would fall into one of two traps previously devised. Other reports of his whereabouts had drifted in from independent overseas agents through their contacts working in two Swiss Alpine chalets. Independent members of the Jewish Mossad had joined the hunt for this criminal, a murderer of Jewish prisoners, their fellow countrymen and women.

Around eleven that evening, a sharp knock rebounded off Bucknell's door. 'Come on in, it's not locked. By that gleam in your eyes Archer, it looks as if something positive and important has come through from headquarters This is the second time tonight they've contacted me. The previous message I was answering on my computer until I heard your knock.'

'It sure has, Senator. I've just decoded this one from Canberra. I think you should read this communiqué before retiring, sir. I've been instructed to await your reply.'

It took the yawning senator a minute to read. 'Right, I'll relay this news to my boss first thing. Other than that, I'll pass the news to Doctor Ross now. First, will you go down to the kitchen and brew some strong coffee. I think we could enjoy drinking something hot to celebrate this good news.' Peter waved the note vibrantly. It fluttered out of his fingers and landed on an ash-smothered leather boot.

An embarrassed, sniggering officer swept past the grubby intruder doused from head to toe in grey soot. He pointed over his left shoulder, said nothing and disappeared.

As the soot-cloaked figure entered his office Bucknell gasped. 'What and where the hell have you been? Here, tread on this paper. Don't step any further.' A large sheet of newsprint spread itself on the carpeted floor, as the figure shook ash everywhere.

Black eyes poked out from under a beanie, followed by even blacker ears. A loud splutter echoed through the room. 'Sorry Mr Bucknell. I couldn't wake Dad or my brothers so I headed in here. Please find me a towel, or I'll wet my pants. Being smart, I overbalanced and fell down the library chimney. That damn roof needs mending. There's a dirty big hole in one of its tiles.' Wiping her soot-coated eyes, Arneka peered through ash-laden eyelashes at the misty figure splitting its sides with laughter.

'How am I going to confront Kendall with you, a scraggy bag of soot?' Bucknell tried to control his mirth while watching a rather bedraggled Miss Baumer trying to clean a pair of smudged cheeks with his damp handtowel. 'Arneka dear, don't move. Stay where you are, while I fetch a bowl of warm water from the bathroom. I'll leave my robe on this chair for you to change into, after you've dried yourself.'

Half an hour later a feminine figure emerged from the senator's quadrant of rooms. Quite unabashed and not a bit embarrassed after her misadventure with broken tiles and a fall down the disused chimney laden with filth of more than century, Arneka shivered. Backing out of his office, she waltzed off to her citadel, five doors down the hall. Wearing only brief panties underneath, on the way her borrowed robe flipped open. Flying through the cool night air, it landed on a guard's head and with her nose pointing skywards, Mata Hari re-entered her room.

Amazed, Bucknell braved the guard's growls and rapped softly on his neighbour's door. 'I'm awake, come in Peter. I recognised your familiar knock. More news?' Kendall looked at his wristwatch. 'If so, it better be good at this time of night.'

'Sort of … it's the best news we've received in twenty-four hours. This communiqué has just come through from Switzerland. No, I'm wrong there.' He yawned. 'It originated in Zurich and ended up in my Canberran office. There have been several positive sightings of …'

Kendall sniggered. 'You may as well have said von Breusch. I don't and cannot think of him as my father. He's just some revolting idiot to me.'

'Fair enough, sport. My secretary informed me of his location earlier. Actually, this communiqué came through from ASIO at four this afternoon. As per usual, red tape ties up everything down there. The moment I hear more of his movements I'll seek you out, even if it's at four in the morning. Otherwise, I'll see you at breakfast. Sleep well on that news, Kendall.' As an afterthought he stated, 'I entertained a strange soot-doused visitor an hour ago. Ask your sister why she delights in frightening the guards in an intrusive manner.'

As his door eased to a close, heavy footsteps in the corridor faded. Kendall couldn't help smirking. 'What mischief has Arneka done now? Good humour goes a long way. Bucknell's effort of mimicking Bjorn sounded weird, yet terrific. He imitated BJ's drawl right down to his last breath.' Still thinking of this wacky portrayal of his sister's staged skit, Kendall climbed back under cold sheets.

Working in conjunction with foreign agencies, a Mossad agent had recently proposed – in the event of von Breusch's capture – a war trial should be held somewhere in Austria, or perhaps Vienna. The latter looked promising. A reliable city, their judiciary would conduct a fair trial without fear of persecuting families still recovering from their experiences in German or foreign prison camps. Beginning in March 1940 until later that year an unknown number of professional tutors, physicians, spies and literary scholars had been systematically tortured by the Gestapo or NCO's. The non-combatants had followed orders to annihilate everyone and anyone who defied, or denigrated Hitler and all those loyal to the Reich.

In those tragic years all throughout Europe, hundreds of defenceless people had accepted Nazi propaganda as gospel. Gestapo thugs persecuted or murdered people who defended their right to freedom, their homes and families. Seldom did an honest German officer need to slither out from under a viper's nest to protect the starving victims of their tyrannical regime.

9

Morning came in with a flurry of rain as predicted the previous evening. Not wishing to spoil her outfit worn yesterday, Arneka chose a navy suit with contrasting accessories.

Unknown to Miss Baumer, her half-brother noticed the collar of her pintucked blouse of lime silk, puckered at the nape of her neck.

'Stand still Arneka, until I fix your collar,' Kendall said standing behind his sister. 'I think your locket chain has caught on this fine fabric. Oh damn, now the clasp's sprung open. Take your pendant off please darling, so I can check it.'

'This would happen now when our federal "watchdogs" are pressuring us to leave. There, it's off. Be careful of my precious photo encased in that tiny gold-plated insert.'

'I will. Now pass me your locket and I'll assess the damage.' He frowned while checking the chain. 'Hey, there's something under this photo. It feels padded. I think it's another miniature.'

'Couldn't be! I've had the locket for yonks. Surely, I'd have noticed it before now. Pass it back to me Kendall. My long nails are stronger than yours.' Arneka freed the heart-shaped support and dug her fingernail under it to expose the photo. 'You're right. That's Erika with my … our …'

'Don't upset yourself sis; I know who you mean. These photos will be invaluable to prove what his evil intent must've been just before he killed your mother. Here comes Peter Bucknell; you should let him have the locket and both those snaps.'

'No, now you've disclosed Erika's secret, I'll treasure those photos. Look, that's Heide's photo. I couldn't possibly part with my locket or them. The Feds might mutilate both my miniatures.'

'What photos?' Bucknell asked joining the couple in discussion. 'I must say they do look interesting. Who are these photos of, Arneka?'

Kendall spoke and pointed to each picture in turn, while noting Senator Bucknell's interest in appraising the photograph of Arneka's real mother, Frau von Breusch.

'I vow your locket and miniatures will be in safe hands. Nothing will happen to the originals. Once our lab technicians have copied them, I promise to return these to you, dear. It'd be advisable to have this small one enlarged. I can assure you they won't be damaged in the process. It might take quite a while before I receive all three miniatures.' She removed the one of her birth-father and passed it to Bucknell who erred on the side of caution. To mention von Breusch, could have a detrimental effect on both these young people. 'Technology these days is marvellous. My colleagues in our photo lab can have the man totally obliterated and your mother's face enlarged to fit a normal frame, if you wish, Arneka?'

'Oh, that would be super, tar muchly Mr Bucknell. I don't want the Nazi's moosh in my locket. Err,' Arneka shuddered, 'it gives me the creeps to think I've worn his smug dial on my chest all this time.' She shivered, contemplating what her mother Erika, would say or think, if she could see this photo of the man they both feared, even hated, bobbing up and down on her breast while on her pushbike.

To bolster her spirits as they watched Peter hurry back to his car, Kendall cuddled his sister. 'My, you do look quite chic today. I thought you looked pretty swish yesterday. Arneka, that outfit would turn any man's head.' Silent in his further appraisal of her, he readjusted the collar of her blouse and gave the nape of her neck a brotherly kiss.

'Do you really think they'll give me, a frump, a second look in their office? I hardly think so Kendall. I usually wear conservative clothes and no make-up, but today I chose to be a little daring … in a flamboyant sort of way.'

'Autumn tones suit your delicate skin colouring and honey blonde hair. I like it swept up on top of your head. It looks terrific.'

Kendall stepped back a little to fully appraise her outfit. *Mm…mm That hairstyle would suit Bridy. I must take a snap of Arneka before we leave for town in the Fed's car.*

'Thanks a ton, brother dear. My real mother and Kurt's wife, Anna, called my hair soft bronze. Good clothes maketh a woman look dignified, even mysterious. Wouldn't you agree Kendall? Sets the men apart from little boys who pass remarks, like you just did.'

Observing her upswept hairstyle, he politely murmured, 'You're a cheeky miss! I must ask Brianna to sweep her hair up like yours. Yes sis, I know that style would suit Brianna.'

'I love you calling me sis. Breaks the formal, tartish word you used yesterday. What's she really look like? You've intrigued me. Now, I'm itching to meet my future sister-in-law.'

'Brianna's pretty, in an Irish sort of way. Her complexion is pale with pinkish cheeks I love to kiss. Her long hair flows down over her shoulders. It's curly, but not over curly. Her face images her beautiful nature.' Both his hands wavered in an hourglass formation. 'Her figure is shapely and slender. Not skinny. She is a lovely girl, one who dotes on yours truly.'

Arneka pouted. 'Now you're making me jealous.' A look of envy displayed her fine features to their best. 'Wish I had a nice feller like you in my life, Ken. Don't frown! Must I call you Kendall all the time? Sounds stuffish, something like Chris would say. In a good humour, his snobbish attitude is tolerable. He can oh, be unpleasant at times.'

Kendall observed how Arneka's beauty imaged her mother's likeness. *Photos never do a person justice. Going by what I heard Kurt say last evening, Arneka is a living replica of Erika. Her quick wit inspires me to believe that under her bravado there's a lonely young woman hiding. One, who may in time blossom into a striking rose. A dynamic force, a terror to be reckoned with when she's angry, she can be charming. In a devil-may-care mood she's a real challenge to everyone she encounters. I know she and Brianna will become the greatest of friends and that includes my mother.*

She nudged the arm closest to her. 'Hey, were you dreaming? I spoke to you doc, and you didn't answer. You omitted to mention Bridy's hair colouring. Is she a copper nut, a ginger top … or what?'

'Auburn, I think mum's hairdresser calls Brianna's long curls, her jewels. Father George refers to her …' Kendall stopped short of saying … as his darling girl.

The priest couldn't help interjecting. 'Brianna's a darlin colleen, if I may be so rude and allowed ta interrupt ya young folks. A lovable girl, with a beautiful nature, is she not me boy?' Turning to Arneka, he smiled. 'Ya might not know this, me dear. This dear man is engaged ta be married soon ta Miss Brianna Skye O'Shea.'

'Yeah, I know that Father. I heard him tell my brother on the plane.' Arneka looked at Kendall with a mischievous grin. 'I hope you don't forget to give me an invite to your wedding, brother dear? If you do … you'll forever regret it.'

'Only if you behave yourself, Miss Baumer,' he affirmed, noting her wry grin broaden accompanied by a cute wink. The scintillating aroma of Arneka's sensuous perfume outshone her mischievous expression. He breathed in deeply to absorb its heady fragrance. It stimulated his senses and created within him a sense of wellbeing.

'Aye, I have a weird idea that the priest thought I was trying to latch onto you. That's why he intervened. Not so me darlin boy,' she sarcastically giggled.

'Sounds even weirder you mimicking him,' Kendall sternly but quietly rebuked her. 'Father George is a close friend of Brianna's parents. The dear man wouldn't hurt anyone, Arneka.'

By the severe glower she received Arneka knew she'd struck a raw nerve. 'Sisters and brothers don't eye one another off,' came her spontaneous reply in a haughty way.

He smiled and together they moved outside to their vehicle. 'Hop in sis, while I hold the umbrella. Please move over so I won't be drowned. Father George doesn't bite.'

Entering the car, she turned towards her half-brother. In a hushed tone she noted, 'You're wrong. All priests hit your pocket.' Looking at Father Brady she smirked. 'No, he won't bite mine.'

'Hush dear, or he'll hear you. Try to be diplomatic, instead of being bombastic, Arneka.'

Smiles broke out all around. Cap in hand and covering his grin, the chauffeur closed their doors and returned to his front seat. Within seconds the vehicle increased its speed and wound along the gravel drive to its inner-city destination. Those in the rear of the car softly conversed.

Kurt Baumer dozed in the front passenger seat. Christian had slotted in beside his father to hold his right hand, so it wouldn't restrict the chauffeur's movements.

Feeling tired, Doctor Ross decided to relax and admire the scenery. Instead, when Arneka touched his sleeve it soon quashed that idea.

'You promised to show me the photo of Brianna this morning. Ken, I want to see it now, not tomorrow. I'm curious to see how she dresses in this sweltering Aussie heat.'

'Okay miss here …' Withdrawing a leather folder from an inner coat pocket he scanned the required items then responded '… this small snap a friend took at our engagement party. The second photo I snapped on that same night over on Jalna, our neighbour's property in Yass.' Resting the wallet on his knee Kendall continued, 'That snap is of Brianna cutting her twenty-first birthday cake. This smaller one I'd taken of my mother in all her fine regalia. Gigi dislikes anyone snapping her photo and she objected strongly.'

Flipping over the snaps Arneka observed some scribbled names and dates. 'Oh, so she's a first of the month babe, aye. November people are usually crafty but nice. Don't know many myself …'

He cut in to say, 'What! Is your birthday in November, Arneka?'

'Hell no!' she exclaimed realising they were in religious company. 'Oops, I didn't mean to swear. Sometimes I do if I'm angry.' Looking at the priest she thought it better not to disturb him. 'Gee, Brianna is a carrot top, a deep redhead.'

The expression on Kendall's face astounded Arneka, who knew he disproved of his fiancée being called both names. Her crudities didn't seem to go down well with him. Knowing her faux pas might be misconstrued as a weapon of war, in a disgruntled way, she suspected this may ruin the chance of becoming close friends. Embarrassed, she refrained from commenting.

Kendall read her thoughts. 'Arneka, you're dancing with fire if you call Brianna a redhead or other fiery tones of auburn.' A glower portrayed his meaning, better than he. 'It's the only time I've ever seen her angry. Someone called her those names on the first day we met, up on the Whitsunday's. She literally breathed fire. Actually, that's how we became acquainted, walking in the rain.'

'What, did you do the same old trick … or a new one?'

'Funnily enough it was an old manoeuvre. Bridy had dropped her handkerchief and I, being the only person around, picked it up out of a puddle. I rinsed her hanky in seawater before giving it back to the most beautiful girl in the world.'

'You did a Sir Walter Raleigh trick to impress her? Well, that's one way to win her over.'

'What, and lose my head. Not likely,' he laughed. 'Being beheaded doesn't appeal to me in the slightest.' Finding their conversation struck a little too close to the bone, Kendall diversified away from this gruesome subject.

'Father George,' Kendall spoke in a demure tone, 'how do you feel now? You were quite nauseated this morning. Those tablets I gave you should've helped. I meant to ask you that, well before ... Arneka and I diverted into puddles.'

'A little better, thank ya me boy.' He sighed. 'I'm not lookin forward ta today, Kendall. It'll be a strain on me tired old brain. But then it's goin ta be hard on us all, I'd be thinkin.'

'Unfortunately, I've not received permission to sit in with you in a medical capacity yet during questioning. They're bound to have a doctor on standby to monitor your vital statistics. Keep thinking positive. It should ease your mind a little, Father George.'

During the remainder of their journey to town silence reigned. Arneka, like everyone else, dreaded being interrogated by the AFP in Canberra. She thought their stringent questioning would unleash memories that were best left in the shadows of time, although she suspected revamping of the dismal dramatic past would be mandatory for the absconded criminal to be apprehended. Since the terrible incident had broken in Yass, only those travelling with her could personally identify the ex-Nazi Captain, Rolf von Breusch.

Father Brady had coined the barrage of inquisitions he'd gone through by the end of the first day as "an awakening of the Devil". The memory of various interrogations hounded him. They dragged up a quagmire of hostilities relating to his past life in German, that he now wished not to recall.

To alleviate some of the gut-wrenching trauma these five would suffer over coming days, the hierarchy did everything humanly possible to assist each person, in this journey of torment.

Doctor Ross insisted on being present when the ASIO officer questioned Arneka about her declaration. Kendall could tell how it wrenched at her heart by the way her fingers gripped the locket around her neck. He felt relieved when the inquisitions concluded.

During the lengthy, stringent interrogations Kurt suffered several minor attacks of angina. Hovering in the wings, Kendall stood by ready to help and console his liberated German friend, and comfort his naturalised Australian family.

10

They were well into the fifth day of answering mind-wrangling questions which produced positive results and interesting facts. Information the Baumer family and Father Brady had supplied welded together a conglomerate of tales, related to von Breusch's years in the Berlin Chancellery. When compiled, this data constituted cruelty beyond belief. It comprised a gigantic bloodcurdling dossier.

Senator Bucknell, along with his federal colleagues and ASIO had literally stirred up a hornet's nest. Their overseas informants who'd delved into the unknown aspects of this Nazi's world brought to life more astounding facets of his clandestine past. Combined results of extensive researches filtered through from the Mossad, German and French sources. All agencies were endeavouring to uncover von Breusch's present location. Hourly, information kept building. Important details came from French collaborators of the defunct Urban Underground, including members of their surviving families. Tied in with this plausible and unexpected data were several letters. An envelope was redirected from an Englishman, a Mr Homer Ellis, an ex-Home office delegate, now retired and living in an alpine town in Switzerland.

All data pertaining to his present whereabouts would inevitably tighten the noose around his neck. The international code, if the absconder was found stated "The rat is caught in the trap". Not so clever, he unwittingly left loopholes for the Australian authorities to delve into regarding his Wehrmacht history and recent past. Out of the Germanic woodwork all manner of poignant titbits wormed its way into foreign files. The majority were indicative of cruelty and atrocities this

Nazi officer had perpetrated against innocent people and everyone who opposed him or unfortunately crossed his path.

Two more letters were, over time, surrendered voluntarily to Mossad in Israel. The initial one appeared quite unexpectedly from a reliable source in Stuttgart. Other scraps of important news, smuggled in by undercover agents from various towns throughout Germany, eventually nosed their way in to this Israeli organisation's files. Names of informants would be withheld until authorities in Europe could apprehend the perpetrator of those horrific war crimes.

Bucknell would neither elaborate nor confirm further details of the current situation after he'd spoken to Doctor Ross the previous night. He now knew Kendall was conversant with most of the unearthed facts. Authentically based on his real father's early years in the Reich, the facts would soon be common knowledge and must, in due course, be revealed to members of his present family.

What Bucknell had heard confirmed his beliefs. Like his superiors, he also knew the data in ASIO's hands would prove invaluable as evidence to incriminate von Breusch. When presented to a panel of court judges in Austria, everything collated thus far, would bulk the prosecutor's briefs enough for him and his legal associates to secure an indictment against the absconder.

Quietly confident that something further may eventuate while searching for conclusive evidence, ASIO ploughed through reams of incoming depositions. All theories put forward from multiple agencies in other northern countries proved that von Breusch, had over time, either visited or resided in their towns and proved his guilt, especially on the Australian home front.

Late on that final day of inquisitions, Bucknell was privately approached by the head of ASIO in Canberra. This request related to his opinion on a more recent problem. 'Peter, before I pursue this matter, I need you to confirm something. In your honest opinion, do you think Kurt Baumer is well enough to cope with another shock?'

'I'm not capable of answering that, sir. Doctor Ross would be the best man to consult. Kendall has been treating him and he's well conversant with his medical history.'

'I see. Well I'll send for him now and confront you both with this problem personally. Then it will save me the trouble of speaking with you separately. Once I've finished, you're free to leave with the Baumer family.' Nodding to the staffer standing by his door, he withdrew an envelope from the top drawer and passed it across the desk to Peter.

'Please give Miss Baumer this with our compliments. The photos her locket contained proved invaluable to our researchers. We've replaced these with authentic replicas. I'm sure she won't mind our office keeping her originals. After the trial everything will be returned intact to you and her family.'

The moment Kendall appeared at his door, Senator Bucknell accepted Arneka's locket. Comfortable, Peter lolled back in the swivel chair to quietly listen to him being interviewed.

On exiting the office Bucknell noticed his face looked quite drawn. His sunken cheeks resembled a man of sixty, instead of twenty-seven. Peter expected this, knowing the seriousness of their current problem.

'Let's collect the Baumers and get out of this morgue. Honestly, I can't fathom their mode of thinking. They know Kurt is ill,' Kendall yawned, 'yet they've insisted on either you or me giving him more shocking news. How the hell do they expect him to cope with it, Peter? I think we should confront him tonight, before he retires. I'll have his medication ready, while you ask the staff to have his bed warmed, so he won't go into shock. That's all we need at this stage.'

On the cessation of their mental ordeal at ASIO headquarters, Kurt was exhausted. Attending numerous interviews had physically and mentally drained him. Even though Arneka and Christian had supported him every day during their recent ordeals, they fared no better.

By mid-afternoon the group of five began their final journey back to their temporary abode at Cooper's Crossing. Feeling elated, Arneka's excitement couldn't be contained. 'Thank God that ordeal is over. Those roughnecks had better not spoil the rest of our time here in Canberra. Tomorrow before we fly out to Yass, I want to see the War Museum. One of my close friends in Yeppoon told me her mother was a prisoner in Auschwitz. I think from forty until close to war's end. Will you come with me, Kendall?' She touched his shoulder. 'Not you Chris. You make me look a fool in front of people when you're overtired or cranky. And it's so undignified in company here in the capital.'

A dignified and staid school teacher, Christian Baumer didn't bother looking back at his sister, sitting directly behind him. Instead, he embraced their father's shivering shoulders.

Looking in the rear-vision mirror, Kendall noticed the diplomat's vehicle was closing in on their chauffeur-driven limousine. 'In five minutes, we should arrive at our destination and I'll welcome relaxing in a hot bath. Chris, you can assist your father from the car. I'll go straight indoors to prepare his medication. Arneka dear, you might help Father George ...'

'Okay, I intended to do it anyway. I'm buggered Kendall, exhausted after running around like a headless chook or an idiot in that damn federal building for days on end.'

Peter Bucknell understood they were due to fly out for Abergeldie next morning. Some news he'd purposely suppressed in case their depositions needed clarifying, or if more vital data had spewed forth through the blower before dawn. In the event of either situation arising all their schedules would be revised. Massive dossiers on von Breusch were compiled and coming in from all corners of the globe.

Immediately dinner concluded Bucknell requested Mr Baumer to accompany Kendall to his office. 'Make yourself comfortable, Kurt. I've asked Doctor Ross to stay during our conversation. I'm also recording everything, if it's okay with you?' He acknowledged Kurt's nod. While talking, he gestured for Kendall to be seated. Peter then nestled into the chair behind his desk. 'The news I must convey to you, I wish wasn't necessary. Nevertheless, I will do so now.'

'By the look on both your faces, it isn't good news I'm about to hear.' An item in his palm quietly slipped on to the desk. The furrowed crease on his prematurely aged brow deepened. The skin on his hollow cheeks had taken on a waxy look. 'I realise sir, what you have to say must be important.'

'No no, we're all friends here. Don't address me as Bucknell or sir. Peter's fine.' A glass of fresh water slid across the desk. 'Feel free to sip that at your leisure, Kurt.'

Kendall picked up the bottle and examined its contents. 'I'm glad to see you've taken my advice, Kurt. In all probability you won't need these Anginine tablets. It's just a precaution.'

'Precaution for what, Doctor Ross?' he vaguely queried. 'I can't think of anything I missed when the federal officers questioned me today and yesterday. I'm quite capable of thinking clearly, although I am a little tired.' Imparting a sigh, Kurt wiped his sweaty brow on his shirt sleeve. Knowing every nook and cranny of Shannon-lea's weatherboard building was air-conditioned, he imagined the queasy feeling in his stomach might be from sheer exhaustion.

'I'll walk you back to your room after this brief interview,' Kendall assured his patient as Senator Bucknell cut in, to distract his thought processes.

'A letter has come into our possession from Germany. An unknown woman posted it to the Jewish Mossad last week. Interns in their organisation forwarded it through to my office, via our diplomatic pouch. The original they've retained. This is an exact copy.'

'What's that to do with me, or my family?' All this official jargon wasn't making sense to Kurt, who, feeling nauseated, condescended to listen.

'Unfortunately, it has a great deal to do with your German family; specifically you, sir.' The diplomat then quietly declared, 'All the delving into your past history by our officers has proved beneficial. My superiors have given me the task to pass this news to you. They thought it better coming from a personal source.' *How pathetic he looks, poor man,* Peter mentally noted as Kurt covered his face with both hands. *Hesitant to disclose the facts, I can't let him remain in limbo a moment longer.* 'I'll be frank with you, Kurt. If I don't explain this letter that's apparently been lost in the corridors of time, my superiors won't be quite so diplomatic.'

'My answer is yes, read it by all means. I consider myself strong under pressure, or I have up until now. Say it in layman's terms so I'll be able to follow you sir. Then allow me time to consider a response, whatever it might be.'

'Right sir, can you please tell me whose writing is on this envelope? Take your time. We must be positive to whom this letter once belonged.'

Kurt reset his glasses on correctly, before attempting to examine the blue envelope. Twice he scanned the writing then solemnly declared, 'It's definitely my sister's scrawl. Yes, I'd know it anywhere.' His mind couldn't assimilate the seriousness of what he'd just read. Silent and with his head bowed he tried to fathom what the letter alluded to, and where this might lead.

'Heide's distinctive script came as a shock. I … I can't believe it. Who did you say the letter was from?' Through misted eyes he couldn't read if the envelope was addressed to him. Wiping his glasses, the pounding in his temples continued to increase and he felt queasy.

'Look at the miniatures clipped to this letter, then comment.' Bucknell passed him the three photographs. 'Kurt, do you recognise the woman on that page?'

'Yes, I do.' He pointed to two snaps, one cut from a larger photo. 'This miniature is of Erika von Breusch and her husband. The smaller one is of my sister I'd taken on a swing in their garden in Hamburg. Heide promised to have it framed for me.'

Mystified, he peered through moist eyelashes at his deceased sister's photo, only his brain failed to comprehend what his mind tried to convey. After the painful fog of memory cleared, he responded, 'Don't you think Heide's beautiful features shine through on that glossed paper? She told me on the morning of her death that he'd tried to rape her numerous times. I can't believe she's gone. That Nazi had touched her breasts in a familiar way. She fended him off, of course. Yes, it occurred just as she's described in this letter to my wife, Anna. I'm positive von Breusch intended to seduce my sister again, either in his car or once they reached Berlin.'

A restricted throat prevented Kurt from continuing. A sip of water eliminated the bitter taste on his mouth and eased its dryness. When able he declared, 'The monster definitely wanted to rape Heide, of that I'm certain.' Reliving the torrid memories of a dismal life, a seething pain gripped his chest. Two minutes lapsed before he could speak again.

'Take your time and stop if you feel breathless, Kurt,' advised Kendall.

'I'm fine, thanks Doctor Ross. Two or three hours before he murdered my sister, she witnessed him killing his wife. Heide told me that she'd actually seen him push Frau von Breusch down their stairs. I have a photo of the incident she took on that terrible day. The snap and negatives are in the suitcase at Shannon-lea.' Kurt took a sip of water to quench his thirst. 'God knows how horrific that sight must've been for her to have witnessed. It devastated her. He tried to murder his wife before, Erika told my sister.'

Kendall looked perplexed over his shortness of breath. A hand gesture warned Kurt to take it steady, or he would collapse. A long pause allowed him time to recuperate a breath.

Relieved, he uttered in a sombre voice, 'That cruel Nazi delighted in berating and whipping all women. I never trusted von Breusch. He is the Devil incarnate. He's a downright evil beast. I hope and pray that the Feds arrest him soon.' Shaken, Mr Baumer momentarily hesitated to regain some composure.

'During the questioning did you declare those facts to the federal interrogator, Kurt?' queried the diplomat, taking notes of this ill man's verbal deposition.

'His wife's murder, yes I did. Not the attempted rapes. Until now I've tried to forget all the horrible memories. The miniatures and that woman's letter brought it all flooding back. Sir, I know the letter you mentioned wasn't meant for von Breusch. Another Reich officer, Major Kassell was a respectable man who'd taken pity on my sister. Viktor became her mentor. He employed Heide as a nursemaid to his children. She lived with his family until she accepted an offer to work in the von Breusch household. Now you know the truth, I'll sign anything ...'

Having listened intently to his patient, Kendall cited in a professional capacity, 'Not tonight, you won't Kurt. The morning will do, after you've rested. Don't you agree, Senator?'

Bucknell nodded. 'Now this signed declaration of mine will substantiate all you've narrated to us. Kurt, tonight I'll photostat your affidavit. You can read and sign it in front of us first thing. You'll never regret giving your honest testimony of your sister's attempted rapes by that criminal. This tape cannot be used in evidence, unfortunately. However, it will be a permanent record of all you've told us this evening.'

'Please listen! I can't afford Arneka and Christian knowing the horrible ordeals he put my sister through by his attempted rapes of her. Sir, there is another stipulation. I refuse to testify against von Breusch if Heide and Erika …' finding it hard to breathe, Kurt gasped for air, '… if their names are dragged through court. Their integrity, even in death, should at all times be respected.'

'That's perfectly in order, Kurt. The authorities wouldn't have it any other way. And I can personally vouch their names and characters will remain unblemished.'

'Actually, I now want to make a full confession to the priest.'

'You wish to see Father Brady now? It's well after ten ...'

'If he'll speak to me, there's something I must confess tonight. Otherwise I won't be able to rest or face court tomorrow. It is important and I can't keep the truth to myself much longer.' Out of Kurt's hearing Bucknell requested Kendall to see if the priest would appreciate being disturbed as such a late hour. Kendall walked from Bucknell's office and approached his room. If Father George accepted the challenge, Kendall knew it would be permissible for Kurt to discuss this ghastly topic in the privacy of the senator's office or in Kurt's room.

In the interim Bucknell stood to leave. 'I'll take this typewriter and the folders to my bedroom. Please stay in here Kurt until Kendall and the priest return. They should be here shortly.'

On his return and worried over Kurt's breathing, Kendall paused to take his pulse. Exceedingly high, he warned him not to overtax himself, especially while in the confessional with the priest.

The instant Father Brady stepped in Bucknell's vacated office Kendall proceeded to dole out the prescribed dosage of tablets to both men. 'Your pulse is fluctuating. It's quite erratic Kurt, so you must rest. If you begin to feel faint or dizzy tell Father George and he'll call me.'

In an indulgent effort to please him, the priest saluted Kendall. 'Sure laddie. This ill man doesn't deserve to suffer more unnecessary pain. I'll be callin ya if he takes another heart turn.'

As Kendall closed the door behind him, the priest insisted, 'Don't kneel down, just sit where ya are Kurt. I'll bless ya first, then tell me what's worryin ya.' In a comforting gesture, he gave the distressed man his blessing. After forming the sign of the cross on his forehead, Father Brady settled his butt on the desktop, opposite Mr Baumer.

'Father, something has plagued me for many years and now I must unburden my soul. I need absolution for knowing, and carrying this unforgivable sin on my conscience.'

'Relax and tell me slowly, in ya own way. Don't hurry. I'm ready ta hear ya confession. Then it'll relieve ya mind and whatever ails ya soul.'

As requested, Kurt unfolded the story of Heide's teen years before and after she began working for Major Kassell. 'The major was my sister's birth-father. Our mother told me about Viktor not long before she died.

I waved farewell to her and my own father, a shoe salesman on the day they embarked on a luxury liner bound for Norway, I think.'

Kurt went on to say how both his parents had drowned in rough seas after the cruise liner ran aground and sank. He concluded by divulging how their deep, sincere love for one another had almost destroyed all their lives. 'Viktor Kassell adored my mother, although she had twice refused to marry him. In her last year my mother regretted her decision.'

A short break, then Kurt told how Kassell idolised Heide, his illegitimate daughter and had taken her in service at fifteen to protect and care for her. Yet he never divulged to whom she belonged. Their secret he kept until his dying day.

'I have not disclosed their relationship until now, Father. Other than that, all I can tell you is von Breusch tried to rape my sister twice, he then murdered her. I can prove that he ordered the "Black Shirts" to shoot Heide. They burned her body and reduced her house to a pile of ashes. At night in my dreams the smug look of those evil men's faces still haunt me. How could I forget such abhorrent and horrible incidents?' Letting the tears flow, his painful throat eased.

The Irish priest lent over and clasped his hands. 'I can understand how ya feel. I have seen what the Nazis were capable of during the war. Please go on, when you can Kurt.' Settling himself in the diplomat's comfortable chair he listened intently.

'What I've told you must remain a confessional vow. Father, I ask you to promise not to tell Arneka or Christian anything of what we've discussed here tonight. They hate and mistrust their birth-father and for tragic reasons.'

'This is a confessional. My vows to our Lord forbid me from sayin a word ta anyone. Why wasn't Major Kassell told about Heide's death? Ah, ya sister was born out of wedlock.'

'Yes, but my mother wasn't a loose-living woman. Their affair was an affair of the heart. They loved one another, she and Viktor. She never broke up his marriage. His previous wife deserted him years earlier, when their second child, a boy, was fourteen. My mother met Viktor at a party and they fell in love, years before he met and married his first wife.'

'One thing's confusin me, Kurt. Why were they not married, after his wife walked out on Kassell?' The priest looked dismayed, trying to think of a favourable reason.

'My father came home from selling shoes and leather gloves in Switzerland. By then my mother was pregnant with Heide. She never mentioned Heide wasn't his child. They discussed the future together and tried to resolve their differences.' Kurt momentarily hesitated to embrace his aching temples. The severe headache was driving him to the brink of insanity. 'They separated. She decided to start life afresh in Munich. I don't think she ever saw Viktor again. My mother considered a clean break would be best for her, our family and both of his children.'

'Ya only crime was lying ta protect Heide. Not fa self-gain, Kurt. I absolve ya of all ya sins.' The Irishman blessed his stole then wrapped it around the ailing man's hands. 'Ya weren't ta blame fa that evil man's indiscretions. I can't see the harm in protectin ya sister from bein raped by von Breusch. I knew of the cruelty he perpetrated against innocent youngsters and women. That man was and still is capable of murder.'

Kurt proceeded to tell the priest that he could never have told Viktor Kassell about this Nazi captain's attempted rapes. He now regretted not attempting to have protected Heide whom he adored. Viktor would've exposed von Breusch's criminal intent. If presented before his peers, Kurt knew von Breusch would have been court-martialled.

Mulling over a conglomeration of problems he grimaced. 'Heide didn't deserve to be murdered by that Nazi, neither did his wife. Father, both of them would still be alive if I hadn't been a coward.'

'Can ya substantiate all the accusations against von Breusch? Kurt, I've known about his treacherous manoeuvres which he conducted in the Chancellery for years. On a recent retreat with other Irish priests, I studied Reich politics. I delved into the dreadful starvation lots of German infants, their parents and elderly people all suffered in wartime.' His in-depth study of Germanic war crimes had consumed multiple lonely hours, a specialty of this gentle-natured Irishman.

'1still have the last letter Heide wrote to my wife. It's in the folder in my room. I didn't think to bring it with me this morning. It deals with an argument he had with his wife over Heide not travelling in his car to

Berlin. I knew von Breusch intended to rape her either there or on the way. Instead, he killed her.'

'No, it's wrong to blame yaself for her death. During me years with the Marquands in France and with the German Urban Underground I saw many lives devastated by rape. Sadly, it became a common occurrence in wartime. Not that I condone rape, I don't. That German maniac, von Breusch, be assured will in time hang, fa what he did ta all those unfortunate people. Criminal acts of murder or depravity destroy all hope of happiness fa everyone who desires to live their lives in peace.'

Kurt's pallid features took on an astounded frown, one of deep concern to the priest. 'Do you mean Father, that you also worked with the Marquand's in France? For over twelve months I worked tirelessly with their underground movement.' He saw the priest nod, yet said nothing.

'I've be thinkin, ya should tell all this ta Mr Bucknell. Not now, but tomorra. If ya keep holding back even one word, it could help this criminal escape the federal officers' net. The authorities here and overseas need every scrap of knowledge ta build a solid case against von Breusch. Something conclusive that will stand up in court. Lies or unverified truths will not do.'

Kurt Baumer explained that he'd told Bucknell everything, except what he knew of his sister's murder. He hoped both his adopted children would be spared from learning the real reason behind Heide's murder. They were also private issues, not revealed unnecessarily. Battling to breathe, he paused to catch another agonising breath.

'Ya can't go on sufferin like this Kurt. I saw ya suck a tablet a while back …'

'I'm fine. Just let me finish. Major Kassell was an honourable officer. He never wanted to ruin either my mother or my sister's life. That's why he didn't divulge to anyone other than to me … about him being my sister's true father.'

George Brady feared this man struggling to breathe might collapse. Excusing himself, he scurried along the hall to find Doctor Ross. Briefly explaining the situation to Kendall and Senator Bucknell, the priest followed Peter back to his office.

Distressed and hallucinating, Kurt fought hard to save himself from falling. Finding it difficult to see anything through blurry eyes, his hands groped for something to hold on to. Instead his fingers slipped off the polished desktop. Floundering to grab something solid Kurt collapsed. This final declaration of disgorging his soul had ripped apart old wounds that for years he'd thrust into the backblocks of his mind.

Kendall rushed to help him. Kurt couldn't speak, just muttered something that neither he nor Bucknell could understand. 'Quick Peter, help me lift him over to your couch. While I loosen his tie, please fetch my medical case. It's on my bed, or should be. Then stand by your phone, in case we need an ambulance.'

In an effort to remain calm, Kendall gave Kurt an injection to lower his blood pressure. 'Don't try to move. That drug will help you to breathe easier. Once we help you back to your room, I'm going to sleep in there. You must try to relax. Forget everything until your breathing stabilises.' He nodded to the priest who appeared at the door. 'Father, ask Arneka to come in here …'

On hearing a muddled confusion of voices, she appeared in the office doorway. Worried over the agitated state of her father she exclaimed, 'Oh Daddy, you look terrible! Here, let me help you to your room. It's only a short walk down this narrow hallway.'

'Yes, I agree with Arneka. I must insist on you resting. Kurt, you're on the borderline of a stroke and a nervous collapse. The drug I injected into your arm will take a minute to work.'

Arneka manoeuvred herself in between her father and Kendall. Slowly she eased him up until he could stand on both feet. 'Come on Dad; take one step at a time. We'll both assist you to walk you down this corridor. A few more metres, easy does it, Dad. You'll be back in your room soon.'

Bucknell took over and Arneka hurried ahead to turn down her father's bed linen. Within seconds she returned with warm towels, hot water and fresh pyjamas. A flip of her hand gestured for everyone to leave his room. She unzipped his fly and removed the belt. Kurt's sweat-soaked trousers dropped to the carpeted floor. A quick sponge bath and dressed only in fresh underpants he sat on the bed.

'Dad, tie the cord on your PJ's, or they'll fall down. I'll tuck the warm sheets around you. Now you should sleep until morning. Kendall will be back soon and he's promised to sleep in here on that rickety stretcher.' Kissing his cheek, she switched the light off. Before leaving her father's room, she opened the bay window. A gentle breeze belled the chiffon curtains.

Lying on his bed, Kendall re-read the scribble on Kurt's medical history. *Most of us blokes are noted for our crook writing. At least this page is legible. It needs a bit of tidying. Damn. I'll do it before morning. By now Kurt will have settled.* Finishing the mug of cool chocolate, he decided to go and recheck his vitals. Kendall laughed. *We professionals are nonentities in this busy world.* Before restacking his overloaded mind, he gave another hearty laugh. Reminiscing over Arneka's outrageous slang, he grabbed his warm jacket, picked up Kurt's chart and closed the door gently, so it wouldn't disturb those asleep in the next room.

In the hall he quietly spoke to Bucknell. They briefly discussed the circumstances leading up to Kurt's collapse. As they talked, Peter took a mental note of Kurt's request not to disclose his sister's repeated attempted rapes by her real father to Arneka.

Agreeing with the senator, Kendall was ready to approach Kurt's room when the priest joined them. 'Do ya think he will be all right? Poor man, his face looked grey when he fell. I think, me boy, Kurt should be sayin a few gratuitous Hail Mary's tonight. This mess is me own fault. I canna forgive meself, fa not advising him ta rest more.'

'I told you before, not to blame yourself. You should go back to bed, Father George. Your medications are on the bedside table. I'll be in to take your blood pressure in a minute.'

Eyes brimming, the elderly Irishman tottered off to his room. Kendall followed every footfall. On leaving Kurt's room Bucknell greeted him. 'Here's a tray of fresh ham and lettuce sandwiches. There's a pot of tea and milk on this tray. You haven't eaten all day, so I thought you needed something substantial in that grumbling gut of yours. Will he be okay, do you think, Kendall?'

'Yes, I just checked Kurt's pulse. I'll nibble these sandwiches then doss down on the stretcher. It looks mighty uncomfortable. I won't sleep tonight, or what's left of it. Thanks for the food. I didn't know you were a handyman in the kitchen. I'm hopeless at boiling water to cook eggs. See you in the morning, Peter.'

'During the night, if Kurt takes another turn call me. I won't be turning in until four or five. Then I'll buzz down to one of the security guards to call an ambulance, if you think one is necessary.'

'He doesn't require hospitalisation. His severe chest pains will ease. I administered his heart medication a moment ago. Kurt's pulse should be normal now. Goodnight, Peter.' His voice dropped a tone. 'I'll read by his bed for a while to constantly monitor his rhythmic pulse.'

'Good, I'll make sure Father Brady is okay. Then it's back to work for me. It's been a long, drawn out and hectic day for us both. You should turn in, before you collapse Kendall.'

Nodding to his sister, who entered her citadel, he interrupted Bucknell. 'I need to finish writing up Kurt's drug chart.' He frowned. *Arneka looks dishevelled, why?* He glanced at his dust-encrusted wristwatch. 'Night has flown and today's ready to turn the corner.'

Relocking his medical bag, Kendall hesitated. 'I'm glad the others have turned in. Now neither Father George nor Kurt will be disturbed. I'll leave his next dose of drugs on your desk. If I doze, you'll know where to find his Anginine tablets. Thanks a ton, Peter.'

'The drugs will be safe in this office. I lock the door every time I leave here.' He thought Kendall looked weary. *Well, I don't doubt he's mentally and physically exhausted. He and the elderly fellers have gone through hell at the hands of their federal inquisitions. To save him worrying, I'll make sure the oldies are checked hourly. That should give Kendall the chance to recoup some stamina. I'm bushed just thinking of how he manages to cope under intense stress.*

'Hey matie, how about I ask one of the guards to rig a bed up in Kurt's room. That way you *will* have some rest. We don't want you collapsing. The guests all rely on you, Doctor.' 'Suits me fine, thanks Peter. I doubt if I'll sleep, just rest on the recliner probably.'

A wearisome gasp petered from Kendall's parched lips and he expelled a long sigh. 'We blokes are noted for sleeping on fence posts at times. A hazard of working in the medical field is lack of sleep. One grows used to it after five or so years of doctoring.'

'It must be providence that both the old codgers are located in conjoined rooms,' Bucknell smirked, not letting on that he'd arranged everyone's quarters. 'There's a tower of paperwork to climb over before I can turn in. It needs to done before the courier drops by in the morning. Now if you'll excuse me, Kendall. I'll see the guards about your bed.'

In the hall Bucknell paused to read a note he'd found on his desk. "Peter, you may not be aware of this. You've been a tower of strength to us all, over the past hectic weeks. It's much appreciated. Will you be flying to Yass with us tomorrow? If so, tuck a note in my medical kit."

Safe to say at the moment yes, I will be. His laughter rollicked around the empty corridor. Peter then grimaced. *That's unless some inconsiderate bastard sends me more indecisive news. The double-ended plugs to Canberra must be working overtime tonight. Oh shit, it's gone one. I better see Kendall first thing. Our political hands would've been crippled without all their valuable depositions.* Peter considered popping in to see how Arneka was coping. With a bitter dawn wind battering the oleander bushes outside his window, he quashed that idea. *With a positive slant on things, I'll check on her welfare before the family awakens for breakfast.*

Overtired, he shuffled down the carpeted hall until he reached their recreation quarters. *I detest some of the guards and their strict, idiotic policies. They strut around like overproud peacocks, poking their noses in everyone's private business.* Minutes after this thought rumbled through his head, Peter spoke to the lieutenant, who ordered two of his strongest guards to carry a solid bed down to Mr Baumer's room for Doctor Ross.

Bucknell dreaded the idea of slogging through reams of paperwork. *It looks as though it's going to be a long stint of sheer drudgery. I hate wading through reams of foreign documents with unconfirmed reports and inconclusive or ambiguous claims that lead nowhere.*

11

Approaching midday, the Cessna touched down on Abergeldie's makeshift runway. As it taxied to a stop Kendall Ross recognised his jeep struggling over furrowed terrain until it pulled to a buck-wrenching halt under one wing of the now stationary aircraft. Seven tired passengers on board showed great pleasure by clapping, after their long ordeals which now had drawn to a close.

This young lady beside me looks haggard and drawn. Father Brady glanced at her drawn features as she stood to exit the seat. Strenuous inquisitions had sucked the life out of all five passengers, ready to step down on firm soil, but none more so than Miss Baumer. He knew her anguish bordered on the brink of despair, although it paled in comparison to his own suffering. Doing the gentlemanly thing he assisted her across the metal air bridge then down eight narrow steps.

Terra firma never felt so good to Arneka. The toes of her navy shoes furrowed deep in the dust beneath her unsteady feet. Tired and feeling languid she resembled an aged woman. Even heavy make-up couldn't hide the dark shadows under her vivid blue eyes.

Bjorn Svensson welcomed the group who exchanged pleasantries. Assisting the pilot to off-load their luggage he frowned at Abergeldie's young boss.

'Hey there sport. You look buggered, Kendall. What the hell have you all been doing down in Canberra? I'll run you and Father Thingamajig over to the homestead. Then I'll return for the others.' Jalna's owner scoffed while lugging their light cases on the jeep's rear seat.

'Take Kurt and his family, along with Father over first, Bjorn. The load you've stashed on this old girl will be the maximum she can handle.

I'm sure Peter won't mind hanging back with me. We'll stand under the wing's shade until the next run home.'

'Righto mate! Once I've helped the elderly gents aboard, we'll see how much room is left for the youngsters. Gee, the priest looks buggered, utterly stitched up. You don't look much better either, sport. Reckon the Feds' tough gruelling must've been harrowing. Well. that's what I think going by the way you all look so exhausted.'

'The strenuous going over they gave us has definitely affected me. Arneka should see Stephen once you reach the homestead BJ. She's really taken their extensive grilling hard. She's a little shaky on her pins, and I fear she might collapse.'

Kendall and his neighbour discussed the flight from Canberra while rearranging an array of cases, parcels and weird things disguised as parcels. Now secured, nothing became flying missiles if the jeep's wheels hit deep ruts or potholes across the home paddock.

Bucknell also took shelter under the wing. The cooling aircraft provided a narrow splinter of shade that gave him some respite from this blistering heat. He nodded as Kendall joined him and rested his foot on the metal wheel brace. Together they watched the disappearing jeep slowly weave in and around rough patches of track, minimising the dust flow.

Discussing the weather, Kendall covered his gaping mouth with a hand. Endeavouring to cool down he constantly fanned his face with a battered Akubra, to repel the blowflies from settling on his eyebrows and nose. He expelled a hefty sigh that drifted in the stifling dry air.

The searing hot wind churned a willy-willy into existence. Fragmented grit buried itself in every unhatched part of the aircraft. Swirling around the men's heads, a sudden updraft ripped their hats off and sent them flying in all directions. Kendall had turned his back the instant he saw a fiery red cloud of turbulence ready to engulf the plane.

Local rain wasn't sufficient to cool patches of parched earth that now resembled the rustic rings of Jupiter. Every skerrick of land cried for relief to quench the scorched ochre-soil. Hand feeding of sheep and cattle was stretched to the limit. The roustabouts were tired of scavenging dry fodder from wayside ditches and tracks to feed their stock and mobs of sheep. Horses contentedly chewed on baled hay or lucerne, home grown and harvested in season.

Nature in all her glory was creating a mystic atmosphere. For some unknown reason the animals detected moisture drifting in on a heavily becalmed air. Westward the clouds burnt off leaving an emblazoned open-ended sky. Flocks of brolgas and long-necked cranes pecked in cracks to keep from starving amid pink-crested galahs and other species of wild birds that created havoc on pastoral lands, now devoid of grass or greenery. Livestock restlessly bellowed under weeping willows for respite from the heat. Their restless hooves dug deep in the earth's hard, barren crust. Wind stirred empty husks of wheat, until their withered leaves and stems rustled under a boiling noonday sun.

'Like me, I bet you're glad to tread on home soil again, Kendall.' Senator Bucknell watched the pilot stretching his cramped legs, prior to the Cessna taking off again.

'You bet I am!' Kendall's modern brogue flicked the loose topsoil, scattering dirt to the wind's mercy. 'I'd hoped Canberra's fine rain might've followed us home. Abergeldie needs a thorough drenching. I must ask Bjorn how the fodder's holding. If our stocks are running low, I better send my crew to scrounge for more hay in Yass tomorrow. We need loads of fodder to keep a mob of a thousand sheep from starving. Dry bales of lucerne and locally grown grains of stockfeed will suffice for a couple of weeks. Maybe longer, if it does rain.'

Stiff from standing he sat in the shade, well out of the sun. The raw heat hit home worse than a furnace blast. Kendall flicked the bluebottle blowfly away after it had bitten his wrist. 'Good, that's now buzzed off. Resting in the shade our sweaty bodies won't fry, in this damn sweltering heat.' He relished the brief sojourn under shelter. Rubbing the itchy and painful swelling on his wrist, he spied a rolling mass gathering overhead. Swirls of black clouds tinged with green drifted across the endless vacuum of blue in quick succession. Neither he nor Peter had noticed the sky darken.

'Praise be for drought-breaking rain. I'll willingly stand in a downpour and sing until my lungs are dry. I'll even toss this old hat of mine in the air then shout Alleluia,' the diplomat crowed, handing Kendall back his rescued Sunday-best Akubra.

He shuddered, saluting the indigo sky. 'You won't be alone there, Peter.' Far beyond the distant treetops growled thunder heralding a vicious storm. Summer hailstorms were common occurrences in productive years. In drought they often turned ferocious, leaving huge gouges in the crazed and powdery soil. Deep furrowed trenches failed to recover, unless their roots were interlocked with strong willows. 'I'm praying for good follow-up rain, or this lot we're about to cop won't even dent the cracked surface.'

By the time they were within ten metres of Abergeldie's homestead it teemed. The jeep's wheels slipped, danced and skated across muddy pockets of drenched earth. These rain-soaked travellers, drunk with delight, revelled in the cool rain dousing their heads. Clear water flushed the surface grime and ingrown dust from slouch hats and summer clothes. It bathed their mud-clad vehicle in Nature's tears. A resplendent feeling of cool air embraced the men's suntanned faces. Languid bodies sweltered as vaporous steam rose to greet their saturated legs.

Keeping a lookout for ruts, Bjorn preferred not to talk while driving. Dust and fly-bitten, sodden eyes addressed his companions. He surmised their eagerness stemmed from being couped up in stuffy rooms down in Canberra. Nevertheless, he silently cursed his jeep's cantankerous mood as its motor spluttered, ready to seize. A monstrous thunderclap drowned the roar of whirling propellers as the aircraft began taxiing down a mud-strewn runway. Even the constant growls of the encompassing storm failed to enhance his abrasive temper.

'When it pours in the outback, Nature sure shows her true colours. Even the atmosphere glows with platelets of sun-gold droplets. Let it rain. Let it drown all our sorrows.' Bucknell flexed his fingers to the tune of a hymn and it imaged their passengers' jubilance.

'Hey sport, we don't need a damn summer flood to wash away the precious layers of unburdened soil. What 'ya reckon, Kendall?'

'No, not at present,' he agreed. 'If heavy rain follows this lot BJ, it'll rip the topsoil to shreds. Then it'll cause even deeper erosion that'll wash the furrowed earth away.'

'We need just enough rain to quell this damn drought. Once the rain ceases it might turn fine,' stated Senator Bucknell whose smiling eyes deflected off his companions' drowned faces, down to his mud-splattered boots. Without warning the jeep bucked. Its wheels dug deep and began to sink in the quagmire of gunk. With great effort, Bjorn turned its tyres until they gripped solid ground, just shy of the home stretch.

'We could do with lots of gentle rain in a day or so. That'd be beaut,' he sprouted anchoring the jeep to a stop beside Abergeldie's back steps and almost ran the priest down.

Without lingering, Kendall placed his all-weather coat over wet shoulders grabbing his suit-folder from the jeep's rear seat. Jumping over the front passenger door he acknowledged their guest. 'Where's my mother, Father George? I thought she'd be here to welcome us home.'

'Nay laddie, she'd be lying down. I think ya colleague has just gone in ta see her.'

'What the hell's she been doing now? I'm not out of her sight for five minutes and she somehow manages to get into strife.'

'Ya mother's fine, me boy,' retorted Father Brady, not wishing to disclose the little he knew of her condition. 'She fainted after seein the Baumer family. Their visit came as a shock.'

This news enraged Kendall even more. He'd specially requested the Canberran interrogator to advise his mother that the Baumer family would be returning with him and Senator Bucknell. Fuming, he cursed himself for not warning Gigi, when he'd spoken to her earlier in the day.

'I'll go straight in, BJ. Fetch my suitcase, please.' Leaning against the jeep Bjorn gave a nonchalant nod. 'Pass me over my medical kit. It's under the cover near your hand.'

On receipt of his bag Kendall hurried inside. One look at his mother confirmed his worst fears. 'How did she get in this state, Stephen? What drug have you given mum?'

'Hey, hold your temper, Kendall. As yet I haven't administered anything to her. Kirsten found your mother slumped over the wheel of her car. Unfortunately, she is suffering from a nervous collapse. I will give her ten mils of Diazepam. This drug and dosage will ease her tension. She'll sleep for four to six hours. Let your mother rest, Kendall. You can check her vitals later.'

'Knowing Mum, she'll fight every drug you give her. Stupid woman usually does.'

Perturbed, Jarvis frowned. 'I'm not in favour of administering large doses of any drug to my patients. I had no option. Gigi literally pounded Kirsten's chest with both her fists. Don't you dare say we should've foreseen her breakdown coming. I did, and this is the outcome.'

'Her collapse hasn't come as a shock to me at all. I expected it long before now.' Standing beside his mother's bed Kendall checked her neck pulse. 'I'll leave. You have my blessing to administer that injection. Her mind seems to be floating in a mysterious realm of dreams. She didn't flinch, not even when my fingers pressed her carotid artery.'

Jarvis pointed the syringe at his colleague. 'Leave now Kendall. Or as heaven is my judge, you'll be the recipient of this drug and not your mother.'

Disgruntled and grumbling Doctor Ross ventured out to imbibe on the back verandah. Not even the stockmen who'd come down to collect their pay could make a dent in Kendall's harsh scowl. He mumbled something incoherent, slammed his untouched beer bottle on the table and stormed off to read in his room.

Misery, plus total exhaustion had overwhelmed Arneka. Alone, and laying on her bed she sobbed piteously. Jarvis, who'd heard her cry of anguish, popped in to assess the reason for her distress. He discussed the cause of Gigi's sudden collapse with Arneka and dulled her anxiety and also settled her nerves. Satisfied she could cope, Stephen left her to commiserate in private.

After a warm shower and feeling refreshed she re-joined the family gathering. With the pressure of guilt lifted, she chose to speak with the priest. Her brother Christian and her father Kurt, were discussing the Svensson's recent spat with their host.

Listening to their conversation it heightened her curiosity. 'Chris, I'm sorry our sudden arrival has upset Mrs Ross to this extent. I didn't think barging in unannounced would cause her enough stress to make her collapse. We can't be blamed. Well, I don't think it was our fault. Kendall should have forewarned his mother that we would be accompanying him home.'

'Listen sis, it will take her time to adjust and accept our family. We're their kin, well, sort of. Don't take her collapse to heart. You're not responsible. I agree someone should've warned her, well before the Cessna landed. That was well over an hour ago.'

'I'm confused how we should address her. To me, it's wrong to call her Gigi. We hardly know the woman.' Annoyed over being chided, Arneka thundered off in a huff.

Brianna, who happened to overhear the discord between Kendall's half-sister and her brother, followed Arneka to her room. 'My dear, you seem lonely and lost. If you have no objections, I want you to share my bedroom. There's plenty of space and we'll get to know one another, without the men interfering in our conversations. Mrs Ross suggested it. Call her Gigi, everyone does. She and Kendall both want you and your family to feel at home here on Abergeldie. So do I Arneka.'

'Oh damn. I always seem to put my foot in it. Can I call you Brianna? Kendall calls you Bridy. I've always wanted a sister. Now I have one in you, Miss O'Shea.'

'Come, I'll help you get settled then we can chat until dinner's ready. Before nightfall, I'll show you around the homestead grounds. In time you'll get used to the way things work on a sheep and cattle station. I'm only learning how to cope with the hectic pace. Can you ride?'

'Me! Heavens no! I couldn't sit straight on a bolted down stone rocking horse, if you paid me a trillion dollars. Astride a bicycle, yes that's more my style.'

'Good,' Brianna responded, quietly confident. 'Then I'll teach you to sit astride his horse while Kendall's overseas. Potchkin is a gentle beast, he doesn't buck and he won't throw you either. Be my pleasure to teach you to ride. Now, let's join the others. By now everyone will be enjoying drinks in the lounge room. An Aussie ritual is to enjoy a cool aperitif before lunch.'

'When is he leaving? Drat it. I'm just getting to know his peculiar mannerisms.' Arneka flicked off the overhead fan as they left her bedroom. 'My brother Chris can be like a stuffed shirt at times, especially if he's annoyed. Kendall doesn't growl over my stupidity or stubbornness. Not like my brother does.'

'You sound quite disappointed. Why Arneka?'

'You hit the nail fair on its head. I am peeved, Bridy. I formed a close bond with your fiancé on the way up from Canberra. I found his quick witticisms quite a challenge. I'll enjoy picking his brains for medical knowledge only.'

A smile broke across the Irish dancer's delicate features. She preferred not to comment on Arneka's brusque manner, nor how she expressed the special name Kendall had given her.

It took hours for Arneka to settle into her new surroundings. Lack of substantial food had caused her nauseousness. She suggested to Brianna that they should step out of doors. Her room lacked fresh air. The ventilation unit was defunct, not broken. Just unconnected. As her squeamishness increased, Arneka found it difficult to breathe in the stifling heat. Relishing a fresh cool breeze, they both leaned over the front verandah railing. Together they watched a deluge of heavy rain sweep in across the parched, grassless paddocks.

'Kendall will be travelling through Switzerland for the first week of his holidays. I think he's booked to continue on through Austria or Germany. He'll be gone over two months. And I will miss him. Still, he deserves this break from studying medicine. He's been working long hours for a pittance. It all depends how his money spins out. I know he's keen on visiting some of his uni friends in Austria or Italy while he's away.'

Arneka's lips creased with a snooty twist. 'Ha, he gave me the impression that you two were tying the knot soon. You're engaged Bridy, so why aren't you going with him?'

'In time, we will marry. He needs this holiday, Arneka. It's his final fling as a single man. On his return home, he's supposed to begin in a medical practice up north. He may even end up working as a locum for Doctor Jarvis, here in Yass. One can't rush these things. The funds here on Abergeldie are restricted at present. Gigi's eager for him to work locally instead of rushing back to city life. At least then we'll both be able to keep my darling on a short leash.'

Arneka laughed. 'Hey, he's not a dog. His mother seems quite pleasant. Do you think she'll object if I call her Gigi? I'd hate to offend her by speaking my mind. Sometimes Kurt reckons I'm disrespectful. No decorum. That's me.' Up shot her arms as if to say take it or leave it, I can't alter my personality.

Together they discussed the idea of Mrs Ross objecting to her suggestion. Mystified, Brianna's peach-flushed cheeks broke into a dimpled smile. 'Of course she won't be offended. She dislikes being taken for granted. Darling, all of us here dispense with formalities. You will grow to love Gigi. I adore her. She'll be a genuine friend to you, if you accept her love. I think of her as my surrogate mum.'

Mm…m I might enjoy staying with Kendall, his family and you. Arneka shooed a blue cattle fly away from her eyes. *It's been a lonely life since our adoptive mother died. Kurt tries his best to console me. I know he misses his wife Anna even more than Christian and I do, although he'll never admit it to anyone, not even to himself.*

Settled on the front swing in the cool night air Brianna discussed what she and Kendall had planned for their wedding. Arneka felt envious and listened intently as her future sister-in-law elaborated on her fiancé's trip overseas, even though Kendall kept denying what day his holiday might begin on. Nothing could be finalised until the AFP permitted him to travel overseas.

The younger women's instant rapport pleased Gigi Ross. Listening to them quietly chatting underneath her window amazed her. Their feminine gestures and friendly camaraderie greatly boosted her morale.

Hearing Brianna speaking in her soft Celtic brogue rekindled her memories of the years she'd spent in her tiny Parisienne apartment. *Yes, they were the best years of my young life. I loved strolling with Pierre through the Montmartre gardens at dusk on mild spring evenings. On hot summer nights we'd drive down to the Quai des Celestins to watch the swans and their signets swaying to a magical tune on the River Seine. Paris in the dim twilight looked resplendent with streetlights glimmering on its rippled waters. Pierre often clutched my hand or kissed my fingertip. Oh, those delightful times still linger in my mind. I adored holding hands with his babes, as their au pair or nanny followed with their pusher close behind us. In my dreams I often see Mercedes dancing to a tune as Raoul sat on their father's balcony to play his flute. I miss his wife Marian as Pierre must do. I wonder what they are all doing now.*

'Now I'm fretting for Pierre, especially as I look upon that glorious sun ready to set over Jalna's fields of sorghum and bronzed ripened wheat,' Gigi whimpered. 'Why hasn't Kendall contacted him? Pierre must be frantic after not having heard from me in weeks. No, I mustn't cry. It will distress the dear man if he hears how lonely I become on these hot nights. I cried myself to sleep last night.' Tossing her blues to the wind, she settled in bed and let the tears flow.

Days of enforced bed rest had a depressing effect on her stressed and impassive mind. Allowed to stand, only to a degree, Gigi still required copious rest to enable her to survive through each day until nightfall. It took her ages to cope with an influx of house guests, some of whom were destined to stay under Abergeldie's roof for an uncharted length of time.

12

On the fourth morning after their arrival, Gigi Ross called the Baumer family to her bedroom. 'I must apologise for being unavailable when you arrived. It seems I made a fool of myself. Sit here by me, Arneka. What a pretty name, like its owner.' Gigi loved lavishing praise on this blushing young woman whose welcoming smile put her at ease. 'I've always wanted to have a daughter. Now my wish has been doubly granted. I'm thrilled in having you two darlings that I can call my own.'

'You really meant what you just said, didn't you Mrs Ross? I think it's us who should be honoured by being classed as part of your precious family.' Mr Baumer acknowledged the nod from both his offspring.

'Please Kurt, don't call me by my married name. I prefer Gigi. Would you mind fetching my son? I'm sure you two sweet young ladies won't mind sitting outside my door. You'll find the tapestry sofa in the hall quite comfortable.'

'Be our pleasure, Gigi,' her female guests trilled in tandem. All laughed.

Brianna held her future sister-in-law's hand as together they discretely left the room, while Kurt hastened to find Kendall.

'What is it, Mother? You should be resting. Not entertaining visitors. I hope …'

'Hush it, Kendall,' she interjected. 'I didn't send for you to chide me.' Gigi then requested something of her son. 'Darling, go to the safe in my office and bring me back the blue velvet pouches. Don't forget the red oblong box. I need all of those. Do it now son. Then I will rest … not before.'

'Stubborn woman, why can't you take our advice? It's sensible. Stephen and I …'

'What's this about me?' barked Doctor Jarvis on entering the room. 'I gave your mother permission to talk with her extended family. I suggest you do what she asks, or I'll do it Kendall,' proclaimed his colleague, standing behind him with a fist raised. Jarvis had resolutely put his point across and intended to stand firm on this matter.

'With both of you snapping at my heels, I haven't much of an option. Don't think I'm a gullible fool Mum, because you know I'm not.' Kendall fumed. Instead of arguing he decided to make a quick exit, rather than prolong their spat, knowing his mother wouldn't rescind and could overexert herself.

Several minutes later he returned with the items she'd requested. 'You'll find everything's there, Mother dear.' He threw the lot on her bed. With a huge sigh escaping through silent lips he snapped, 'Now what?'

'Be a dear, fetch Mr Baumer and his son. Brianna and his daughter are outside. That includes Father Brady and all our *personal* house guests. I wish to speak with everyone now.'

'Would you like me to call the relieving shearers and our stockmen in as well, Mother?'

This sarcasm sounded out of place for her quiet, reserved son. She could tell Kendall was annoyed by the way he had pronounced *mother*. Livid, he stormed from the room.

Doctor Jarvis followed him into the main hall. 'Kendall, one moment please. Don't take your frustrations or your anger out on your mother. She's not the one at fault here. You are.'

'Oh, come off it, Stephen. My mother's being obstinate and you damn well know it.'

'You're wrong. I approached Gigi. Not her me. Can't you see there's a lot of hurt in both your families? You're not the only one whose anger and damnable memories have buried your life in sorrow. The melding of your two families will be, I think, a predominate part of the healing process.' Jarvis tapped his left arm and the hand remained stationary.

'I won't ever apologise to her. In her grouchy mood, she would probably snarl at me. I suppose you are right,' he responded curtly to his colleague, whose glower made Kendall clamp up. 'Why did you agree with her, Stephen? Because you knew damn well I was right.'

'It's my prerogative to state what I considered to be honest. Wake up to yourself, before it's too late. By helping your mother, you'll help the Baumer's to cope with their tragic past. Then your own grief will dissipate. You only have yourself to blame for abusing your ill mother. In time you will be able to alleviate all this unnecessary anguish and misery.'

What can I do but agree with this knowledgeable man, who has offered me a job in his practice once my dreaded nemesis is either behind bars or hanged for his evil deeds? Jarvis understands my frustration while trying to control my delusional mother. We both know most of her health problems are compounded by grief.

'Now you're being sensible. Gigi needs every skerrick of love and understanding you can give her to pull through this because a nervous breakdown is a devastating illness, in case you Doctor, can't remember from your lectures. I for one intend to see her retain her sanity, with or without your help.' Having expressed his opinion dominantly, Jarvis hurried along the main hall until he reached her bedroom. With one hand embracing the brass doorknob, he paused to discipline the pup. Amber growled and with her teeth bared, she snarled and refused to budge, or let him re-enter her mistresses' room.

A sharp whistle from Kendall brought her to heel. 'Amber, you know better than to stop a person from entering your mistresses' room. You're a good girl for protecting Mum. Let's go for walkies. Sit and stay until I fix this lead to your collar.' She obeyed him.

With Stephen's latest lecture on his mind, Kendall headed to gather the clan. Still fuming, he followed Brianna back his mother's sanctuary where the Baumer family were assembled. With the two chairs occupied by Kurt and his son, Gigi gave the bed two taps.

'Arneka, you can sit here near me. Brianna, sit on the other side of the bed. I promised my pugnacious and obstinate doctors not to talk for long.' She glared at both her physicians, also present. 'I ask you all to be quiet for a moment. Then you can say what you think.'

Silence reigned in the room as each guest complied with her wishes. 'Good, now this is for you, Arneka. Open this first. You can show your menfolk its contents later.' Gigi passed her the box. The two velvet pouches lay untouched by her knee. 'Undo the clip. Be careful not to drop the box, or its contents will become disarranged.'

Cautiously Arneka unclipped the silver clasp. The lid sprang open to reveal a glittering array of jewellery and a diamond necklace. 'Gigi. No, I mean Christian, look at these stones.'

He peered over her shoulder. 'I don't believe it, sis. This is incredible. I can't believe my eyes. The sparkle of those small diamonds in this light is dazzling. Arneka, show Mrs Ross the miniatures in your locket.'

Instead of undoing the gold chain, Arneka slipped it over her head. A flick of the locket's clasp exposed two sepia snaps. Her fingernail prised the heart-shaped bracket loose. A second gold heart remained attached to the locket's case. 'That woman in the miniature is Erika, our mother and she's wearing this necklace. Look, now we know the centre, elongated stone is an emerald. The six gemstones down each side are small diamonds. We've always wondered what they all were. Now we know.'

While Gigi perused the miniature photographs, Arneka looked up and whispered to her brother, 'Should we tell her whose photo is underneath our mother's one, Chris?'

Not quite knowing what to say, Christian replied softly, 'No, what good would it do. Let her enjoy this special moment with you and Miss O'Shea. Why put the ill woman through more agony. We now know whose image it is. No, forget it sis.'

Their minds absorbed by the emerald jewellery, not one of the women had noticed Kendall coax his male companions to quietly gather by the end of his mother's double bed.

'Arneka, pass me back my jewellery box, please,' Gigi pointed. 'Let your father and brother hold the necklace. Then they can see how the diamonds glimmer under lamplight.' By giving orders it made her feel superior. The thrill of being in control made all their dramatic traumas worthwhile. This treasured moment of drawing both families together produced the desired effect of closure. A brief pause gave Kurt Baumer time he needed to cope with his horrific past. His suffering both in Hamburg and on their trek to freedom would forever be listed on the annals of time. He preferred to leave it buried with the phantom of evil.

Soon, God willing, they may all settle into a normal lifestyle with my family while they are holidaying here with us on Abergeldie. Gigi sighed and she saw tears in Arneka's eyes. *From now on, I vow to accept the future and whatever lays ahead for this, my conjoined family.*

'They are emeralds,' Kurt concurred. 'I've never seen this necklace before. Heide told me how magnificent the pendant looked on Erika's slender neck. From memory, I think my sister mentioned something about a ring …' he ceased speaking, when their hostess politely intruded.

'Excuse me, Kurt.' Gigi smiled at his daughter. 'Please dear, tease the silk cords of both those velvet pouches apart. Inside you'll find what I think your father was referring to.'

She cautiously tipped the contents out onto her unsteady palm. 'The earrings were Erika's and that was her engagement ring!' Arneka gasped with delight.

'Fasten the necklace around Arneka's neck, son. You're the nearest.' Gigi watched him comply with her directive.

'Lift up your hair please sis, then I can see what I'm doing. Ouch!' he squawked.

'What's wrong, Kendall? Did the clasp bite your finger?' Arneka sniggered. Slightly twisting her head, she held her long saffron curls up enough for him to see the chain.

'No, not exactly!' He sucked his index finger. 'Something made it tingle. I reckon it was static electricity. Keep still please, Arneka. Ah, there the clasp is now fastened.'

The five men standing by the open window felt a gush of warm air breeze past them. The force had sucked the chiffon curtains towards the wire gauze. Released of enforced tension, the soft lilac fabric belled back into the room. Astounded, they looked at one another. Enraptured by the sparkling gems, the three women noticed not a thing.

'I think it's time we let the ladies enjoy their quiet sojourn, without us interrupting. Wouldn't you agree, gentlemen?' Doctor Jarvis focused on the door.

A communal "Yes" was said as they quietly exited the room, leaving their counterparts to dream.

In the lounge room Kendall opened up by saying, 'Hell was I glad to get out of there. Did you all feel a spooky atmosphere the instant that necklace touched Arneka's flesh?' He gave a fearful shudder. 'It felt as though an electric shock had passed right through my body. I know you all sensed something weird. The shock showed on your faces.'

'Yes, your face imaged mine,' stated Kurt. 'How could such a weird feeling affect all of us at the same time, Father George? It struck me as though a ghostly apparition, or cold fingers had touched my shoulder. Then I felt a cool breeze brush past my face,' he said, leaning his bottom on a barstool.

'A similar feeling overwhelmed me,' declared Stephen Jarvis.

'I agree with ya all. A creepy feelin passed right through me. It's peculiar how or why it occurred. In me time as parish priest in County Clare, I saw weird things happenin at our church at home. I've heard of priests exorcising people who'd been inflicted by evil spirits.' Father Brady began quaking in his size six shoes. Unfamiliar with sorcery, he preferred to abstain from commenting on the occult. Uncertain what to say, he couldn't think why an apparition could have caused such confusion in his hostess's room.

'Whiskies coming up all round. I think we can all do with a drop of that delectable nectar,' proposed Kendall. 'You all know what I think.' He uncorked the cut glass tantalus and selected five whisky glasses from the bar fridge. 'I'm sure the women were oblivious of that weird event which occurred in there a moment ago. I also felt a really odd sensation ripple down my spine. Like everyone here, I think it may have had something to do with Erika. I know that statement sounds absurd. But it does put a positive slant on the incident.'

The furrow on Kurt's forehead deepened. 'It seemed as if Erika had returned to bless her daughter, the moment you secured the catch of her gold chain on Arneka's neck.'

'Yes, I also felt a tingling sensation travel down my spine. What's your opinion Father?'

'Well it sounds logical, me boy. How can we be sure what caused the strange phenomena? Cold shivers ran up and down my back. More of an unpleasant feelin I'd be thinkin,' the priest declared, still focusing on the uncanniness of the situation.

Trying to fathom the reason, Christian Baumer shook his head in disbelief, thinking of the unstable feeling he'd felt as a wisp of cold air breezed past him. Seeking an answer, he looked at his father, now seated on a lounge covered with autumn leaves fabric. 'The perception I got wasn't unpleasant. Dad, it seemed more of a gentle feeling that lasted a split second. I think Erika's spirit, bound in tragedy all these years, may've suddenly escaped its bonds. Whatever it was, we can be certain it's gone.' Chris rendered his version while sipping a whisky-dry.

Sitting on the bed in Gigi's room, Arneka fiddled with her diamond and emerald necklace until she made herself comfortable. She was thrilled over the gift of her mother's jewellery from their host, although she disapproved of Brianna's envious glares.

Alert of her quandary, Gigi refrained from commenting. Uninspired by her lack of enthusiasm, she pensively appraised her grimace, until an idea was born in her mind. 'On the dressing table, beside my clock you'll find a trinket box. Be a dear and pass it to me Brianna.' Once she received it, Gigi gestured for her to return to the bed. 'Darling, I was going to give you this on your wedding day. Now, I want to see your face light up with joy as you lift its tortoise shell, silver-embossed lid.'

'This is your black opal ring on a gold chain. I can't accept such a precious gift.'

'Why not, Brianna?' Taken aback she queried, rather puzzled over her refusal. 'It's my gift for you to wear on your wedding day. Darling, turn the opal in different directions then you'll see fire flashes of brilliant red to deep purple. They meld in with the pale apricot streaks, hidden in the opal's fiery lustre.'

'Look how the colours shimmer against my fingernail, Arneka. I shan't take the ring off until dinner. I mean until lunch is ready. Would you mind if I ask you to keep this ring in your jewellery case, Gigi? I shan't lose or misplace it until I'm dressed in my wedding gown.'

Her nod indicated no. 'Darling girl, one never can be sure when our life runs short. That's one lesson I have learned during the past few weeks. Wear the ring with pride, Brianna.' A brief tap on both the impulsive gigglers' laps indicated she wished to doze. 'Later, much later, I may have another trinket to give you. Only time will tell.'

'I … I hardly know how to say thank you,' Brianna stammered. 'I didn't expect anything. And I am delighted and I will adore this magnificent gift.' In a genuine effort to express her feelings Brianna embraced her elder. A flamingo kiss left an imprint on her florid cheeks.

'You've already proved your love for me and my son. I could not wish for two more precious darlings than you two giggling youngsters. You've doubled my happiness, in this droll life of mine. I call it "The birth of wisdom". Now leave me to rest.' Gigi glanced towards the door and nodded. 'Brianna, if you can stop twittering a second, ask Kendall to come in here, there's a dear. Ask Kirsten to do so. I think she's talking to Harriet, our cook in the kitchen.'

About to trot to her room Arneka coughed. 'One moment please. Mrs Ross, I think you should keep the emerald necklace. It rightfully belongs to you. Not me.'

'No dear, you're wrong on both accounts. Under such stringent rules that govern all our lives, to keep these gems would be an injustice. They were never mine. They belonged to your mother, Arneka. I know that now. I want you to enjoy their beauty while you're still young. Old fogies like me don't go anywhere to don extravagant trinkets,' Gigi sighed. 'I wore the emerald necklace to a gala performance in Milan once only. I'm ecstatic to think it's where it belongs … around your dainty neck.'

A long gasp emitted as she held her throbbing head. 'Now please fetch my son, Brianna. I won't keep him long. Darling, I know you both wish to discuss your wedding and his overseas holiday. Perhaps you'll find time this afternoon, or later tonight to sort out your differences.'

13

The instant Kendall entered her room he was aware of his mother's confused state of mind. Her eyes lacked spontaneity: they'd lost their fiery spark. Exertion had wilted her spirit and left a vacant expression on her drawn, though still beautiful features.

'You wanted me, Mum. Why, may I ask? Overdone it this time, I'd say by the look of those flushed cheeks. Why won't you listen to Stephen and me? We're trying to help you.'

'Really, it's him I want. Not you. I feel a bit fain … t.' Her voice faded into obscurity. Everything swirled, curtained in black, her mind lacked the power to accept speech. Suddenly the vortex of colliding thunder ceased in her inert brain. Her head slumped forward. A blank fixation clouded her eyes. They stared through a misty haze at Kendall. She could neither see nor hear her son, who spoke to his colleague as he entered the room.

'We'll do this quicker by working together, Kendall,' Jarvis affirmed, selecting a cluster of instruments from his medical case.

A pen-torch shone in the darkened room. The negative response to its flash behind her eyes indicated to Jarvis that Gigi had lapsed into a semi-comatose state. Methodically both the physicians assessed their patient's condition.

Her head lay in a distorted position. It lolled to one side and remained stationary. Briefly her eyes fluttered. A minute later Gigi recovered consciousness.

'In my opinion your mother is verging on a stroke Kendall,' Stephen whispered reviewing her chart. 'We must be more stringent with visitors from this week on.' Standing back to allow the son to examine his mother, Jarvis smiled as her eyes flickered. Blood flowed to her pale cheeks and the lifeless fingers began to twitch.

The old hand at medicine straightened, to quietly confer with his junior. 'You must agree; a stroke is possible. When is anyone's guess. No wonder, with the stress she's undergone this month. I really think hospital is the safest place for your mother.'

'No … no Stephen, not … not that, not hospital,' her voice faintly ebbed and her breathing shallowed. Battling to stay calm she muttered. 'I can't afford to be in hospital, not now …'

'Mum, don't be foolish. You'll overexert yourself again.' Kendall stroked her forehead to relieve her worry. The wistful smile expressed her desire to remain under Abergeldie's roof, which he couldn't ignore. 'Fine. I'll give into your demands, just this once. Why must you keep fighting the important decisions we make to stabilise your health problems? You know why we try our damnedest to make the right medical choices.'

Jarvis knew she detested being in hospital: so did her son. Quietly they conferred how to restabilise her condition. After some deliberation, they reached a unanimous decision. It related to a specific matter that must be done to prevent her mental collapse.

'My suggestion is to acquire a full-time nurse, a woman who is capable of administering all manner of drugs we suggest,' stated Jarvis, mindful of the consequences. 'We can arrange for a live-in nurse to care for your mother. You know the woman I mean.'

'Sylvia. She's trustworthy and an extremely capable nurse.'

'I agree. Hey man, why look further than your own back door?' Jarvis tested Gigi's reflexes. Both her knees and ankles responded as normal. Drawing up the drug in a new syringe, Stephen flicked it to expel the air.

'What do you mean exactly? Perhaps I'm dense, but I don't quite follow you.'

'Let's discuss it outside once I'm finished here. I've administered this injection to your mother's bottom so she should relax and probably sleep for some hours. I'll keep monitoring her vitals half-hourly until she fully responds. Her prognosis looks promising.'

'Fine by me,' Kendall growled stepping from the room. Stephen followed, quietly closing the door. 'Sorry, this trauma has affected my brain. I can't think clearly. What did you mean by saying look no further than the back door?'

'Gigi will require round-the-clock specialling,' Stephen Jarvis responded to his nod. 'I'll call Sylvia Benson while you approach Arneka. It will keep her from becoming maudlin. She is a qualified nurse, Kendall. She'll be doing the job she's trained for, I'm sure.'

'You've solved a difficult problem. Stupid me! She'll be ideal.' His hand brushed a wisp of dark hair from his temple. 'I'm gratified with the solution. Mum will be thrilled, if Arneka agrees to be her daytime nurse.'

'Sylvia works nights in the local respite centre. Or she did until she began specialising in Goulbourn Hospital. Another problem solved, if she accepts the offer to be your mother's nurse.'

'Cost's no object,' Kendall replied, eager to quell his appetite. 'I didn't eat breakfast or lunch. I'm not fond of pig's ears or kidneys sautéed in red wine. I'd rather tuck into a rare to medium steak marinated in wine, smothered in mouth-watering gravy.'

'I'll attend to Gigi's medications, before she drifts off. It'll be friendlier if you call Sylvia. You know her better than me, Kendall.' Jarvis scoffed. Eyeing his wristwatch, he stepped back towards his patient's room. 'You will probably catch Nurse Benson at home now.'

After speaking to her on the hall extension, Kendall hurried back to his mother's room. Keeping a low profile, he observed Jarvis administer the injection of Diazepam into his mother's thigh. 'All set. Sylvia's off duty for a whole week. She'll be here at six tonight.' A crass smile flashed crossed Kendall's smug face. 'Stephen you old codger, we work well together, you and I as a team.'

Casually glancing down, he noticed a slight smile embroider his mother's lips. Although she understood, she couldn't speak. Gigi winked at Jarvis who smiled. Both her physicians smirked at Nurse Baumer who'd just entered the bedroom.

Arneka fluffed up her top pillow then resettled it in a comfortable position under Gigi's neck. 'Don't try to talk Mrs Ross. I want you to know how honoured I feel, being your full-time day nurse. If you need me or anything at all, ring this china bell. It belonged to my nursemaid in Germany. Heide was my aunt, my father's sister, whom we as kids called Button. It seems eons ago now. I'll tell you her sad story one day. If I don't, Chris or my father will fill you in with all the sordid details of our horrid life in Hamburg. Not until you're giving me cheek and sparking on all four's again. My aim now is to make you feel well again.' Arneka kissed her fiery-hot temples. She then placed a damp compress on her forehead to keep it cool.

A hoarse whisper emitted through parched lips. 'Thank you, Arneka. Call me Gigi, please dear. I hate Missus. It's not applicable to me now. I feel thirsty … a dry mouth.'

'Suck this crushed ice through a handkerchief and let me do the talking.' Arneka's gentle fingers cradled Gigi's stressed neck, while her other hand held another ice-cold cloth to her lips. 'I am fortunate. In my short time on earth I've had four mothers. My birth mother, Erika. Then Anna, Kurt's darling wife and his beautifully natured sister, Heide. Not many people I know can boast of all that love in one lifetime, let alone in almost thirty-odd years.'

'How sweet of you to think of me as your mother, Arneka. Are you that old? You really don't look it,' she uttered in a soft drawl. Battling to breathe and feeling exhausted she paused for a breath until the grogginess passed. She grasped her temples which throbbed constantly.

'You shouldn't talk, Gigi. If your son, or his quack mate out there hears us gabbing, we'll both be in strife.' Sister Baumer put a finger to her lips. 'As of now, I think of your son as my full-blooded brother. I hope he doesn't reprimand me for not following doctor's orders.'

His mother's parched lips formed a crescent smile. And her eyes beamed with adoration.

Placing another cool compress on her patient's fevered brow, Arneka sat on the bed and held her sweaty hand. With gratuitous tears ready to flow, she embraced the limp fingers and silently prayed until Gigi drifted off in a peaceful slumber.

Quietly and mentally recalling the traumas of previous weeks, she heaved a reluctant sigh. *I know now we were destined to become one family. And I couldn't ask for a sweeter mother than you, Gigi. You're a very special person and precious to me. Now our family is complete.* Arneka gasped with relief. *If only you could hear my genuine heartfelt cry.*

Silently Gigi acknowledged it, while drifting into the realm of lost dreams.

Sitting at the desk in her room, Doctor Jarvis revised and wrote up all her medications. Arneka, on a tea-break, had joined Kendall in the kitchen. For longer than ten minutes they talked of the overseas holiday, the trip of a lifetime which he'd planned for months. Their future together as a family remained forgotten in the backblocks of his numb mind.

'Before you leave for Austria or Switzerland, think of me as a 'kangaroo charm' and tuck me in your suit pocket. I can't afford to travel overseas. Not on my lousy salary. Money isn't easy to come by, and nurses aren't privileged like your lot are, Kendall.'

'Who's having who on here? Working thirty-six-hour straight shifts was the only way I could boost my wages. As an intern I'd score fewer hours, if I was lucky. It's not all beer and skittles you know, so don't kid yourself, Arneka. At the moment I'm awaiting to hear positive news of two positions. One is working with Stephen in Yass. The second job is in North Queensland, if it eventuates. Whether I get a reply from there, is uncertain.'

'Gigi mentioned you would be a wealthy landowner soon. Is it true?'

'Well yes, only when we sell Abergeldie. This property and its landholdings are on the agent's books. Mum and I intend to split the sale money equally. Don't say a word to her, Arneka. I've yet to broach the subject with both my darlings. I think Bridy has some idea of my devious ploys.'

'Bit underhand, isn't it? Gigi led me to believe that she loves living on the land. Everyone here thinks she's the cat's whiskers. It would kill her to leave Abergeldie and reside overseas somewhere.' Arneka relished sipping her second Bundy rum and coke.

Kendall laughed. 'To the contrary, Mum reckons if she stays here the worry of managing this huge property will shorten her life. Once she would never have admitted that. Now it's different. A sad fact of life!' Huge tears welled in his eyes. 'She's gone through hell and won't even consider remaining here. Bitter memories constantly toss her frail mind into a vortex of despair. His evil can't be forgotten overnight. It's like living in a black hole with no escape.'

In a thinker's pose on the lounge, he tried to dispel the sad memories from a dysfunctional brain. *I regret having to witness my mother shivering from her heart down to her feet in fear. My greatest fear is for her sanity. Her health was wrecked by that Nazi contriver, a man born of evil. Mum spent the best years of her life caring for those she loved. Worry has caused her to age prematurely. Now she resembles a woman of seventy. I think she's torn between anguish and the desire to live free of fear. I hope she relinquishes the sentimental crap he created because of his jealousy and brutality. I'll never forgive him for the torment and pain he caused my mother. My pain and anguish are nondescript, compared to hers.*

Arneka glanced up from reading *The Lancet*, a medical journal and frowned. 'Were you dreaming, Kendall? You muttered some indistinguishable garbage, quite abstract as if your mind was lingering somewhere in the past.'

'No, not really. Just thinking of a positive flash and how to resolve our dilemma. I'll go overseas as planned. On my return I'll sell this mausoleum and all its landholdings. My mother will not prevent me from putting Abergeldie on the first stable market. I'll be damned if I'll let that Nazi beast ruin all our lives. My mother's desire to die will not be fulfilled. This is a battle of wits. My strong willpower will endure against her weak contentious one. Arneka, I *will* strive to win this controversy, even if she keeps grumbling or sulks for days on end.'

14

The night nurse began her shift at eleven then Arneka returned to the kitchen. Not having eaten with the family at six, she was feeling hungry and a bit nauseous by eleven. In preference to the heavier meal put aside for her, she munched on a cold lamb chop and bacon. Even at this late hour she caught Kendall walking into the kitchen to make a ham sandwich.

'You're a fibber, sport. You assured me you were turning in yonks ago. What are you doing up at this hour? It's gone eleven.'

'I couldn't sleep with all the day's happenings rattling through my brain. Do you want to bunk down? If so, I'll hear if Mum calls.'

'No need to Kendall. The nurse you employed has just arrived, so I've given her my case notes to read. Shit, I forgot to enclose your mum's temperature chart. I'm not bunking down until I fetch that damn chart. I may not even then. It's such a hot night, I might go for a trot up the drive.'

'What! Ride a horse in the dark at this hour? You're cuckoo, sis.'

'No, I'm not. Just restless. In a tick, after I return from speaking to Sylvia, I want to discuss something with you.'

'And I you, Arneka. It is important. It's seldom we have time to talk. While it's quiet we can sit on the front verandah swing. We won't disturb mum, if we keep our voices low.'

On her return she seemed lost in a world of her own. Once he'd finished glossing over Sylvia's notes Kendall replaced them in their folder. A slight hesitation ensued to remind her not to forget to check the drug chart, because Jarvis had upgraded his mother's nerve medication. He asked her to leave Gigi's chart on the hall table for him to peruse. Kendall then followed Arneka out to the verandah.

'What do you think, after breezing through your mum's chart? Sylvia will browse through my notes before she makes a coffee.' He nodded. 'She also has all my general assessments I'd done today.' Arneka pouted. 'Aye Kendall, she seems quite a nice person. Knows her stuff by the way she spoke.'

A huge yawn invaded his ageless face. 'Yeah, she sure does. I'm appreciative of her offer to stay here for a week longer than scheduled. It'll free you. Besides, it'll give you time to spend with Brianna.' Kendall yawned again. 'I should have the damn accounts finished before dawn. Until now, I've shelved the bookwork because I detest totalling five rows of figures in one hit on these accountancy pages. Keeping everything up to date is an utter pain. I'm not a damn mathematician.'

'Not my cuppa either. I'm not a bookworm and I think bookwork sucks. Still, I suppose some idiot has to keep all of Abergeldie's records up to date.'

'Thanks for calling me an idiot,' he laughed. 'Arneka, I ain't no mug. Having to keep tax records and farm produce itemised accounts creditable is a tedious business. Bookwork is a necessary evil, a thankless job.'

They joked over the hazards of working in their medical spheres. Kendall thought his half-sister a comical wag. Some of the sarcastic comments she made about doctors she'd worked with in Queensland sounded unreal.

Once their frivolity subdued, and back in his office, Kendall broached the subject of his mother's health. This was accomplished while trying to total eight columns of figures. 'You may not realise this Arneka; mum thinks it's terrific having you here. Since her illness began, she is often quite cantankerous, especially if she's overtired. She complains of loneliness at times.'

'How's your mother going to cope with all this bookwork while you're overseas? I can't imagine her ever leaving here. She loves living on this huge property. This place is her life's blood. Brianna told me tonight, they both love riding Abergeldie's horses across your high paddocks on cool summer mornings.'

Accepting his nod in agreement of what she just said, Arneka held her empty glass aloft. Agreeing to a small whisky over ice, his fingers indicated an inch.

'I think mum will cope far better than she did before he absconded with thousands of our wool clip dosh. I hate that beastly Nazi mongrel, because of the savage beltings he gave Mum and myself with the buckle end of his leather belt. In a furious temper, he thrashed my three-year-old bottom, for a slight misdemeanour. When angry, he'd smash everything within an arms-length. That violent temper will be his downfall, Arneka.'

'Yes. I hate that Nazi swine. I was forced to live with his evil eyes peering at me, in my infancy. It must've been sheer hell for you and Gigi, before he absconded from here. Neither Chris nor I trusted him, as kids of three and four. Chris hates him more than I do.'

Downing her drink in one hit, Arneka crusaded on. 'Kendall, I know what it's like to see the mother you love being cut to pieces by a leather strap. He battered my mother senseless. We'd left with Button, I mean Heide, not long before he murdered our mother, Erika.'

Unable to control her emotions Arneka sank to the floor. Tucking her head between her knees she sobbed inconsolably. Unable to cry, or express her innermost feelings since the devastating news of von Breusch's atrocious and evil past broke, she now felt relief.

'There, there darling, let the tears flow.' Kneeling alongside his sister, Kendall cuddled her. 'Arneka, all those pent-up emotions of yours need releasing. My shoulder's a solid one for you to cry on, if needed.'

'It's been a terrible struggle to keep sane. I never wanted to recall those horrid memories. The week of endless interrogations by the Feds dug up memories I'd almost forgotten.'

Kendall's finger teased the damp curl from her cheek. 'I can't imagine the anguish and torment you and Chris suffered in Hamburg, nor how he wielded his cross-strap to strip away your dignity. The shocking beltings that swine gave you and your brother were similar to the ones he lashed out to me. He mistreated me unmercifully. Mum fared no better. He belted her with something solid until she lay unconscious at his feet up in our top paddock. Lying on sodden ground, he covered her with shrubbery and left her to die. If our native backtracker hadn't found Mum that night she would've frozen to death before morning. I remember running my fingers over the welts, blackish-blue bruises on her arm and neck. No father should ...' His mouth clamped shut, realising he should not have mentioned 'father'.

The word repulsed both he and Arneka. *How cruel and uncaring of me not to think of her feelings.* To rid his mouth of the bitter taste Kendall ran his tongue over his teeth. It repelled him to even think of mentioning their avenger by name. Nobody in their household or over on Jalna could bring themselves to speak of Abergeldie's ex-employee, without defaming the Nazi.

'That swine's never been anything of the kind to us. Kurt's the only father we have ever known in our lives. And he has been since our escape from Germany.' The same strand of saffron hair flicked down to grace her eyes. Tossing her head, the filament settled back in place above an eyebrow. Sniffing, a solitary tear blemished her flushed cheek. 'No one can imagine the hell we went through, before we sought asylum in this beautiful country. I love living here in Australia. In Queensland on fine sunny days I often wade in puddles or gather driftwood on the beach just below our house. Sometimes I ride my pushbike or walk for kilometres along the coast road or through forests and enjoy a swim in a pool under a waterfall, surrounded by ferns.'

Kendall listened to his sister rambling on about how she enjoyed the little pleasures of life in Yeppoon. Reminiscing of an unhappy childhood, his mind deflected to numerous times when he'd tried to escape a beating over some stupid childhood prank. *The night he forced my head down our toilet after he'd done a piddle, I gulped in mouthfuls of unflushed water. Afraid of drowning, I lashed out with my fist and struck him in the groin. He backed off and in agony let me go. Being immersed in piddle, even for a second stifled my breathing. Gigi never raised her hand to me in anger. An intolerant man, he used to wield his leather cross-strap on my bottom. In my childish mind and in pain mum tried to stop the bleeding by bathing me in warm salted water. Then she retied my jarmy pants and I cried myself to sleep. So I know how much pain and anger Arneka has suffered. I'll never forgive that inhumane beast from Hamburg.*

Embracing his sister, he asked, 'Are you feeling better now, sis? That good cry has eased some of the tension. Many a time I've suffered from his nasty handiwork, Arneka. That creep delighted in taking his temper out on me in front of my mother. Gigi couldn't do a thing to save me from being bashed senseless. She was helpless. As I said, he's a useless idiot not worthy of consideration as a cattleman's bootlace. You should think of him as I do as Max …'

Angrily she rebelled. 'Don't ever call me Max, like you just did. In case you didn't know it Kendall, that beastly man's middle name is Maximillian.'

'Oops. I stuffed that up, well and truly. The simplistic way you phrase things reminded me of my mentor at uni, a professor. Maxine Bolton's attitude to Nazism equalled ours, sis.'

He delighted in calling her "Sis". 'Think of him as I do Arneka. I'm here because of a tiny seed implanted in my mother's vagina. I don't and never will consider myself a product of his loins. You and I originated from a sperm that fertilised an egg impregnated in our own mother's womb.'

Deeply touched by his sentimental, though true words, an occasional sob from Arneka rendered her incapable of speaking. In an effort to gather her thoughts, she endeavoured to forget the past. In silent reflection the torment in her mind eased to a tolerable level. Contented in his peaceful arms her stress eased. The support from her newly-found sibling was deeply appreciated, yet it had come quite unexpectedly.

'Once you're composed, try to express your feelings aloud. Then you'll be able to cope with every problem you encounter. Remember, we all love you dearly Arneka. Ah,' his index finger wavered, 'keep in mind I do, but as a doting brother.'

Appreciative of his joke she solemnly decreed, 'Oh, I know that. Kurt says our families were destined to be reunited. We became naturalised Australians not long after landing in Melbourne five years or so ago. The Australian authorities won't send us back to Germany, will they? I will never go back to confront all those horrid memories that haunted us there.'

Sipping the dregs of his now cold cocoa Kendall shook his head. 'Listen to me Arneka, because you're naturalised citizens, our government will never deport you or your family. Well, not unless one of you does something illegal. Something really offensive against our federal laws or policies, and I can't see that happening. If anything, they should award your family medals for helping to capture the Nazi war criminal. Locked behind bars in prison is where he belongs. Better still, they should hang the bastard. He's a murderer, and a woman hater. The sooner they capture and string the mongrel up the easier I'll breathe. In clink he won't be able to harm you, us or any other innocent person ever again.'

This genuine appraisal from Kendall, boosted her spirit. It also bolstered her ego and gave her the courage to face the future and whatever it threw in their direction.

'Now darling, we must be up early and it's gone twelve. Off to bed with you, Madam.'

She giggled. 'Not yet, I'm not. A madam, I mean.' Her quavering tone and dry satire sent his misery drifting to the night wind and set their humour ablaze with silent mirth.

He enjoyed her sardonic humour. He now knew Arneka wielded the power to challenge any man and win hand's down. And he revelled in their friendly camaraderie.

Downing several gulps of whisky Arneka stretched then sighed. 'It's been an exhausting day.' Feeling a bit tiddly and through glazed eyes she appraised his drowsy eyes. 'Kendall, it's time you turned in. If you sit there much longer fiddling with those damn accounts, you'll fall off the perch.' His eyes rolled, and he ignored her witticism as she sauntered back to her room.

Yawning, she crawled under the covers of her warm bed. Tomorrow was another day.

15

Seeking solace in a place of tranquillity, Kendall found the key to his father's office, or den. Flopping down on the chair, lacking its original graffiti, he pushed a red leather ledger across the desk which allowed enough space for folded arms to support his head. Lost in the fantasy world of dreaming came a vision of two babes, their faces smeared with blood and tears. He pictured a tall man in a grey uniform, a cap bearing a skull and what seemed like crossed swords. A convoy of gypsy caravans laden with bales of straw; children huddled beneath hessian bags. A young woman in tears, her bronze hair braided, vivid storms, people fleeing from terror, a grey submersible tossing on a turbulent sea. A sharp clap of thunder disturbed the dreamer. And a calamitous racket up Abergeldie's drive caused its occupants to jump out of their beds in amazement, and in an awful hurry.

'Bloody hell', he gasped, leaving his father's den, 'that sounded like gunfire. Who the blazes would be making such a damn racket at six? Must be a rustler trying to steal our sheep or cattle again.' Hurriedly he washed and donned clothes before a hot morning sun burnt off the remaining clouds, much to his despair. On leaving the bathroom, Kendall confronted a dark figure wielding a gun of some description. Not a rifle.

A federal guard, a Frenchman by birth, pointed towards the front verandah door, just past his mother's room. 'It's okay Kendall, we have the matter under control. Now you, your family and your guests are safe. Check on your mother and fiancée. Her loud screams alerted my men that something was afoot, either here or up the drive.'

Shocked and ready to question the reason, Kendall blurted in a daze, 'What the hell's happened Rob. Has a bungling intruder tried to harm my mother? I've not long come from her room, and Mum was asleep. Brianna and I were about to enjoy breakfast in the kitchen with my sister. I want to know what's just happened. Tell me, Rob.'

'Patrolling the area close to your front gates, Tim my offsider, discovered a youth hiding behind the azalea hedge. No, amid a clump of Rhododendrons. He fired a shot to maim, not kill him. The kid then ran towards this house. We apprehended the lout and he's now in custody. A local copper nabbed him. Later today I'll drive you into Yass. You'll probably identify the kid, Kendall. We couldn't get a word of sense out of the stupid twit.'

'Shit, this is all we need.' Anxious to check on his mother, Kendall, clad in denim jeans, checked cotton shirt and with his leather chaps dangling over one arm, cursed. 'Rob, I was ready to head down to the stables. We'll be mustering once a fresh crew of circuit shearers arrive at ten. Will you give Bjorn a buzz and tell him what's occurred? I know with federal restrictions still in force, we can't dial out from here.'

'Okay, I'll do it now, before we get bogged down with red tape from Canberra. I'm being honest in saying this; we expected some sort of intervention. Forewarned, we devised a plan to nab the youth and his old man, who we suspected may've been in cahoots with von Breusch. That Nazi bastard couldn't escape from here, not without some local bastard in his pay. His departure occurred late one night, as you know, so the Nazi would've needed some assistance. Over a period of four weeks my men and I have analysed and studied his recent escapades in and around the Cross in north Sydney. Well prepared, we alerted their security blokes; if seen in a local brothel they were to apprehend the absconder.'

Kendall laughed. Something else lay undetected besides what he and Bjorn suspected of his father. 'Rob, who'd be insane enough to put their lives in danger to assist him to abscond from here? Like you, I'm damn sure he must've paid a local bloke to cover his tracks before heading north. It's an eight-hour drive in good weather from Yass to Sydney. I've done that journey often enough myself, more so over the last twelve months.'

'A German, an ex-Nazi lieutenant with an Austrian affiliation has been living right under your noses, here in the Southern Highlands. We came across their association by accident. Stumbled right onto it through one of our undercover agents. He resides in Bowral, and often frequents his cousin's place in Yass. I'm forbidden by law to reveal either of their names. Suffice to say they're both in custody. Tomorrow sometime before nine, two of my officers and Charlene my secretary, will accompany the two traitors down to Canberra. As of now, I think it might be by helicopter. I'll know the precise time and more details tonight. Everything we've discussed here must remain confidential Kendall, as per usual.'

'Phew, what a relief, Robert. Thanks a lot. Bjorn and his crew are expecting my blokes to meet his relieving shearers in his lower paddocks at nine. I'll be in with mum for ten minutes if you need me. Then I'm riding my horse Potchkin over the lower paddock to Jalna.'

Grabbing a warm jacket off the doorknob, Kendall heard his mother snoring. Not sure whether to enter Brianna's room, he was tempted to leave, until Arneka called him.

'Come on in, don't stand there and gawk. Surely you've seen a woman in her knickers and bra before? How's your mum? I heard her screams and rushed to her aid. She was scared witless, a bit shaken and shocked by the gunfire. What are you doing on this fine morning?'

A little embarrassed he blushed, frowned and spluttered, 'Mum's asleep. Or she was until I poked my head around her door a second ago. It's you I want to speak with Arneka. I need your advice on a minor problem. A favour. I almost asked it last night. Perhaps, we can discuss it after breakfast. I …'

She interjected, 'No, what's wrong with asking it now? I'm decent, park your bum on my bed. I'll straighten it later.' Arneka gave the sheet a firm tap.

'Stephen and I think you're a responsible and capable nurse.' Kendall laughed at the way she curtsied. 'I want to thank you for boosting Mum's morale. We medicos can do only so much where she's concerned. I hope you have a few uncluttered and dignified gags ready to amuse her while I'm overseas. Better you than our staff trying to keep her sane. My mother can be cantankerous and downright rude in a rotten mood, and quite arrogant at times.'

'As a qualified quack, you haven't lost your sense of humour. A lot of the people in our profession are devoid of humorous frivolity, and they lack courtesy. I find most of your breed prudes, with swollen heads *and* everything else.' Having off-loaded her honest opinion, Arneka rendered a hearty chuckle. 'Talking of your mother, with Bridy's help I'll make her toe the line, and keep her sane by telling her respectable gags. Or we'll wear her out with our nonsensical nattering.'

'Take my advice, Arneka. Don't test her patience, especially if she's grouchy. She's been known to make a protagonist's ears ache with a tongue lashing.'

'Ruhr, not mine she won't, Kendall.' Arneka giggled at his clownish downturned mouth.

His enlarged eyes flickered as the buffoonish cheeks broke into a smile. 'All joking aside, I've agreed to take this holiday under duress as you, my dear sister, know. While I'm in Paris or travelling through Europe I'd appreciate you specialling my mother in a paid capacity. Don't argue, I insist on paying you the regulated fee. It will alleviate a lot of worry. I'm sure you won't object to staying under Abergeldie's roof for another six weeks. My offer includes your family, Arneka.'

'I'll have to ask Kurt and Chris first. You know I can't answer for them. Kendall, I'll stay here and work without wages. My duties will repay your mother for her generous hospitality to us all, although my brother's overpowering and argumentative attitude won't improve.'

'Wages or nothing. I can hire a trained nurse. Brianna might return with Kurt, if he goes home earlier than expected. Be a pleasurable break for her. She'll be terrific company for Chris. If you agree to my proposal, I insist on paying for their fares.'

'Codswallop, you're not going to waste your hard-earned dosh on Chris. He can pay his own damn fare to Yeppoon. Besides, what you've proposed is blackmail.'

A stern glare from Kendall warned her to desist. He disapproved of her unjust criticism of Christian. And he totally disapproved of the dominant attitude she displayed. Shaking his head, he frowned. *This dominance seems so unlike Arneka. Well, it serves me right for sticking my nose in where it wasn't wanted. I must know if she's willing to special mum, before I finalise my holiday travel plans overseas. Arneka's indecisive mood leaves a lot to be desired.*

'Poor Bridy,' she scoffed over his last suggestion. 'Don't inflict Chris on your fiancée. My brother can arrange for extended leave from the principal of his college. He's not scheduled to recommence teaching until next February. Kurt's a free agent, he'll please himself. That's if this new lot of miserable federal turds agree, Kendall.'

'Arneka, what a shocking thing to call those dedicated federal officers. They've proved their worth tenfold, by working double shifts to protect everyone here.' The harsh look he received would sole the toughest workman's hob-nailed boots. 'Amber runs her little legs stiff to protect my mother and your family. The pup follows your father everywhere he goes in this house as well as down to the shearing shed. Kurt's interest lays in watching them scour the wool. The raw smell of lanoline oil that lingers in its aged floorboards helps to clear everyone's sinuses.'

Climbing into her brief shorts, Arneka looked down her nose at Kendall. Discreetly, he turned away. With her tee-shirt on back-to-front and inside out she looked hilarious. He gave a half-hearted snigger. Indignant, she scowled and then smiled at him.

'Kurt will stay here with me while you're overseas. Then I can monitor his heart murmurs twice daily. The rigmarole we were all subjected to in Canberra was ridiculous. Completely unnecessary. The Fed's lengthy inquisitions have broken his spirit. I shouldn't have to tell you how stonkered he feels after a short walk. Kendall, I refuse to let his lung capacity deteriorate until he's unable to breathe without an oxygen mask, or without a tube up his nose, while you're galivanting around Switzerland or Austria for six weeks. You know how my father's health has deteriorated in less than a week.'

'I monitor his vitals hourly. So do you, Arneka. Don't snipe at me either. I can't spare the time to hold his hand or take his pulse twenty-hours a day.' Kendall was furious. 'Approach your menfolk and get back to me tomorrow. As it stands, I'm due to fly out for Paris a week tomorrow. All my previous travel plans are now defunct because of Mum's relapse. Peter informed me last night that he's rescheduled my flight and my accommodation in Switzerland. He also mentioned that my renewed passport and documents will be finalised in two days.' Kendall didn't elaborate on the latest news of his father's possible encampment overseas.

'Oh, has Mr Bucknell flown down to Canberra again?' Arneka grizzled while plaiting her hair.

'Peter rang me from home. It's an old weatherboard cottage, located five kilometres past Jalna. His road is third on the left and then it swings right to Coolalie. Although he frequently travels to Canberra by plane, he's a lonely and reserved man. I wish he could meet a woman who appreciates carte blanche cooking and superb wine. I doubt if he'll find his ideal woman here in Yass.'

'Is his property bigger than Abergeldie?'

'Not quite, his rams run wild in twelve-hectare paddocks. His ewes and their newborns are housed in a barn during the winter months. The lambs will eat chaff and fodder from your hand. The overseer and groomers keep his stables spotless. His indoor staff work long shifts to keep the house immaculate. Peter's a perfectionist, he dislikes slovenly workers. The gardener's a repatriated soldier who doubles as his nightwatchman. Pastoralists hereabouts are harvesting their crops of sorghum, sheathing wheat now and their massive threshing machines have stripped the stalks bare. Some property managers get the casual labourers to scythe and thresh their crops manually. Other farmers breed their steers to kill. The cows they call milkers are bred for home use.'

Fascinated with how Arneka had braided her hair distracted Kendall. Secured with fine hairpins her plaits draped over the nape of her neck. 'You'll find Peter a generous person, once you get to know him, sis. Since the absconder left here, we've become firm friends.'

'Is Peter married?' A wistful twinkle glistened in her eye.

'No, he's a widower. A barrister, a QC, that's a Queen's Counsel. He's a fully-fledged accountant with a degree in political science.' Amused by his sister's inquisitiveness, he smirked and readjusted his reading glasses. 'You'll appreciate his spontaneous sense of humour. It's scintillatingly wicked, like yours. He's not what you'd expect of a well conversant and busy diplomat. Peter's a sincere and humble man who loves tennis. He enjoys swimming on steamy summer nights in the river that flows through his property.'

Arneka hadn't seen a glimmer of humour in the surly-looking senator's face. Their only contact had developed in a professional capacity while in Canberra.

16

4am next day: On this glorious summer's morn, Kurt Baumer paused on approaching Senator Bucknell's office. He fiddled with the letter in his hand, hesitant whether to disclose its contents. 'Oh damn, why not. He'll find out …'

On leaving his room, Kendall looked blankly at Kurt, standing not an elbow away. 'I didn't expect to see you up this early, Kurt. What can I do for you?'

'Kendall, can you spare me a moment? There's something important I need to talk over with you. Can we go somewhere quiet so we won't be disturbed? Perhaps your room?'

A nod beckoned Kurt to follow him to his father's den. 'We'll talk in there. It's quiet and no one is allowed in here. As a restricted area, we're forbidden, by our federal guests, to enter this office. What's the letter in your hand? It looks official, Kurt.'

'No, it's something I found tucked under the lining of my suitcase. My wife put this letter there in our caravan. The gypsies saved me from being murdered by Nazi deserters. I remember Heide tucking this envelope in Arneka's change bag in my car, not long before the Nazis murdered her. Anna must've put this envelope in my case. On reading its contents you will be horrified, Kendall. It's written in Heide's distinctive script. At home I have legal documents written by my deceased sister towards the end of forty-four.'

On scrolling through the mind-blowing handwritten text, Kendall closed his eyes. A jittery hand held the signed letter. 'It's inconceivable to believe all these sordid details describing the attempted rapes of your sister by von Breusch could be true. Kurt, I can't apologise enough for what that mongrel did to Heide. No wonder Arneka and Chris hate him.'

'Listen to me. You and your family were not responsible for that criminal's evilness or his cowardness. Never condemn or blame yourself, Kendall. I could never hate a living creature, not in my lifetime. Nor do I wish to now. My patience is expended whenever I think of the atrocities that beastly man was responsible for, while in service for the Reich. I've seen many declarations written by men, women and children, mostly of Jewish origin. Some of their crippled hand penned letters would make you, a grown man, cry. I have deposited a list of the known atrocities his men committed in my Yeppoon bank. This copy comprises a hundred or more names of escapees, people maimed or persecuted by the Nazis in Berlin. Shocking incidences that occurred in the latter war years, signed depositions witnessed by myself and members of the Marquands working undercover in Paris and Germany. French partisans who conscripted their countrymen to rid their forests of Nazi collaborators and infiltrators. People who'd lost their entire families in the ghettos, Hitler's thugs destroyed … not liberated, in Poland and Austria.'

Capable of speaking Kendall touched the distressed man's shoulder. 'Kurt, I think you should take this letter down to Mum's office. Colonel Montsard will welcome this news. He and our federal officers need every scrape of information to put and keep this Nazi wretch behind bars. I'd sooner sanction his neck to be throttled by the strongest rope to hang the bastard.'

Incapable of answering he saluted Kendall. Without saying a word and clutching Heide's letter in a shaky hand, Kurt Baumer hurried off down the main hall.

Discussing Kendall's offer for her family to stay on Abergeldie for a couple more weeks Arneka approached him before breakfast the following morning. 'Kurt assured me just now that he's willing to remain here while you're away. Chris reckons its fine by him, Kendall.'

'That's terrific, I know he won't regret his decision. It will also relieve Mum's anxiety and ease my conscience. Now I can finalise my travel arrangements and set the date for my flight to France. This news might divert Bridy's attention away from constantly nagging me to pack enough warm clothes, as it'll be freezing in Paris and Austria over Christmas. There is a bleak forecast from the new year right through until March with blizzards sweeping in from the North Sea. The Scandinavian countries are also in for heavy snowfalls and Iceland will become a frozen paradise. I hope my friends in Paris have an ample supply of warm blankets. I detest having to wear cumbersome winter overcoats. I'm like you in that respect, Arneka. Living in Queensland you wouldn't need winter clothes, woollen scarves, fur-lined jackets and gloves.'

'I only have one thick coat in my wardrobe. Why don't you take Brianna with you? A holiday would do her good and ease the tension between you both. After all she has suffered the indignity of being traumatised by our problems with … Well you know who I mean. Besides Kendall, she must miss her own folks back home in County Clare.'

'Brianna wants me to relax and enjoy this holiday with my uni friends without her tagging along. The "silly-wag-pack" she calls us; jokingly of course. Bridy reckons I'll settle into a new practice better after this break and once we're married. I love that girl, Arneka. To me she's every man's dream.'

'She worships you. I've seen it in her eyes, how she admires your carefree manner and how you can switch a terrifying incident into a humorous one. She told me so yesterday.' Nose held high Arneka giggled. 'I doubt if any man could ever tame me. Chris criticises the way I muddle things, instead of being frank. He can be a stuff-shirted arsehole at times. And he thinks he's always right when criticising me over some insignificant misdemeanour. He reckons I'm headstrong, with the defiant manner of a mischief making vixen. Kendall, I hate being hounded by him and those two watchdogs. The bludgers are working in your mum's office. I'm a born freedom hunter, not a lazy loafer who doesn't care a damn about people I love or admire.'

'One day Miss, I'll remind you of all those words.'

Her spontaneous laugh heightened Kendall's miserable mood. A good gag and bit of frivolity always worked wonders to brighten a dismal day. They laughed together about Arneka's idea of being tied down to one man. She didn't believe in marriage. In a fit of exuberance, she bet Kendall ten dollars that it would never happen to her. He accepted the bet. Poignant in attitude, her jest came as no surprise to this man, who loved working hard on the land to make ends meet.

'Now we should eat, I've a lot of work today. I'll either be mustering or working in the top paddocks with Bjorn's crew. Then we'll be mending the fences in the home paddock. My blokes should be down here by noon. Arneka, work never ends for people on the land, or men and boys herding sheep or drafting steers on their cattle stations.'

'So I believe Kendall. Struth, Sylvia's due to leave in five minutes. If I gobble down this brekky of eggs and bacon on rye, I'll be ready in time to take over from her. How's your mum on this sparkling morn? Gigi looked a bit peaky when I popped in to check her pulse at six.'

Kendall looked over his new tortoise shell rimmed spectacles. *The Daily Telegraph* newspaper flopped down on an empty plate. 'She's okay. Had a bit of rough night, so Stephen told me just now on the blower. He promised to drop by around ten. I think he wants to revise her medication. The dosage of Diazepam wasn't enough to stabilise Mum last night. This morning he'll probably prescribe a different drug, prior to leaving for his practice in Yass.'

Kendall wrote a note on his outdated prescription pad and passed it to Arneka. 'Should you need me before lunch contact Jalna on this number. Kirsten will send one of their roustabouts up to fetch me. She knows where we'll be working until twelve. Keep in mind, all phone calls in and out of this homestead are still being monitored by our watchdogs. Ask them first. They'll probably approve and connect you to Jalna on their phone.'

Arneka's mouth twisted in an obscure angle. *What the hell is he carrying on about? I'm not an idiot. He waffles on like a plum-duffed fairy and thinks he knows everything. He's worse than Chris when he's in a stinking, rotten mood.*

The barrier of ice between them dissipated to a degree, after her apology. 'Did you know Kendall your mother thinks of Brianna and me as her daughters? Gigi told us that yesterday.' Inciting mischief, Arneka continued to torment him by adding, in good humour, 'Your mum wanted a girl, instead she got you, a squawking kid. What a disaster for her.' As Arneka walked past Kendall, she gave him a tap on the bum. Laughingly he reciprocated by pinching hers.

Arneka must have the last word. 'Slowly, very slowly you're growing on me, mate.' They cracked a smile together and then went their own ways to do what must be done.

The next few days were crucial regarding Gigi's health. Gradually she regained the capacity to stand unaided without any permanent damage to either her brain or limbs. The risk of having a severe stroke now seemed unlikely.

It took weeks before she regained the use of both her arms. From her knee down the left leg there remained a slight tingling feeling. This debilitating episode had proved a warning for all concerned, including her regular physician, Doctor Jarvis. It also forewarned Kendall.

Stubborn and set in her ways, Gigi insisted on her daily riding escapades up through the top paddocks. Illogical in decision, it went against Jarvis and her son's advice. Her stubborn attitude of defiance caused resentment between her and Kendall, to the extent that she refused to take or accept any of her daily medications.

With only she and Kendall at the breakfast table he approached his mother. 'I refuse to tolerate this childish behaviour and these stupid ideas of yours any longer. Either you accept Stephen's advice to take your prescribed drugs, or I'll take Brianna back to Ireland.'

'Oh, don't be silly. What a stupid thing to say. It's you who is being childish, not me. I have the right to reject anything that I feel will affect my thinking. I've suffered enough being bullied by that beastly man whom I hate extensively. Don't you keep bullying me, Kendall.'

'No Mum, you listen to me. You forget that being born in the British Isles I have dual citizenship. I'm a fully qualified physician and I can buy a private practice in Ireland. My doctorate in medicine should be sufficient for me to register for a licence to practice there.'

'Kendall, how can you say such a cruel thing to me?' Several minutes of serious thinking brought Gigi to her senses. 'Uhr, I suppose you are right,' she snarled. 'Well, you won't stop me riding Wallaby Downs. He gives me a good reason to live and a lot more pleasure than you do.'

'Nobody asked you not to ride. We only wish you'd take a more sensible approach to the sport. No galloping, like you did yesterday. I saw you. You forget, during this dry spell dust squalls can be seen from the woolsheds.' Leaning over the back of his mother's chair, he added, 'Otherwise, I'll make certain none of our horses remain on this property. This is the last time I'll warn you. And I'm serious, Mum.'

'Changed your mind about selling too, I hope?' she snarled sarcastically. 'Seeing you're so damnably adamant so am I. As of now, I utterly refuse to sign those papers ...'

Ignoring her caustic remarks, Kendall laughed. 'That's where you're wrong. If you recall Mother dear, I countersigned the affidavit in your presence. Bjorn then witnessed both our signatures. All those documents are still in my possession.' An icy glare warned Gigi not to interrupt. 'I have no intention of aggravating or fighting with you, Mum. The suggestion to sell actually came from you. Not me.' Kendall omitted to say, my idea you willingly accepted while convalescing.

'Yes, he did. I remember Bjorn signing something. Excuse me,' Gigi said curtly, 'I prefer the company of Wallaby Downs to you. Especially when you're in such a pig-headed, obstinate mood. Calm down or you will have a stroke Kendall. Not me.'

Spearing a glare in her son's direction, Gigi reefed a key off its hook on the back verandah. The door shuddered on slamming. She hurried down to the stables. Its saddle room housed a collection of saddles, whips, rugs and horseshoes, plus a huge variety of riding regalia.

I'll teach that obstinate son of mine to treat me with contempt. Kendall can go overseas and stay there, for all I care. Why he is so dominate and bombastic is beyond my comprehension at times. Blast him and his holiday. I'll ask Jamie, our aboriginal backtracker to saddle Wallaby Downs. He's a pleasant youngster. And he does know how to treat a lady with respect.

17

Around six the following morning, accompanied by his manager, Kendall rode up towards the home paddock. 'Thanks Tom, for bringing those top fences to my notice. Until we inspected them yesterday, I hadn't realised that quite a number of their posts were in such deplorable disrepair. I bet Bjorn isn't aware his lot on the boundary fence are almost rotted through.'

'Well, the ones lower down aren't much better. Want me to tell him when I go over this morning, after tucker?'

'No, where I'm concerned breakfast can wait, the fences are more important. Once I've assessed the ones we're heading for, I'll see Bjorn myself. I particularly wanted to check those lower ones now, before the heat of day rides in. All boundaries must be completely sound, so that when a buyer does appear on our horizon, he won't be put off by their unstable appearance. And that could be as early as next week, even sooner.' Kendall didn't let on about the tentative nibble regarding the sale of Abergeldie that looked promising.

Stationary but still astride their horses, the men continued their conversation. 'Righto Boss, that means I better get the men to doin them today. I have ordered iron posts and reels of barbed and fine wire. Treated iron won't rot and it'll stand the test of time. I've also allowed extra posts for them lower fences. Everything should arrive here about noon.'

'I'll leave a crossed cheque for the load in our electricity box for the driver immediately his truck pulls in here. If I'm not around he knows where to collect it on the back verandah.' His old battered Akubra jumped into action and Kendall swiped a nest of blue cattle flies away from nose and eyes. 'I can't leave here next week without all the bills finalised. If you need anything before then Tom, give me a buzz.'

'Sure will, Boss,' Marden replied patting his dappled grey's withers. His stallion whinnied and its tail swished violently. Eventually the pesky flies took flight. 'Be seeing ya mate. I better get going or BJ will be bellowing in me ear. The old fart gits quite agro if I keep him waiting fa long.' With a parting nod Abergeldie's manager manoeuvred his horse around and headed towards Jalna's lower fence markers.

After a review of the remaining nearby fences, Kendall rode back to the homestead where he tethered Potchkin to a verandah post. Wiping his muddy boots on a metal shoe-scraper set in cement, he removed his bluchers. Setting them down inside the back door, he tossed his "old-fave" on a wicker chair. It missed. 'Blast that hat, it should've landed on the table. Well, it can't fall any further, now it's on the floor. Someone will pick it up or I'll collect it on my way down to the stables.'

Tucking into a hot breakfast, he paused to sign the cheque and left it alongside a note for Arneka, telling her where he would be, if Gigi or Kurt needed him. With the paperwork done, Kendall invited Robért to join him on the back verandah, the coolest spot out of this blistering heat. He and the Frenchman sipped their wine while discussing Abergeldie's proposed sale. Kendall seldom drank beer during the day, unless it was excessively hot and never this early.

'Kendall, will you thank Mr Baumer for loaning us his sister's letter? It's been a boon to our top blokes in this investigation. Kurt and his family must've gone through hell, while escaping the Nazis in Hamburg. Do you remember seeing black or grey inked numbers under your father's armpit? The six or seven numbers have intrigued me. All German officers were tattooed on, or before their graduation from the military academy in Berlin. I could be wrong about Berlin. I spoke to Senator Bucknell a moment ago on the blower. He's on the ball this morning. Up bright and early ready to pack, I suppose.'

'Peter's leaving at noon. He'll be a passenger on the Cessna that delivers your electrical equipment. It's due to take off again for Canberra on the earthen strip in our top paddock.'

'Don't ask. I've arranged for one of my crew to give him Kurt's letter. Peter received it an hour ago. He promised to hand-deliver it to my superior officer, as soon as he hits town. You know the letter will be in safe hands. I took the precaution of running off several duplicates. One for Kurt, the other I'll keep on file for you to browse through at your discretion. Kendall, the letter's text has really shocked me. The absconded ex-Nazi won't be able to wriggle his way out of all accusations levelled at him. That's if we catch the bastard. It's blokes of his calibre who delight in torching every scrap of evidence, well before an agent can get his hands on it. The fingers of contention all point in his direction.'

'What a weird expression. Still, I caught your drift.' Kendall watched a strong gust of wind send red dust swirling over the top paddocks. It formed a willy-willy then spiralled down towards the shearing sheds, transforming a clear sky into a gritty brown mass. Without warning another cloud of ochre-dust spiralled out of control towards heaven. Kendall coughed. After copping a mouthful of fine grit, he spluttered.

'This afternoon's going to be a scorcher by the look of those crimson clouds against a hazy sky,' he declared, eyeing the great expanse of red as they reached his tethered horse.

'Yes, it'll be a sizzler by dusk,' agreed the Frenchman, 'and a long afternoon. I'll see you later, Kendall. Do you want me to ask your cook to send your men up a cask of fresh water, while they're drilling postholes to replace the broken fences? Fire a shot and my blokes will know where you're working.'

Kendall scanned the steamy horizon. 'Phew, I should change into something cooler. These thick jodhpurs will cook my crutch. There's a light pair of strides in my saddle bag, plus a cotton shirt. I'll swap these hot clothes on reaching the high paddocks. The Cessna should be landing soon to deliver your replacement secretary, and to collect that irreparable radio transmitter, as well as its sole passenger, Peter Bucknell.'

Colonel Montsard flipped one hand to prevent a marsh fly from landing on his eyes. 'I'm allergic to your Aussie creatures. They not only sting but they also leave huge welts on my skin. Kendall, before you ride up to those top paddocks, I might need a lift. There's an important parcel of documents to return to our archivists that must leave here as soon as possible for Canberra.'

'Fetch it and I'll be your courier. My saddle bag won't be overloaded with a bit more stuffed in its pouch. Rob, please let Mum know I won't be down until four or thereabouts. She might worry if I'm not here by dusk to sooth her fevered brow.'

This age-old gag no longer held momentum, and the colonel laughed. On his return he threw a small brown parcel to Kendall, who caught it in the jeep. With a broad grin, Robért Montsard waved and turned to walk indoors, well out of the ravaging heat.

Driving down from the plane Kendall almost missed his manager riding towards the house. 'I would've missed you if the damn rust bucket's engine hadn't seized. I'm about to join you and your crew, Tom. How are the traps we set yesterday? Catch any foxes? The new iron rust-inhibited fence posts are due to arrive at three.'

'Boss, you better tell BJ ta watch out for them pesky foxes. With all the traps we've set he might snare one, if he's lucky. I'll get our blokes ta do the baited claws before they start work tomorra around six. Well before the sun rides in, if you agree?'

'That'll be okay, Tom. I'm off to Jalna across the lower paddocks.' Kendall broke off speaking to dislodge a fly sniffing his nose. 'How many ewes and rams down this week? We've lost thirty in the past two months. If the foxes or vixens keep tearing our ewes to shreds and the southern eagles keep feeding on their flesh, the newborns will die.'

Tom Marden broke the dreaded news to his boss. 'Makes six all up, so far. Probably find more dying lambs ta shoot today. Up until now, the vixens haven't killed or mutilated our rams. Three ewes down in lambing yesterdey, plus a newborn. Can't afford to lose many at this damn rate, that's what I reckon.'

'No property can sustain huge losses for long. Set more traps if you must, Tom. We'll pull in all the hounds to scout the watch. I'll alert Bjorn and with the fences mended, he can run Jalna's mob with Abergeldie's in these lower paddocks. The new lambs are all branded with our irons. There won't be any confusion or hiccups with our sheep running around or falling into crevasses in Wave Rock now it's fenced nor in the sown paddocks. This old crate is sturdy enough to carry fifty to a hundred bales of dry fodder to restock the hay shed. Don't forget to relock its doors, once you've fed the lambs and their ewes.' Kendall pulled on the reins and flicking them, dug his heels into Potchkin's ribs. This sharp jab made the stallion lope towards Jalna's sloped, grazing paddocks.

'On leaving ya hay shed. I'll blast one shot in the air,' Tom yelled as his boss rode away. 'Be seeing you shortly, matie.' Tom Marden swung his horse in the direction of Abergeldie's stables to round-up the muckers, roustabouts and all of their available station hands. With this new job under way, Tom led his horse to a willow tree while sweating on his crew to load more iron fencing posts on a disused grader, well past its prime. A rust bucket, it served the purpose.

By the end of that week all broken fences conjoining Abergeldie to Jalna's paddocks were replaced and stable. Kendall's men couldn't have accomplished this huge job without Bjorn's crew and other workers they'd scrounged from neighbouring properties.

Sprinkled rain fell to briefly quell the heat and it barely laid the dust. By noon the blistering heat soared, which stifled every living thing within range of this hell-fired inferno. As predicted, by nightfall in swept the wet with vengeance. It poured relentlessly for three days, until the station hands' common refrains were in full vogue. "The sweltering weather's gone. Now this bloody lot's trying to create a damn sea at our doors." Their cries echoed across and up to the top paddocks. "So much for a sunburnt country, our plains are drowning in mud".

Two days on, good news for a change afforded the residents of Abergeldie some relief from stress. The interested buyer had fronted to purchase the property and all its holdings.

This morning the prospective owner had arranged to drop by Abergeldie for a final run-through of its potential. Sheep, paddocks and woolsheds would all be scrutinised. Also, on the buyer's agenda were inspections of the shearing shed and its capacity to bale. During their initial visit he and his accountant had missed inspecting the outhouses, due to a previous commitment.

Trying to manoeuvre the jeep around deep puddles of congealed mud, Kendall found it almost impossible to keep all four wheels aligned. He cursed the rough section of track. With the woolsheds in sight he managed to steer clear of corrugated ruts and potholes that could topple an unstable vehicle.

Spotting a jagged rock, he shuddered. 'Hang on Lanky, we're going to hit …' Intending to say hard, Kendall gritted his teeth as the vehicle bucked and struck another deep rut. Cluttered with debris and flotsam, by unclogging the revving engine it spluttered to life. Run-off hadn't drained, nor did it hinder its driver trying to see through a mud-splattered windscreen.

With teeth still chattering and knuckles white from gripping the wheel, he managed to pull hard on lock. This enabled the jeep to form a tight turn to escape another muddy furrow.

With sweated brow, he savagely acknowledged to his top shearer, 'Struth Lanky I didn't expect to miss that last pothole. I'll tackle Tom Marden this arvo. His crew can bring their grader up here to shift the mess from these clogged furrows. Tony Smithers won't be landing in on this track, thank God. I dislike the bloke. He's a fusspot and his pedantic attitude is really irritating.'

'Oh yeah, I'd forgotten he's the prospective buyer. Most probably he'll drive in through ya side gate. I went over that track at six this mornin, its stable. A bit muddy in patches. Where wouldn't be after this sudden downpour. Good we finished them fences before the deluge rode in, aye Kendall.'

'Yeah. My thoughts exactly. Dame Fortune at long last is blessing us sheepmen. Why she hasn't heard our pleas for rain is a landholder's guess. I, for one am glad she's taken pity and spared us all from starving. The crops of barley, sorghum and lucerne will benefit from this downpour, as long as it doesn't wreak havoc and end up deluging the sewn wheatfields.'

'Ya right there, Boss. I've been thinking along the same track as you.'

'Good steady rain is better by far than a deluge. This sudden burst of sunshine will dry the mud. It'd be impossible to tackle another big job in unstable weather. Ironic, isn't it ...' Kendall bit his tongue as the jeep bucked on striking a huge, unseen boulder. Until they hit solid turf Kendall couldn't answer his 'top gun'.

The well-weathered elderly shearer, bred and dyed in the wool, understood. As the mud-trail snaked and twisted Lanky saw his reflection in sun-flushed puddles.

Finding it difficult to see through a mud-splattered windscreen, Kendall shielded his eyes from the glare. Steering clear of another boggy quagmire he responded to Lance Hancock's last query. 'What did you mean, Lanky? Were you referring to the sudden onset of rain, or what? I don't have a crystal ball, and I'm not a follower of the occult to decipher your wacky notions.'

'I meant the rainsquall. If we tackle this rough mud-soaked stretch soon, it shouldn't deteriorate before winter. More heavy rain and that last section will be impassable, with the exception of heavy vehicles.' His boss nodded in agreement.

Lanky eyed Kendall suspiciously. *He looks puzzled. His mind seems distracted by some unusual problem. His mum's illness probably. BJ and his missus will be frequent visitors to Abergeldie while he's tripping over yonder in this huge world.*

Uncertain what to say, he casually said, 'What did ya mean about bein ironic, Boss?'

'Last week we prayed for the wet to arrive. Now we're cursing it. After this good drop of rain, it might show up the faults on this home run. It should give the local turfmen enough time to fill these potholes well before Withers takes over. Tony seemed undecided about shelling out more than 200 000 buckaroos for our homestead, the shearing sheds and a mob of well over 10 000 sheep, the last time I spoke to him.'

'Lookin forward ta ya holiday, Kendall? Bet you'll have a beaut time on them snow fields. Wonder ya girl doesn't want to climb in ya pocket.' With the woolsheds in view, Lanky gave a jubilant gasp. His bum ached from being jostled every time the jeep hit a pothole.

'Aye Kendall, she's a smasher your girl. Thought ya mum was a good looker, but I reckon she beats Gigi's pretty smile and her xuberance.' He wasn't sure how to spell or say the word. Not an illiterate, he left school at thirteen. Self-taught, Lance often mispronounced words.

'Keep your eyes off Bridy while I'm overseas,' Kendall sniggered. 'Don't let my mother hear you say what you just said. Mum's the boss-cocky of this spread. Or she thinks she is.'

'Only joking, Kendall. Who'd want ta look at an old codger like me. Past them capers. In me younger days, I adored ya mum. That's when you were a stick of a kid. Considered her me best girl then.'

'Right, now let me concentrate.' Kendall focused on the closest shed. Cruising to a stop outside it, his boot pressed hard on the jeep's brake pedal. 'Tony Withers isn't here yet. Good, it'll give us time to recheck the combs, shears and other equipment. Everything must be in working order, or he'll withdraw the sale.'

'Yeah, I'll show ya which blade's been shortin Boss. I hid it, so no bastard would get lectrocuted. Did I say it right? I'll grab the broken blade as we pass the breezeway.'

'A lot different to the old days, when men sheared with manual clippers. Wouldn't you agree, Lanky?'

'Suppose so. Makes no difference ta me. I can fleece a sheep faster than you can spit, and quicker than any of them modern gadgets. Cleaner clip, I reckon.'

Jumping over the jeep's passenger door, this tough bushman with his shirtsleeves rolled, dusted a mud-splattered hat against his patched riding breeches. Their tattered hems clung around wrinkled ankles. Boots once of crisp, shiny leather now looked shabby, worn in by time. Lanky waited for Kendall to alight before treading the boards. Thumping up six steps like a lame draught horse, he followed his boss through open doors, swinging to the wind's tune. Wide oak floorboards, trampled well from years of wear, groaned with every step he took. This well-weathered shearing shed had taken on the persona of its present visitor, whose gnarled fingers of over seven decades could still shear five to six hundred sheep in a day.

Kendall had just finished rechecking the combs and clipping gear, when he heard a truck's brakes screeching to a stop, on the mud-impacted track. Walking across the breezeway, its loose floorboards squeaked under his boots. Disturbed, the ironbark planks impregnated with lanolin released their strong aroma. Sniffing the scented air laden with natural oils he sneezed. The potent smell of dried dung drifting up from the disturbed boards had accosted his nostrils. No amount of cleaning could rid the timber struts of such an indelicate aroma.

Hard on his heels, Lanky also sniffed the oil-drenched air. It aroused the shearer's senses and stimulated his longing to click the manual shears still clutched in one hand.

Kendall looked back at the loosened planks and remarked, 'Those wonky boards should be hammered back in place. Lanky, you can supervise the two young roustabouts on that job. They're keen lads and neither are afraid of bending their backs to do a full day's work. Don't walk back this way. Keep clear of those rotten timbers. You better get the loafers to go over this entire floor, Lanky. We could easily lose this sale, the state they're in at present.'

The roar of another vehicle's engine seizing intruded on Kendall's thoughts and disrupted the still morning silence.

'Looks like the clumsy agent's brought Withers back for another looksee. Shit, what's gone wrong now?' Kendall uttered observing the smug duo exit their Citron.

'Hey Kendall, ain't they the same blokes who trudged through ya homestead yesterdey? I saw them when ya mother went down ta feed Wallaby Downs. I hope that Withers bloke keeps me on here. I'm too friggen old now to pick up me bluey and hump it anymore. Besides, I feel part of this place now, after working here fa thirty years. It weren't called Abergeldie then. The old crank who owned ya homestead changed its name from Huntsville ta that.'

With their minds tracking in different directions, Kendall winced. *Withers has probably returned to give our shearing sheds the final going over. It's odd. I'd already signed the release papers including that last report on this entire property. What devious scheme has he and his sidekick hatched now? Withers haggled over the final sale price. We came to an amicable decision or so he led me to believe. I don't trust the Yank. He could have another property lined up, and hasn't the guts to confront me. Oh well, I'll soon know the truth.*

'Ah Boss, da ya think that old coot will buy this whole shebang? I don't trust the bugger, he's out ta rob ya. Get his hooks inta everything ya and ya mum own. I hear he's more interested in her than paying a good whack for this property, its sheds and them bunkhouses.'

Reconsidering the proposal of five hundred grand, Kendall began to think twice about off-loading Abergeldie to a man whose shrewd reputation lacked the scruples of an honest buyer. Withers' delaying tactics in signing the final papers of sale were irritating and not what Kendall expected. Well not after sighting the man's credentials which seemed to be above board.

He apologised to his top gun, 'Sorry Lanky, I didn't hear what you just said. Something about keeping you on here, I think. Never fear, I have approached Withers about keeping the staff on. You're a bonza crew, and if the man has one iota of sense, he'll agree.'

Removing his sunglasses Kendall focused on the stocky, prospective buyer as he cruised in across the unsound floorboards. He held his breath when the timbers didn't move or creak with each footfall as they'd previously done. Releasing a sigh, he greeted Withers.

'Good morning gentlemen,' Abergeldie's owner said, removing his hat. 'My top gun, whom I think you've already met, Lance Hancock.' Copping a mean look from Lanky who hated his real name, Kendall smiled.

'You only just caught me. I'm about to head back to the house. Everything's fine, I hope?'

'Hi there, Doctor Ross.' The Yank's slow drawl could be heard a mile away. 'Why wouldn't it be?' Frowning, Withers used his broad-brimmed Stetson to brush the dried mud off his trousers. 'Just passing, I wanted to take a gander at the property down the strip a bit. I thought we'd drop in here on the way.'

Kendall held his breath and wondered what lay behind this arrogant American's return visit. His shifty gait accompanied by a languid roll of one shoulder warned Kendall to observe and study this man's peculiar mannerisms.

'Looking for a plot of good earth hereabouts for the son.' Withers scanned the first shed's exterior. 'The property I've just gone over isn't structurally sound. My boy's new to the land, so it'll give him a chance to see how a prospering ranch is run. How many sheep to the acre, Kendall?' Tony, looking up towards the rear top paddocks.

'Of my mob, I run eight to the acre, in old terms. That's about standard in good times, Tony.' Kendall looked down at the narrow-combed shears in his hand. His finger carefully probed the sharpness of its blades, after the recent clip.

Most shearers preferred their own electric shears. Here on Abergeldie and likewise on Jalna, the local men insisted on using manual ones. Their overseers knew the combs, blades and shears were reliable and they would get an even cut. Damage would be apparent after each fleece settled on the lanolin-impregnated, huge oblong bench. In extenuating cases both these property managers would relent, letting the drifters use their own gear.

'Mind if we have another wander around these sheds, Doctor?' the agent queried, on his client's persistent demands.

Amused by their effrontery, Kendall shuddered as another floorboard groaned. 'Be my guests. Take your time, there's no hurry.' He knew he'd said the wrong thing. Just shy of the festive season and the way Withers dithered, it could be new year before he decided.

While Tony scouted around this shed, the land agent took note of its condition. Standing in the breezeway Lanky caught Kendall's attention with a low whistle. He pointed to the jalopy parked in shade adjacent to the woolshed. 'Ya reckon our jeep is a heap, get a gander at that old bomb bout ta cark it.' Lanky scowled on giving the agent's rust bucket the once-over. 'Suppose it's better than nothing, aye boss.'

'Well, our ute's on loan to us. Bjorn did offer his four wheeler to Withers. His is off the road, and needs a new accelerator. Withers will probably purchase a new vehicle if he takes over Abergeldie. Five hundred grand is nothing to a man in his position.' Kendall knew the vehicle below where he now stood wasn't roadworthy, and not fit to run over a rough bush track.

'I wasn't prepared to lay out fourteen grand for an expensive utility when selling. Besides I'll need a new vehicle myself, before I begin working in the practice up north.'

'Okay boss, if ya don't want me no longer, I better be off.' Lanky coughed as the strong aroma of sheep urine wafted up from between two damaged timbers.

'Please stay Lanky. Just let me do the talking. I'll need you to back me up, if I forget or miss something important. A devious bod, Withers is likely to sound me out, after scrutinising the manifest. He thinks I'm trying to diddle him cash-wise. He's a miserable turd.'

Kendall stubbed the toe of his blucher on an unstable board and flinched. He caught the grey-haired roughie glare as his boot prodded the wonky timber. They looked at one another, raised an eyebrow and sauntered through the breezeway, back into the shed.

Glancing over his shoulder, Kendall watched the well-weathered buyer wearing a frown walk from the rear shed. His gaze then deflected over to the agent taking notes as they both approached him.

'Tony, is there anything specific worrying you about Abergeldie? Not the price I hope?'

'No, everything seems okay,' Withers confirmed. 'I've just been going over the prospects of your property.' The man's gruff voice sounded like sandpaper scraping on rough cement.

Kendall with his professional expertise smiled. *Withers should see a specialist.* Not at liberty to say a word he looked at his fingernails. Still, its raspy sound grated on his nerves.

'Dams, how many?' the Yank, sparse on words queried. 'Only sighted two so far.'

'Three all up,' replied Kendall sharply. 'Top one should be on the way to finding its depth after this rain. An artesian bore with a good flow in below it, with a second on the drawing board, so to speak. One of our lower dams will eventually tap into that bore.' Observing the man in tailored navy slacks with blue overshirt handling the combs, Kendall realised they'd previously covered this subject. He frowned at Tony Withers displaying his bulldog glower.

'Bjorn tells me most of your mob have been running on his ground in recent times. Why? No contamination on Abergeldie, I take it!'

Kendall knew this question would be lobbed in his direction. Prepared for it, he had the answer ready and seized upon the moment to express his opinion.

'Simple Tony, I'm off overseas in two days. Well, put it this way; I suppose you've heard about Mum's illness.' Withers nodded in reply. 'Now she needs copious rest to recuperate. That sir, is the sole reason I have condescended to put Abergeldie on sale.'

'Mm...mm!' the grump muttered and stamped his boot on the breezeway floor. 'These wooden floorboards are solid. Your property's fine, suits my needs. It's good.' Withers seemed to be distracted by something. His wily eyes surveyed everything in this second shed and way beyond.

This was fortunate as Kendall's face had flushed crimson as he trod the boards. It was the first time he could recollect telling a blatant lie in his twenty-seven years. Give or take a few days.

Fiddling with the shears, Tony questioned him again. 'Shearers including drifters ... how many and are they top class?'

Lanky, who'd remained quiet, shot an angry glare at their arrogant intruder and jumped in before his boss could answer. 'Better believe it. We only employ the best workers here.'

'Correct,' agreed Kendall, feeling the pressure. 'Thanks Lance. Eight to ten locals come in before the season breaks plus my top man here and two lads in training. All my men do an excellent job. You're welcome to ask Bjorn, if you doubt my word.' Narked at him rehashing these questions Kendall quietly fumed. 'Svensson will vouch for my employees. They're fully trained shearers, the best crew this side of the range. Our women running the homestead cannot be beaten for their culinary and domestic skills.'

'Thought he said as much,' agreed the buyer. 'Good! What about protective paddocks?'

Tired of all this idiotic quizzing, Kendall over-politely retorted, 'Three or four with plenty of shade in the warm paddocks, to protect our newborns. We lamb off shears here, so does Svensson. Any … more … queries Tony? Or haven't my responses satisfied you?'

There, that should shut him up, Kendall grimaced and his furrowed brow deepened. Stuff him! Is he going to buy Abergeldie or not? Ready to explode, he gave Withers and his sidekick a courteous smile tinged with sarcasm. *What does this wretched bloke want? Does he know something about our recent trouble? Bjorn wouldn't betray a confidence. Never! What's the idiot fishing for? This business deal I discussed with his agent long before Withers first stepped a boot on this property.*

'Your property's fine; bonza in fact, as you Aussies say,' declared the intended buyer in his cocksure manner. 'Yeah, I'll accept the asking price, you've quoted Doctor.'

This smart-arsed bout of Australian slang sounded odd coming from an ignoramus, who revelled in exhibiting some knowledge of his adopted culture.

Wiping the sweat off his sun-bronzed brow, the American continued in an arrogant, loud voice, 'I couldn't do better pal. Be in town tomorrow and sign. Money's ready the moment you put pen to paper.' Withers sanctioned the purchase with a bushman's seal, a handshake. The estate agent stood back and glared at him. Words couldn't express his dislike for Kendall. And sporting a staged grin he walked out of the breezeway, down a few steps to his own car. The rusty dilapidated rattle trap on loan to Withers was now out of the shade, its leather seats baking under a sweltering sun.

Serious faced, Kendall reciprocated in-kind. Yet he inwardly sniggered over the ridiculous way both men had behaved. With a decent sized cheque awaiting him, he didn't give a damn.

18

Bored stiff with their pretentious shenanigans, Lanky wandered off and left them in the shearing shed. Gazing over the foothills, he pointed to a distant clump of bushes.

'Hey boss, look, ain't that Wallaby Downs, ya mother's horse? What the hell is he doin out ere on his own.'

'The gelding wouldn't be out here on his own. It means Mum's been thrown. I'd bet my last buck on that.' Cutting dead his buyer, who was ready to say something, Kendall spoke in the sternest tone he could muster from a dry throat. 'Tony, drive back to the homestead and see if my mother's there. If she's not, call Bjorn.' Confused and in a hurry, Kendall sang out while cutting through the breezeway. 'Tell him I said to bring one of the floats over the top way. He'll know what I mean. Lanky, you come with me.' Jumping into the jeep Kendall hollowed. 'Tony, you better fetch several of my men to ride out as well. I can't drive this thing far and not through huge boulders.' The engine purred and it fired on all cylinders on command.

Lanky clamped the battered army hat tight on his noggin. A puff of wind then lifted his slouch hat and swirled it high, never to be seen again. He'd seen Kendall's Akubra dance to the wind's tune the previous day.

Annoyed, he tapped into his thoughts. 'What would ya mum be doin out ere at this hour?'

'Buggered if I know, Lanky.' Kendall pressed the accelerator flat to the floor. He cursed his mother's stupidity. 'She promised not to go riding after yesterday's heavy rain. And what does she do when my back is turned? Bloody well saddled her horse.'

'Look boss! There ain't no saddle on Wallaby's withers. She wouldn't ride him bareback, would she?'

'My mother has been known to do lots of idiotic and stupid things. She's not going to hear the last of this, I assure you of that, Lanky. Gigi's a damn fool. She's likely to try anything once. Well, this is the last time she *will* ride Wallaby Downs.'

'Ya wouldn't shoot the poor horse, would ya? T'aint his fault,' the shearer yelled, gripping the door until the taut skin of his knuckles whitened under the strain.

'Good mind to,' sprouted Kendall in anger. 'No, from now on her gelding, plus all our personal horses will be stabled over on Jalna. Only you men will be allowed to keep your horses on this property, until the sale's finalised. The draught horses are far too big for Mum to sit astride. Besides she wouldn't dare ride them.' The vehicle bucked and bit into bedrock. 'Hang on, I'll have to back up, otherwise this damn thing won't budge. Blast, I didn't see that boulder. But then I didn't see her blasted horse wasn't saddled either.'

'Struth! That was a near miss. Now this heap's spluttering well and good.'

'Damn!' Kendall blurted, his temper rising. 'We'll have to hoof it from here on, Lanky. We must be somewhere near where Mum was thrown. Tether her horse to a tree, while I climb down this rock face.' With his top gun out of hearing, Kendall lashed out with one fist. 'She'd have to be an idiot to have ridden to the Bluff. It's dangerous in the best of weathers.'

'Hey Kendall, Wallaby Downs is lame. He's thrown a shoe,' Lanky shouted down to him after checking all four hooves.

'There'll be someone here soon. Tether him Lanky then scout around up there. Keep calling cooee, and if Mum's somewhere close, she's bound to hear you.'

'Cooee! Cooee!' the old woolly bellowed. He hollered a third time. Listening, he heard an echo rebounding off Wave Rock's overhang. A faint cry of distress arose from below. Hancock put his ear to the ground and heard a whispering whine. He knew the voice.

'Kendall, look below ya. I think ya mum's wedged between them rocks.' Cupping his hands Lanky called over the bluff. 'Want me ta ferret around the base of this big one, boss?'

'I'm almost down to the base of this sheer drop now.' Kendall couldn't afford to look up. If he became disorientated or lost his footing it could be calamitous. Trying to jostle his feet in narrow crevices he could overbalance without a rope or some support. The hazardous rock face restricted his movements. Extremely difficult to negotiate, if distracted he'd fall metres.

'Stay where you are until our crew arrive. They shouldn't be long now, Lanky.' Five metres lower Kendall's boot touched bedrock. 'Thank goodness I'm down safely. What a hell of a climb that was. I'll never do it again, not without a harness.'

A familiar voice rose to greet him. 'What delayed you, may I ask?' Gigi, chewing a blade of blue tussock laughed while perched on a shelf of sandstone rock. The wave formation towering overhead shaded her from the intense heat.

'You're a stupid woman. I could've broken my neck or lost my life climbing down that rock face. What I can't fathom is how you haven't killed yourself,' shouted Kendall watching her spit out a saliva-chewed mess. 'Mum, are you hurt? Although I don't for the life of me know why I bother to ask.'

'No, I've only twisted my ankle. And no better for your rudeness. Kendall, don't you dare blame my horse. Wallaby Downs shied when a snake slithered across his path. What would you care if I were bitten? He put his life at risk to save mine, if you must know.' Crabby, her obnoxious manner didn't alleviate the pressure of an ugly situation. Blatantly she bellowed, 'Help me to climb down from here, son. And don't keep harping at me. I'm not a damn idiot either. Where's your horse? It'd be impossible for you to walk hatless out to Wave Rock in this blistering heat.'

'Broken down and stuck in the mud, thanks to you.' Kendall moved over to lift his mother free of the rock ledge. He hesitated when her riding boot lashed out to kick him in the groin. It missed and clobbered a bird in flight.

'What!' she snapped; astounded she tried to jump. Looking down at the sheer drop of four metres Gigi froze. 'Have you gone mad?'

'Bjorn's jeep is bogged. It's completely buggered, if you must know. How did you know I'd be here? Don't answer.' Up went his hands. 'Give me a look at that foot.' Kendall removed her boot then surveyed the swollen ankle. In the intense heat he felt like choking his mother.

'Move your ankle. Good. Now do it again. Nothing's broken, but it'll keep you out of my way for a while. Stay off it,' he snarled abruptly. 'I'll carry you the long way until we reach the top of this bluff.' Tossing her bodily over his shoulder, Kendall whacked her bottom. 'With such an empty head, your brains might well be in your bum for the little good they are.'

Squawking, she tried to wriggle free from this precarious position. Her efforts proved useless. Kendall's firm grip of her torso restricted any movement. Gloatingly, he let her keep squealing like a snared rabbit. Her fists pounded his back, and he took no notice. Out of puff she relented.

Everyone standing on top of this giant monolith smirked over the fracas rising from below. Bjorn, who'd just arrived in his horse float, thought the debacle hilarious. Yet none of his or Kendall's men were game to utter a word to his mother.

A couple of rescuers assisted Bjorn and three of his roustabouts to secure Wallaby Downs in his horse float. Tethered well, he couldn't move one hoof.

On sighting a bedraggled creature smothered in mud, Bjorn turned away and sniggered. Answering a hoy, he guided Gigi over to the float. Kendall took over to settle his mother on its front seat. 'Now stay there. Don't you move, or I *will* throttle you Mother. I find these idiotic escapades of yours intolerable. They are beyond a joke and so are you. Your idiotic stunt could've cost me my life. I have made up my mind to have you certified, if you do this again. Think over what I just told you, because I mean it.'

Spotting Bjorn's grinning face she pouted. 'What do you think is so damn funny Mr Svensson? Fetch my saddle, its behind that mound of flotsam I think you call my son.' Feeling triumphant over her successful escapade, she glorified in the idea of humiliating Kendall.

'Ah, nothing's wrong Gigi,' Bjorn sniggered, 'Kendall's cursing struck me as funny. You know, he really should try to curb his temper. It doesn't do for a doctor to fly off the handle.'

'No, I suppose it doesn't. And if it will appease your curiosity, I didn't fall. I needed a break away from that dishevelled bit of flotsam. I disapprove of him poking his snotty nose into everything I do. Now will you drive me home, please Bjorn.'

Hearing her snide comments, Kendall smiled. He and Lanky doubled with two of the other men astride their horses. The jeep remained stuck in the mud. Later some wanderer or the new landowner might redeem it for scrap.

Gigi displayed a prominent pout. Silent and still pouting, she couldn't express her depth of anger. *I refuse to sell Abergeldie now. Kendall better not try to make me kowtow down to his whims. He humiliated me in front of these men, some of whom I don't know or recognise.*

The instant his float pulled up near Abergelie's back verandah door, Bjorn assisted his passenger to climb down. Gigi's bruised shoulders and bum throbbed. She couldn't move her aching arms. The stiffness of her bruised spine, shoulders and neck burned in the hot noonday sun.

'Can you manage to climb those five steps, Gigi? Or will I call Mrs Graham? I just saw her walk inside that door. Unfortunately, the men are waiting for me to unlock the float. We'll make sure Wallaby Downs is safe in his stall and I'll feed and groom him.'

With her head held high Gigi pushed past him. 'No, I'm quite capable of walking through to the kitchen. You go, Bjorn. I won't fall again, thank you. If you see my arrogant son, tell him I'm resting in my room and do not wish to be disturbed.'

A tad uncertain how to reply to Gigi's abrupt response and her rudeness, he hung back until he saw Harriett Graham standing by her, like a hen hovering over a clutch of chicks.

Now I'll unlock the float and release the soft ties on my horses' fetlocks. Shit, where did I put the friggen keys? Found the bastards. They were in the left pocket of his tattered jeans.

Free to return the horse rug to Abergeldie's stables, Bjorn walked through to its saddle room, where a farrier was applying his skills to shoeing another of their horses. 'How's his hind hoof healed, Brett? I disposed of the rough-edged lid he trod on last week. The vet reckons Sky-high could've got tetanus from that bit of rusty tin. The antibiotic injection he jabbed in his withers has prevented any infection. It actually saved the bay gelding's life.'

Settling on a chair Gigi hadn't realised that Kendall had followed her into the kitchen with a shotgun in his hand. With the catch on lock he slammed it down in front of his mother.

'Why the rifle?' Gigi challenged him. 'You're not going to shoot my horse?'

'A man ought to shoot you for being so stupid. You realise Mother that you could've died out there in that extreme heat today. If Lanky hadn't twigged to what you'd done I would be in Yass tomorrow, paying for your funeral. He saw your horse and pointed him out to me. I knew you had done something foolish. Don't apologise Mum, you've broken your promise to me.'

'Tell me, why the gun?'

'If you must know, I left this rifle in the shed last night. Luckily I didn't use the damn thing to shoot you.' Kendall looked at Bjorn, who'd just come indoors. Instead of interfering, he clammed up till Kendall had finished chastising his mother. 'From now on all our own horses will be stabled on Jalna. I'll leave the old hack here for you to ride. He won't buck or move six feet in a millennium. That's all you deserve to ride.'

'Kendall…'

'Zip it, Mother. I've had enough of you and your tantrums for one day.' Re-dressing her sutured ankle, he barked, 'I'm going for a cool shower. Bjorn, please see the horses are in their floats and taken over to Jalna before dark. My crew will help your blokes load our four thoroughbreds. I know they'll be well fed, watered and exercised over there. Otherwise I might do something drastic, before I leave here on Friday.'

Considering her son's rudeness unwarranted and in a huff, Gigi asked Bjorn to assist her into the lounge room. There she made herself comfortable on the cretonne sofa decorated with a citrus floral pattern.

Kendall hadn't walked six paces when a car horn tooted up the drive. Striding towards the front door, he heard Peter Bucknell speak to someone.

'I'm glad you're home Kendall. Here.' The diplomat handed him an official looking envelope. 'I received this by courier before leaving Canberra this morning. Thought you should read its contents first. Then you can give it to your mum to peruse the contents.'

'A government seal? It does look ominous. More rotten news, I suppose?'

'Don't jump to conclusions. Open the envelope first, before you make a damn fool of yourself. This might cheer you up, seeing you're in a grouchy mood.'

'Well, don't ask the reason. It's complicated and beyond your wildest imagination.' Fumbling, Kendall couldn't rip the waxed medallion apart. It relented finally. He gasped on reading the text. 'Hell, I don't know whether to believe my eyes.' Another look dispelled all doubt. 'Everything's here, done, completed. How come, Peter?'

'Go through each item. Once you've followed the directives, sign this receipt in front of me.' The diplomat pointed to the chit in his hand. 'Then I'll countersign it.'

'Come through to my office. Please, don't say a word to Mum or the others about this. I'm annoyed with my mother, and I intend to make her pay for what she did today.' While signing the document, Kendall briefly explained the situation at the Bluff.

Instead of siding with him, Bucknell neither condoned Kendall's anger nor his frustration over his mother's stupidity. The diplomat thought his snide remarks unwarranted.

'Your mother is always in strife where you're concerned, Kendall. I think it's high time you took off for Europe and left her in peace. You don't realise the treasure you have in Gigi. Wish I had someone to love me like she does you,' Peter adamantly stated, leafing through his wallet for a misplaced item. 'This envelope has a grump's name written on the front. Yours, if I'm not mistaken. Its contents must be signed by you and your fiancée, grouch.'

Skipping through the papers, he laughed. 'I'm not a grouch, just frustrated. Everything here looks in order. I don't know how you or Father George managed this. This is the best news we've received in weeks. My passport?' Kendall flicked through the documents in his hand.

'I think you'll find it along with Gigi's inside the envelope. I felt their hard covers as I passed that manila envelope to you. In service, our fingers, like our minds are well-trained.'

'How right you are.' Kendall withdrew the passports from their protective leather sheaths. 'This photo doesn't look at all like me. Mum's is even worse. She looks older than her years. Thank goodness you or some top brass have spelled her names correctly. She'd hate anyone to find out her real Christian and middle ones. They're a well-kept secret between us.'

'I've yet to see a reasonable passport photo. Mine looks repulsive, unlike yours matie.' Bucknell frowned; his terseness so unlike Kendall's casual manner. 'Don't keep your mother in suspense. She doesn't deserve you tormenting her.' Senator Bucknell smiled as together they casually approached the kitchen.

'Want a cuppa, Peter? I've just boiled the kettle.' Gigi posed with the teapot on her well-padded hip. 'Harriet will be the pourer.'

'No thanks Gigi. I can't stay long. I wouldn't mind a swig of the best stuff, before I head back to town. My office awaits this impatient senator.'

'Peter, I'd appreciate you staying for dinner. There's plenty of food and I have something of a personal nature to ask you, once we've eaten.'

'Suppose I could,' drawled the diplomat. 'Yes, why not. I was only going into Yass to have dinner and buy food. Thanks, I will stay.' Over his spectacles, he looked directly at her son. 'Gigi, you may retract your offer, after reading the letter I gave your son on my arrival.'

'Spoilsport,' sniggered Kendall, sporting a grin. 'I'll make your whisky quarter strength in future by watering it down, you old bushwhacker.'

'Come son, don't keep taunting me. Include all our guests in your little gag.' Limping, she sidled up to Kendall. 'I saw the official envelope. I ought to know what one looks like by now.'

'Bjorn, will you fetch Brianna. I think she's in her room. Then we all should sojourn to the lounge room. With the leather poof under your bandaged ankle Mum, you should be comfortable.' Kendall walked to the bar while Peter helped Gigi hobble in to partake of drinks with her guests.

Impatiently waiting for his fiancée to appear, Kendall poured drinks all round.

'May I call Kirsten? She won't want to miss this excitement,' Bjorn whispered to Kendall in Brianna's hearing.

'Please do. Use the hall phone BJ. Where's Father George tonight, Mum?'

'He's lying down. Father's not well tonight. I'll go and ask him while you attend to your guests,' remarked Harriett before scurrying up the hall to his room.

Idle chatter and speculation continued until Kirsten showed. Settling his bottom on the arm of Brianna's lounge chair, Kendall withdrew each paper from the envelope, one by one.

'Stay seated, Mother and hold this.' He passed her the big envelope, with a smaller one inside on her lap. 'Open the small one and be careful. What does the inscription read?'

Cautiously she fumbled with the seal. Tearing the flap with gusto it parted. She read the wording then gasped with excitement. 'My full name! Everything here is inscribed in my maiden name. This means my annulment has finally come through. I wondered why …'

'Why I made you sign *all* those papers weeks ago. Yes, Mum. Now Father George will tell you his news. If you will?' Kendall nodded to Brianna's family priest.

'Gigi, ya no longer a married woman in the eyes of the church. This is the official Vatican seal.' He passed the official document to her. 'The Holy Father has granted ya His blessing with dispensation. No longer will anyone call …'

Her hand gesture hastily foiled what he intended to say. 'Don't spoil this moment please, Father. Now, I would like your blessing. This is terrific news. News I never dreamed of hearing in my lifetime.'

He complied and blessed not only Gigi, but also the entire gathering, which included Brianna, their cooks and the federal officers. He saluted Colonel Montsard, who entered the room as he spoke.

Feeling thirsty, the priest settled on the sofa to drink his whisky. Nobody minded and everyone joined in the festivities. Kendall, with Brianna's help, cut his belated birthday cake.

Half an hour later he whisked Brianna out to the front swing, in time to hear a vehicle rumbling down the drive. It stopped in line with its steps. Kendall assisted Mr Baumer from the car. Neither he nor his passengers could fathom why a commotion echoed from within the house.

Kendall explained the reason for their frivolity and suggested Kurt's family might like to join the happy throng. He was puzzled to see Arneka on the verge of tears. Without a word she darted past Kendall and her father then headed for her room.

'Leave her be, darling.' Brianna pleaded. 'I have an idea why she's crying. Her tears are tears of relief. She never envisaged that Gigi would receive her annulment. I think she feared the opposite actually.' Bridy's womanly intuition proved to be correct.

'Until now, I haven't realised how emotionally disturbed Arneka is, or how she must be reliving the hell of yesteryears. All these weeks she's battled through, yet only once has she confided in me.'

Feeling his own emotions bubble to the surface, Kendall composed himself. Feeling calmer, he kissed Brianna's forehead. 'Excuse me for a moment darling. I'll go and see what's wrong. If I give Arneka a big hug, it might help to relieve her anxiety.'

Brianna never answered. She understood her fiancé's compassion for his distressed sister. Slowly unclasping her fingers with Kendall's and with damp eyes, she watched him go.

'Who is it?' Arneka replied to the quiet rap on her door. 'Is that you, Brianna?'

'No. It's me, Kendall. Open the door sis and I promise not to stay long.' Confronted by the distressed expression on his sister's face, his eyes smarted and his heart ached. 'I'd like to sit with you for a while. I won't talk if you don't want me to.'

'I'll be okay. A spur of the moment thing. How did Gigi take the news? I feared …'

'Hush. And let me cuddle you. Mum's fine. In fact, she's ecstatic over her news. It's you Arneka, I'm worried about.'

'Kendall, I feared you would be unable to alter your surname to your mother's maiden name. Chris and I thought we'd fared horribly enough by that beastly man's evilness. Kurt and Anna's adoption of us as babies saved us both from being alone or wandering the streets with nowhere to go. Honoured, we took their family name. You, a doctor could've been branded illegitimate and haunted by the bastard's surname. Even though his last one here was fictitious.'

'Darling, I should've approached you earlier and we may have clarified this problem. I failed in my responsibility to you, your family and my mother until I received something positive to tell you all. Don't you see? It would've killed Mum if I had boosted her morale and increased her hopes for a better life without being tagged with his Germanic moniker.'

Kendall tossed his sister a hand towel. 'Dry your eyes, then we'll go down to the lounge room and celebrate Mum's good news otherwise she may think she's upset you.' A brotherly kiss on the cheek encouraged Arneka to follow his advice. 'Don't let her see you've been crying. Wash your face, put on a party dress and a bright lipstick. I'll be in the hall.'

Choosing a new tube, she engraved her lips with fuchsia red. She dabbed a little rouge on her pale cheeks and rubbed it in, to hide the blotches on her skin. Quickly running a comb though her hair, now falling loose over her naked shoulders, it boosted a sagging morale.

She tucked the lime cotton blouse in her navy skirt with a matching belt clipped at her waist. Buckling her white sandals and with a defiant air, she joined Kendall in the main hall.

'Do I look presentable now, brother mine?'

'You look adorable,' he said bending the crook of his arm for her to accept. Together they sauntered down towards the lounge room; there humorous jokes were in vogue.

'I appreciated our talk and I'll accept your sensible advice. It put everything in a new prospective. Tell me Kendall, how did the sale of Abergeldie go? Did the man buy it?'

'Well, I didn't have to drop my price. He intends paying me in full before I go overseas, Bjorn just told me. I have a feeling he might renege. I don't trust Smithers.'

'Shit I hope the old buzzard doesn't pull out of the sale. If he stays true to his word does it mean we'll have to leave here soon?' A dropped bottom lip manifested Arneka's fear.

'No dear, it doesn't. Tony Smithers has granted us three months grace, or as long as we need, provided we keep the property viable. He's not in a hurry and never mentioned a precise date. I would never have consented to sell otherwise.'

A contented smile encroached Arneka's striking features. Silently she conceded Kendall was right. *This news should put both his and Gigi's minds at rest.*

'I haven't told Mum yet. I'll let you tell her. That should brighten you Arneka and make both of you contented. Come and talk to Mr Bucknell. Peter's staying for dinner and you two have a lot in common. Do him good to chat with a wise old bird like you.'

'Old, or wise, Kendall?' she laughed and together they joined the merry throng.

19

4am Friday: The day of his overseas adventure had arrived at last. Kendall arose with the dawn, then showered, dressed and finished packing his port and skiing gear ready to leave at seven with Kirsten. Knowing she must be in Canberra for an appointment, he couldn't afford to delay her. Finished breakfast, he looked as the girls appeared at the kitchen door. He stood to greet his fiancée with a hug. Arneka turned her cheek for a kiss, while Harriet poured their teas.

'We caught you sneaking off without saying goodbye.'

'Course not, Arneka. Is Mum awake yet? If not, don't disturb her. I'll pop in on my way out to the car.'

Having heard his voice, Gigi crept into the kitchen. 'You called, son? I wouldn't let you go overseas without seeing you off. Kendall, give my love to Pierre. I believe I said some shocking things to him last November.' She nodded, retying her ming blue dressing-gown.

'Don't we all know it. You really hurt Pierre's feelings, and insulted him Mother. I'm not sure he'll forgive you, unless …'

'Unless what? Come on Kendall; don't keep me in suspense. What mischief are you lot brewing now?' Gigi seemed a tad curt over not being called to say goodbye.

Kendall refused to commit himself. It gave him great pleasure to witness his mother linger in suspense for a change.

'I gather this matter concerns Pierre and myself,' decreed Gigi, wiping her hands on the hand towel. 'I'll wager you're up to no good son.'

'For us to know and you to find out … ole girl,' Kendall grinned. On the pretext of leaving the table, he coerced Brianna into joining him in the bedroom, where he secured the door.

'Darling, I'm going to miss you awfully. It's still not too late for me to pull out. Perhaps you will change your mind and follow me on a later flight?'

'I've made up both our minds. You're going on your own. Enjoy your holiday, Kendall. This is your last fling as a bachelor. You may not get another chance if our plans are thwarted by the trial. You've not said a word …'

'No,' he interrupted her. 'Although I think Mum suspects some mischief's in the wind. Darling, this is our secret.' He waved something from under the flap of his overcoat pocket.

'Don't forget to ask "you know who" to find a solution to our dilemma. He's bound to think of how to accomplish it, and accept our offer.'

'Ha and who's suggestion was it, Bridy? Stand still and let me cuddle you.' Drawing her close Kendall inhaled her delicate French perfume. 'If you keep your promise, I may buy you a bottle of exquisite perfume from that chic boutique on the Champs Elysées in Paris. Oh sweetie, I'm going to miss you awfully. I'll be lonely without you for six weeks, especially over the festive season. I promise to ring you on Christmas Eve, around seven our time, Brianna.'

'I'll probably weep buckets, while you're away. Six weeks sounds an eternity.'

His index finger touched her lips. 'Shush, don't spoil our last few moments together.' As he cuddled her, he hummed their song, the one chosen for their wedding. *I love you truly, truly dear.*

'You'll be too busy keeping my mother occupied to miss me,' Kendall sighed. 'Write often. Your letters will console me on miserably cold days in Paris. If I'm in Austria or on the continent, Pierre or his daughter Mercedes will forward your letters to me.'

'Darling, I'm sorry you didn't physically love me last night, when we were alone in my room.'

'Don't start that again,' Kendall sighed. 'My love for you, Brianna is far deeper than a mild flirtation to deliberately make you pregnant with an illegitimate babe. Most methods aren't infallible. Precautions, other than abstaining, are not always effective. Once we're married, it'll be my pleasure to sleep with you. You know my views on the topic. I can't begin to imagine the pain and sadness my mother went through, before my illegitimate birth.'

'We love one another, Kendall. Surely that makes a difference. You're not like him and never could be. You're sending my passions wild, holding me this close. My dearest wish would come true, if you could defer your travel plans until the new year.'

Kendall pulled away an arms-length. 'Sweetheart no, don't put yourself or me through hell again. I'd be less of a man not to want to love you physically. I adore you. It's natural, when a couple are in love. Perhaps it's a good thing I'm leaving, or I might weaken.'

'Oh, Ken …' A knock disturbed their tranquillity.

'Kendall, I have your flight tickets. You left them on the table. Where will I put the folder son?'

'On my suit-folder in the hall, please Mother.'

Gigi gauged by him calling her "mother" that she may've intruded on their privacy. While putting the tickets on the smaller of his two cases a smile embellished her face. Little did she realise that smile would soon turn to a grimace.

A second intrusion on their privacy came almost instantly from Father George. 'Excuse me, Mrs Svensson's car has arrived fa ya Kendall. Ya mother is talking ta her and she's in a hurry.'

'Tell Kirsten I'm coming now, please Father,' Kendall called. Lowering his voice, he smiled down into Brianna's moist eyes. 'This is not goodbye my darling, because my love for you goes beyond the paranormal. I promise to ring you every second night Aussie time. You have the number where I'll be staying, Bridy. Keep in mind both those countries are also on different times. One more kiss and I must fly. No tears.'

'Have you got it, Ken?' queried Brianna sniffing.

'It's safe here in my suitcoat pocket. Don't forget to take the photo. It's important and in this good light it should be a beauty. Don't let Mum get her hands on the negatives or the prints. Knowing her, she'll probably destroy them.'

'Will and won't do. Look after yourself, darling. It's going to be dreadfully lonely here without you Ken. I'll miss you until the seas run dry.' Releasing her hand, she kissed his cheek.

Brianna dried her eyes. She promised not to weep as tears should be for happier times. She understood what Kendall meant and waved goodbye from the verandah. She and Arneka pretended to catch the kisses he'd blown from Kirsten's car.

'Kendall, do you have everything?' He nodded. 'Have a safe trip, both of you.' Gigi tapped his grey serge sleeve. 'I'll be like Brianna, sweating on your calls thrice a week.'

'Don't be so cocksure, Mum. After we leave here you might dance to a different tune. Get this thing moving, Kirsty,' he said, with a sniggering smirk.

'Why! What mischief are you up to now?' Gigi called as the vehicle slowly ascended their drive. Amber jumped with fright as the car cruised to a stop alongside the cedar tree. She barked and feeling relieved after a wee, hightailed it back to her mistress.

Tempted to see what was wrong, Gigi hesitated when Kendall's hand waved from the front passenger window. The item clutched in his fingers fluttered in the breeze.

'Mother, do you remember a certain letter you wrote to Pierre? I'm sure he'll love this. Ha, and you haven't missed it.'

'You didn't! You couldn't have found my *billet-doux*, Kendall. I'll kill you. You'll keep. And don't think you have heard the last of this nonsense.'

'Gigi, turn and face me. Wow! What a good shot.' Brianna clicked the shutter button twice. 'He did and he has. You shouldn't throw love letters in your bedroom bin. They could be incriminating evidence.' Brianna grinned and Arneka giggled.

'You conniving pair of monkeys! That was ages ago and it wasn't meant to be read,' Gigi bellowed. No way could she see the humorous side of their little charade.

'We know. Arneka and I found it. Kendall's the mailman now, delivering your precious declaration of love to Paris,' Brianna trilled as her compatriot in crime roared laughing.

Bewildered, rather than dumbfounded Gigi was incapable of retrieving her letter or the incident. They'd stolen her letter of love and an apology. What could she say? Anger failed her.

Kurt stood between Father George and Christian; they split their sides laughing. All of the guests were in on this prank. They enjoyed it far more than those driving off.

Gigi turned to reprimand her future daughter-in-law and Arneka. The pranksters had flown, taking her camera with them.

Until Kendall rang each day Gigi found it impossible to settle. Since him leaving almost two weeks earlier, she felt desolate and lost. Swathed in misery her days dragged into nights.

Bored beyond reason, Gigi cajoled her cook into leaving earlier than expected to attend their hair appointments in Yass. When finished in the salon, they completed their Christmas shopping. Together they delved into every jewellery and giftshop, then visited elegant clothing boutiques until their feet ached.

Abergeldie remained relatively quiet all morning, with only the rich smell of freshly baked pastries pervading the kitchen.

20

5am Saturday 29 December: Bjorn arrived at the homestead to collect Father George and the Baumer men for their day trip to Cootamundra. After the men had attached Abergeldie's horse floats, Bjorn took the wheel. He left Lanky to drive his single float across to Jalna.

His manager Bernie Staples, hitched the double float to their jeep's tow-ball. Once this job was completed all three vehicles set off for Coota's pre-Christmas yearling sale. Bjorn intended to boost his stud with the progeny of proven stayers with superb bloodlines, rather than wait for the January sale.

Understandably, this was the first insight into horse wrangling for Abergeldie's two male guests. Elated over the prospect of naming one of the horses, Christian propped his butt against the saleyard fence. His interest lay in studying Bjorn's purchase of three fillies.

Initially he'd chosen a roan, followed by a sorrel and bay. Next horse in the ring was a dapple-grey colt. This finalised Jalna's buy. Cheques having changed hands the yearlings were secured in their floats. With this task completed, the men knocked down a few beers in a local watering hole.

Before leaving Red Kelly's Blue-boy Pub, Bjorn called home. After checking the floats and horses the men began their journey home.

Arneka and Brianna, the sole remaining guests on Abergeldie enjoyed their well-earned freedom. They gathered apricot rosebuds, their petals glistening with dew, from the courtyard garden leading to the homestead's long drive. A rainbow arching across their valley indicated the rain was dissipating. The sun looked like breaking through a patch of blue.

A brief trip into town and with a new expected guest due shortly, they'd prepared Kendall's room ahead of time.

Waiting for Kirsten's vehicle to churn the mud-puddled drive to a flurry, the mischievous pair settled on the verandah swing, while discussing Brianna's future. With the prospects of her wedding day closing in, she asked Arneka to be her bridesmaid.

Arneka couldn't imagine herself dressed in a frock of ankle-length chiffon; she felt like Cinders. With little or no finances to bulk her purse, she wondered whose pocket this money would spring from. The elaborate gown Brianna had chosen lay well beyond her paltry means.

'I'm a nobody Bridy; a person without a wealthy parent to buy me things. I'm almost a pauper who defies my past. Why would you want me to take part in your wedding, with my Nazi family history?'

'Don't say another word Arneka! By belittling yourself, you're letting Kendall and Christian down. Remember, they're both your brothers now. You should never denigrate yourself. We love you and sincerely care.' Brianna tried to allay her fears by explaining this in a gentle manner. 'Kendall intends to buy your gown. He told me that before leaving for Paris. This morning his mother backed his suggestion.'

'Weddings are expensive at any time, without adding my frock and flowers to their bill. Kurt will …'

'No dear, Kurt won't be offended. It's you who will hurt his feelings and theirs. The matter is settled. I don't want to hear you denigrate either yourself or them again.'

'Well, I probably sounded ungrateful, but I'm not. Enough said. Who am I to argue with a beautiful and loving bride who will be my sister-in-law? I'm honoured that you have decided to include me in your wedding plans. Even to think strangers have taken our family to heart is unbelievable and indeed surreal to me.'

What Arneka had just declared stunned Brianna. *How could this sad woman ever consider her family being strangers to my fiancé. Kendall would be insulted if he'd heard her say it.*

'True friendship is what love builds on. By helping and being here for one another, we help ourselves. It would devastate Gigi to hear you say what you just said to me.'

'I didn't mean to be offensive. Our life has been one long struggle to find where we belong. Some people frown on our past history. Having a war criminal's blood flowing through Chris and my veins repulses me. Being a Jewess would be fine. A Nazi: no way.'

'Wake up to yourself, Arneka,' Brianna scolded. 'None of what happened in Germany was your fault. None of us are perfect. Tragedy brings families together in unity, like it's done now.'

'Suppose! I fear if anyone discovers he was our birth …' A bitter and harsh swallow hurt her throat. 'I can't say or think of his family relationship to us, Brianna.'

'Don't be upset. What did Mr Bucknell tell you? He promised nothing will ever be circulated about the tragic situation that occurred here on Abergeldie.'

Arneka couldn't respond. There was no logical answer. The truth of their past in Hamburg was frightening enough and she realised the past must be forgotten. She would learn to forget the horrific times she and Chris had suffered due to their father's cruelty. It wasn't easy, considering their adopted father had and would suffer every time the name of von Breusch emerged. Her most devastating fear lay in the chance he may return to Yass one night and murder her family while they slept.

Brianna tried to calm Arneka by confessing, 'As an outsider, I cannot understand how any of you feel. Imagine your trauma yes, but understand it, never. How could anyone who hasn't lived through your wartime dramas and agonising past comprehend them.'

Similar to Arneka, she also feared if their situation became public knowledge, it would destroy all their lives and everyone in a close relationship with themselves.

From what she'd heard Peter confirm to Kendall, she understood little of what ASIO and the federal agencies were up against. Brianna surmised once they pinpointed the location where the criminal might be hiding, they would crucify the demon.

Brianna knew Kendall seemed to think there was an on-sight warrant out for his father's arrest. They lived in hope that the federal police might catch him and soon.

'Well Bridy, I for one am sick of thinking what I want to say every time I open my mouth. I'm scared I will slip and say something I shouldn't. And I'm tired of being constantly warned about doing it. Chris can be quite belligerent and nasty if I say something wrong to insult him, or make my father angry.'

'Arneka, some people have a tendency to run off at the mouth that's all. None of us would ever divulge anything to the press, or publicly. We all have a lot to lose.'

'Does Gigi know the full facts of what we suspect?'

'She knows nothing of what I've disclosed to you and we must keep it that way. Peter informed Gigi when this criminal's known whereabouts are proven, the federal police or ASIO will work from there. Personally, I don't think any of you will need to appear in court.'

'You're wrong,' Arneka sniffed. 'Kurt has promised to testify at the initial hearing. And if he has, you can bet Father George will also.'

Brianna lent forward and looking into Arneka's eyes she declared, 'Can't you see why? They both lived and worked under and against the Nazi criminal regime. They were crippled by lack of power. And they've also seen how evil the officer we've been discussing was and still must be. A person of his calibre doesn't change. He fooled everyone, including those who loved and trusted him, like your mother Erika and Gigi. Even your nanny, who you have mentioned to me, suffered and died by his hands.'

'Heide, yes he did fool her, only not for long so it seems. Chris heard Daddy mumble in his sleep recently, that Heide was manhandled by von Breusch. We assumed Kurt meant the Nazi wanted to rape her. If it's true, the idea is revolting.'

'You should've talked to Kendall about your fears. He could help you to mourn your losses. Have any of you ever had professional counselling, Arneka?'

'No, bearing his soul to the Feds and Peter Bucknell, besides Father George has helped Kurt. Chris is a private person, he would hate to express his feeling to a stranger, unless he's forced to. Me, I have told Kendall some of my traumas, but not everything. Not of my low esteem or my insecure fears. Just talking this out with you has made me see there's a brighter future beyond the dark horizon.'

'Well that's a start. You can come and talk to me or Kendall whenever you're stressed.'

21

Relishing the cool night air on the swing after their talk, Brianna enjoyed the glorious carnelian sunset ready to disappear into obscurity. A whirlwind channelled skywards some distance up the earthen drive. As its swirl escalated, Brianna focused on a dust-swirl hovering above Abergeldie's front gates.

'Looks like Kirsten's car is ready to pull into the drive. She won't stay long; I'll help her guest in with the suitcases.' Walking down the steps Brianna turned to further address Arneka. 'Would you mind assisting the guest through to Kendall's room? Poor darling's bound to be tired after such a long journey. I'll fetch fresh towels out of the warming cupboard. With this chilly wind and the weather closing in again, they'll both be frozen.'

Kirsten pulled her Monaro to a stop alongside the front steps, only she didn't alight. 'Wow! It's good to be home. Here darling, your hat and coat. See you later tonight. Look after our special guest, Arneka. One very tired and treasured friend I've placed in your care.' Kissing her passenger on both cheeks, Kirsten waved a farewell to all and headed for Jalna.

'Hope your flight wasn't rough, sir? I've done it twice from Ireland and it never seemed to end.' Taking their guest's hand luggage Brianna remarked, 'My, this suitcase is heavy, feels like you have a ton of bricks in it. Oh, do take care on these steps, they're wet and the wood is slippery.'

Carrying the heavier suitcase Arneka followed them up the steps where she whispered to their guest, 'Hey, won't somebody be surprised to see you?'

'You have not said a word? Still our secret, Miss Baumer?'

'I'm sure Gigi hasn't a clue that we're expecting another guest,' she smiled. 'She and her cook should be home shortly. Unless they're rummaging through bargain sales tables, and that's more than likely at this time of the festive season.'

'Oh Brianna, have you heard from Kendall? We had a wonderful talk about your future wedding. I will tell you everything, once I have changed my winter clothes into something cool and comfortable.'

'We receive a phone call from him every other night.'

'We picked the apricot rosebuds and a few sprigs of baby's breath that you requested in your letter, just before it rained,' called Arneka strutting ahead of them.

Abergeldie's newest guest had no sooner freshened and changed into summer clothes when Gigi's vehicle hit a furrow at the drive entrance. A flurry of red mud sprayed the tea-tinted, white gates. Doused in red they complimented the vivid sunset breaking through a series of white-feathered swirls.

Arneka addressed their guest in his doorway. 'Brianna or I will comment on the weather; that will be your cue to come out.'

Their guest nodded and stepped back into Kendall's room. The door remained ajar.

The instant Gigi walked into the main hall she deposited an armful of parcels on her bed. A quick shower and dressed in light clothes she joined the women in Abergeldie's kitchen.

Settling her bottom on a stool she confirmed, in a weary tone. 'Well, if that's Christmas shopping, Harriet I never want ever to do it again. Town's hectic. Everyone's rushing around looking for bargains. Fetch me a cold drink please, there's a dear.'

'Okay Gigi.' Kicking off her shoes the cook nodded. Passing over the glass of iced orange juice, Harriet caught sight of the roses in a bucket near the first coolroom. Quickly she stepped in front of the wire door leading to the back verandah to block Gigi's view.

'You're sweating Gigi. Why don't you have a cool bath or shower? Dinner will be later than usual seeing we've just come home. The lamb roast and veggies won't be ready until around seven.'

'A cool drink is all I need. There's nothing better than an ice-cold orange juice and a good soak in a tub of cool water to ease sore feet.'

'Don't forget, Kendall promised to ring you round six-thirty,' Brianna reminded her. 'You better not stay in the bath for long Gigi. Sometimes it's hard to get through from Ebensee or Salzburg. The Austrian phone lines will be very busy tonight.'

'Won't everywhere be busy on the eve of Christmas Eve? You're right, Brianna. I'll have a quick bath. Pour me a glass of my favourite Heaven's Dew please darling. One scotch won't hurt on this festive occasion.' Finished giving orders and clutching her glass Gigi hurried to her bathroom. The phone rang and she paused to listen.

'Arneka, it's Kirsten. I'll tell Gigi to take this call on the hall extension.'

'We're clanging pots and dishes in the kitchen. Gigi won't hear you here. Hang on, I'll ask the Feds to tell her.' The federal officer on duty transferred her call. He warned Mrs Svensson to keep her conversation short.

'That's Gigi out of the way,' Arneka giggled. 'For how long, is anyone's guess.'

Just as Harriet announced dinner was served, Brianna walked up the front hall where she casually remarked aloud, 'Arneka, the sun's setting before it sinks below the horizon. And it looks as if the rain has gone.' Standing beside Kendall's door she nodded to their guest, who sneaked passed her in the hall. 'Stand behind the kitchen door and I'll cough if I hear Gigi coming down the hall.'

Silent and with eyes misted, the guest anxiously stood silent for what seemed like ages. With his heart thumping and fingers grasping the cellophane wrapped flowers, he counted the minutes until Gigi appeared at the door.

'Kirsten and Bjorn are on their way here for dinner, Harriet.' Gigi danced an excited jig on her entry into the cook's domain. 'Please set two more places at the dining table.'

'Thought they'd be over, so I've done it already,' Harriet replied with a beaming grin.

'A bath makes one feel refreshed, and it's cooled me down a little. Phew! It's been hot today.' Gigi accepted the crystal goblet from Arneka. 'I'm going to enjoy this whisky and then I might finish wrapping my presents …'

'Why not sip your drink now, while you're resting. Put your feet up on this kitchen stool and relax, Gigi.' Arneka slipped the footstool in place. 'You said they were aching, so having them elevated will ease those swollen little toes.'

Removing her boss's slippers Harriet smiled. 'There, you've earned that rest after all the trotting around we did today in Yass. The village streets were chock-a-block with sightseers and festive buyers.'

'Oh Harriet, I'm absolutely stitched up, buckled with tiredness.' Unable to move from the chair, Abergeldie's owner elegantly sipped her delectable drink, relaxed and closed her eyes.

Slowly their mystery guest eased out from behind the door. A pair of sock-clad feet glided silently two paces across the room. Quiet and on edge, he placed his hand on Gigi's shoulder, as Brianna redeemed her drink.

Her eyes reflected on his clean-shaven chin and as her gaze settled on his smiling eyes Gigi emitted an excited gasp, 'Pierre! How on earth …'

Passively, he smiled down into her eyes. 'Hush, my *chérie,'* he stated in a passive tone. 'Let me see your beautiful smile. It seems to be sweeter this eve than the last time we met.'

Endowing her temples with kisses, he presented Gigi with the bouquet of roses. Quietly and without ceremony, Pierre held her free hand and escorted Gigi to his room, where they could talk in peace.

Entwined in blissful rapture he cocooned his darling in his arms. Bound in quiescent love their lips melded in love as they embraced. Words were unnecessary.

He placed a finger on her lips and binding them in silence, he announced, 'Long have I craved this moment, my chérie. You to me are my all. I've waited a lifetime for these precious moments alone with you, my treasured Gigi.'

'I never dreamed our love would find a way. After one failed marriage I …'

'Chérie, I ask one thing of our love and you. Never must you mention that Nazi again. Wash the infidel from your pure heart. For should you not, our union is doomed to fail.'

Allowing Gigi time to assimilate his demand so tenderly phased, Pierre rested his lips on her hot forehead. A finger again held her lips to ransom.

'I have not come all this way to embrace the past. Although something of mine I will tell you later. Embarking on our future together is my sole purpose to be here with you. I do however, have one request for you to do in Paris.'

'What might that be, Pierre?' Her curiousness increased dramatically. She instinctively grew suspicious and tried to decipher his reason for an operetta with her in its leading role. *This seems so unlike Pierre to make demands of me. What does he have in mind? Something to do with his new ballet troupe, I suspect, by the sensual tone of his deep, serene voice.*

'I suspect that you are doubting my motives Gigi. I would never deceive you, ma petite. I propose one more ballet for you to perform. The stage and venue will be apparent should you accept my offer …'

Disappointed, Gigi pulled away from his arms. Retying the robe's sash around her waist she declared, 'Never, my ballet days are over. Finished! Where once I breathed, slept, ate and worked with and for you to perform on stage, I now cannot.'

'Can I say this, s'il vous plait …'

'Sorry Pierre, you've had your say and now can I have mine? Sadly, I confess those erotic dances are now pipedreams that must be locked in our memories. Like all our majestic performances, a ballerina's dreams cease with the finale, and in her prime. Mine closed the moment you left me in Sevelen, after our final tour of Switzerland in forty-five.'

Wisely Pierre let his ex-protégée complete speaking before he disagreed. 'Not so! My request is for another finale, with your presence on centre stage. One, I daresay you will not turn down, nor require an explanation for, Gigi.'

Giving her a roguish smile, he gently but cautiously added, 'What you chose to do in the past performance proved correct. Who am I to pass judgement on you, or anything you may have done? The past is gone. The twists of life, however, demand change. Shall I go on, Gigi?'

'I'm intrigued, yes please do, Pierre. I'm interested to hear more of your fantasising.'

'The venue shall be in Paris. The scene is set with you, my prima donna dressed in an ankle-length gown of magnolia satin. Your low-cut bodice will gently cover your naked shoulders. One Cattleya orchid tinged with pink, or a missal with a spray of miniature roses you will carry in gloved hands. A crown of orchids will wreath the chignon atop your beautiful head. No jewellery will adorn your sweet neck or dangle from petite ears.'

'Pierre, the scene you've described is of a costume I wore when I danced the principal role in La Bohème, a wedding scene at Covent Garden. Am I not right?' She found her answer in his deep-brown pools afloat with endearing love.

'Would you believe, I still have my pastel pink satin pointe shoes? They're dangling over the dressing table mirror in my room.'

'*Oui* I do believe it and you. Gigi you will look as grand on our day, as you did on the night you brought the house down. You made this old fool proud of you then. And you will do so again in our dual performance on our special day.'

Holding both her hands he gently crushed them to his chest. Cocooned in ecstasy, their eyes focused only on each other. This was their realm of dreamtime, of their togetherness.

'My chérie, I will stand by the altar. Then as the Bridal March begins, I shall turn to see my bride walk down the aisle decorated with your favourite bouquet of flowers.'

'You've asked me to marry you. How delightfully put. Although I disagree …'

'Why Gigi, does the idea of marriage repel you?' Dismay clearly showed on Pierre's sad, but as yet unlined features. His lower jaw was agape. Tousling his dark hair, greying at the temple, he feared her response.

'No darling, I'll be honoured to become your wife and I graciously accept your sweet proposal.' She beamed as Pierre kissed her cheeks which sealed their vow. 'I disagree with the graceful gown you described. Not a style befitting a lady over fifty.'

'Perhaps my idea is an old-fashioned Frenchman's choice. Well, what might your preference be ma chérie?'

'Sweetest man, I shan't tell you what I shall wear. You will have to be patient. I believe it's a bride's prerogative to keep her future husband in suspense. All I will say, I'll carry a silk-covered missal plus an orchid spray. A pillbox hat will adorn my upswept hair; as it is now.'

Swathed in his loving arms, Gigi stood on the tips of her toes and kissed Pierre's not yet receding hairline. 'Darling, I'm thrilled that you want me to marry you. Your proposal attests that my awful son has passed on my private letter …'

Pierre interceded, '*Oui*, I will discuss the *billet-doux* with you tonight after dinner. Not at present. Your guests may overhear us as they pass by to go elsewhere.'

His wry smile intrigued Gigi. *What has my deceitful son said to Pierre? I feel betrayed by Kendall stealing my billet-doux. Oh well. It's over and done.*

While entrenched in thought, Pierre placed something from his pocket on the bedroom dresser. Withdrawing a two-carat diamond solitaire he turned to slip the ring on her finger.

'Gigi, keep your hand still,' Pierre politely demanded of her, 'this ring signifies a lifetime of endearing love together. Chérie, wear it proudly with my undying love.'

'This diamond is huge and so magnificent I can't believe the gemstone is real,' Gigi declared. Agog with excitement she shook her head in disbelief of the gem's grandeur.

'Chérie, you can admire that ring later ...'

A familiar knock disrupted their precious moment. 'Come in Harriet. We're just talking.'

'Sorry to disturb you, Gigi dear. Doctor Jarvis, Senator Bucknell and his lady friend have just arrived. Dinner is ready to serve.'

'Have Kirsten and Bjorn also arrived? Consoled in ecstasy, I didn't hear their cars outside or footsteps in the hall,' she uttered all of a dither.

'*Oui*, I believe they have Gigi. It would be rude of you not to greet your guests. My dear, you are their hostess.'

'Allow me a second to get suitably gowned, please darling. I won't be a minute. Once I've changed, I'll sneak back in here to your room and you can carry me down the hall. Then in your suave accent, you, Pierre can announce our betrothal,' she said. Giving him a quick peck on the cheek she darted back to her room.

True to her word Gigi hurriedly returned. Her apricot, full-length gown of cotton swept the parquetry hall floor as she glided towards him.

'You look exquisite chérie,' Pierre paused to relish her delicate aroma. 'Your *parfum* is delectable, my gift to you last November on Abergeldie. Remember how you ignored my plea to be calm. You were ill from stress caused by that evil man, who no longer will hurt you Gigi, because I shall protect you in Paris. Mercedes sends you her love as does Raoul. This episode caused by that man will kill you, if you stay in Australia. The police here cannot be everywhere at once. Until the German is captured, and he soon will be, you and Kendall can't remain in this house.' He acknowledged her frown while kissing Gigi's forehead.

'My darling, you forget we have lots of protection. Amber will attack anyone who tries to hurt me. Then there's our two-legged watchdogs, Arneka calls the federal agents who will enjoy dinner at our table tonight, and keep their guns ready to shoot any intruders. We're safer here than in your bank building in Montmartre and in all of France.'

'How can I argue with you on that score, my dear? You were right and I was wrong to interfere in your private business. I apologise for suffering the guilt of your fears, Gigi.'

Gently kissing her shoulder, Pierre let the door close of its own accord. He stood silent for a moment before returning to speak with her guests in the lounge room. 'Kirsten, I am really concerned by your friend's abrupt attitude and her health. We cannot talk here, let's go out to the front garden. It is peaceful there and no one will disturb us.' She agreed.

22

7pm Christmas Eve: While being carried down the hall Gigi began to panic. Feeling guilty, she pleaded with Pierre to phone his family of their engagement before confronting her friends.

'Chérie, I have Kendall's blessing. Later you can phone Mercedes and Raoul. They will be pleased to hear of our betrothal. Now is not the time to ring Paris,' Pierre assured her.

With Gigi still cradled in his arms, the Frenchman paused under the lounge archway. The room was vacated. Not a person was in sight.

'Where is everyone?' moaned Gigi disappointedly. Astounded by the silence, her tone faded to a whisper. 'It seems strange; I can't imagine where they might all be. The breezeway. Yes, that's where they'll be on a sweltering hot night.'

Pierre smiled as he eased her feet to the floor. Ready to step over to the carpet, they hesitated as voices echoed through the room. Friendly faces popped out from under curtains, from behind the bar and furnishing, plus the kitchen and its servery.

'We fooled you, Gigi.' The crowd of ten cheered. 'And you never suspected a thing. We kept you in suspense, not knowing what might occur. Your son and I arranged this little soiree without your knowledge.'

'You, Pierre,' she began to accuse him and faltered, 'no you're all guilty of setting me up. It's wonderful to have all our kind friends welcoming us. Not forgetting my extended family, Brianna, the Baumer's and Father George.'

'Kirsty and myself, I also hope so. We've been part of your family for longer than I care to remember.'

'Well, that goes without saying Bjorn. You and Kirsten are my dearest friends. I love you both more than I would a sister or brother. I didn't expect this pleasant surprise, thank you all.' Gigi looked in Pierre's smiling eyes and instantly knew he and her son were in cahoots as the instigators of her surprise party.

'Enough of your speeches old girl,' Bjorn garishly lauded. 'How about letting your better half have his say?'

The Frenchman stood and nodded. 'Mesdames, no I prefer to ignore formalities. Dear friends, I have the pleasure of presenting to you my future wife, Madame Bouvier, my fiancée, whom I am proud to say, has just accepted my proposal of marriage.'

Enraptured, he swept Gigi into his arms and kissed her long and lovingly. As their radiant eyes met, their happiness shone stronger than a divine guiding light.

Everyone gathered around the ecstatic couple to congratulate them. Shaking hands with Pierre, the men praised his choice of a wife. As the *le femmes* hugged Gigi, she flaunted her magnificent and greatly-prized diamond ring.

'I hate to spoil your little *tête-à-têtes.* Dinner is served. Please take your places at the table, or it will spoil,' Harriet sprouted, flipping the oven cloth. 'The honoured couple are seated together at the end of this table. Name tags are for all of Abergeldie's guests.'

'Surely you and Rose aren't eating in the kitchen, Harriet?' queried Gigi, her eyes rescanning the settings.

'Trust me! Wouldn't you just know it? With all this excitement, I forgot to set our places, plus another for Helen. She'll be over in a minute to help me serve.' Seeing the humorous side of her blunder, Harriet laughed and then returned to decorate the sweets in the kitchen.

'The table looks beautiful, decked with festive bonbons and flowers. I'm pleased you used my best gold-edged china and those fine crystal goblets. To whom do I owe the honour, apart from our treasured cooks?'

'Both of us,' heralded Brianna beaming. 'Arneka and I drove Rose to town early this morning to buy fresh goodies. On our return we decorated the table. Harriet and Jalna's cook, Helen did most of the cooking. Her daughter Jasmine iced all the cakes.'

'Bridy and I had fun icing your engagement cake, Gigi and Pierre. The Christmas cake we found a challenge. Jasmine scrolled your names and the interlocked double hearts on it. I must confess, I mucked it up a little, before Jas took over,' Arneka pragmatically chirped up with a hearty grin.

'Really, to me both cakes look professionally iced. The entire house looks magnificently decorated with Christmas baubles and streamers strung above every doorway. Not forgetting fresh flowers everywhere. I noticed you pair pinched some of my prized roses, carnations and greenery to put among the baby's breath.'

'I asked Brianna to gather the roses from your garden, my dear. Gigi; I could not manage my luggage, the demijohn of champagne and flowers.' A gracious kiss touched her cheek. 'A *perfumed fleur* by any other name is ere so sweet as you, mon chérie,' quoted Pierre, the quote a shade out of context.

Casting an eye in Bjorn's direction, and up standing he nodded. 'May I ask Monsieur Svensson to do the honours with the champagne that is in front of him?' The massive bottle was strategically placed.

'This bottle hails from the Province of Champagne in France. I carried it to Australia in my luggage. The champagne is specially to celebrate this precious moment of our betrothal.' Having declared his speech, Pierre concluded, 'To my future wife, and a precious jewel in my life. I salute my glass to you, Gigi.' He raised his goblet aloft.

All upheld their glasses and chorused, 'A wonderful life, Gigi and Pierre.'

'May all their troubles ... well, it's rather late for that I suppose?' The burring of the hall phone saved Bjorn's hide and his witticism went unheeded.

'Gigi, I think in all probability that might be Kendall. Shall I go or will you?' queried Brianna, on tender hooks.

'I will, excuse me all. Can't leave my son dangling on the line from Austria. Besides, I think he deserves a reprimand from an engaged, not an enraged mother.' Discarding her napkin to the four winds Gigi hurriedly vacated the chair.

'If you don't mind,' Brianna clasped her hand as she passed, 'I would like to speak to Kendall, after you've finished speaking to him, please Gigi.'

'Naturally my dear, you're welcome.' Her brevity of words indicated her son was to about to cop a mouthful of chastisements. Her indignation reached no bounds.

A brief tranquillity embellished the room in silence, as Abergeldie's matriarch silently listened to the strains of the distant voice. One she loved. Not a note of discord entered her conversation to Kendall.

Satisfied her son was safe and missing her and his precious twins, as he'd just referred to the pair of gigglers, Gigi frowned at the instigator of the rising ruckus. She then nodded for Arneka's co-conspirator, Brianna to take the phone.

'Gigi, what did Kendall have to say? We're all ears.'

'And so you shall remain, Arneka my dear,' her reply was curt. 'My private discussion isn't important. What is important is Kendall sent you his love and best wishes for the festive season, besides his greetings to all present. Please charge your glasses to my son, Doctor Kendall …' Befitting with her frivolous mood, Gigi held her guests in suspense, then quietly added, 'That is my secret. You will all learn our new surname in due course. Sorry to disappoint you folks. Don't you say a word, Bjorn.' She cautioned him to be silent, by a wave of her index finger.

A hushed murmur emitted from the guests. All smiled for reasons of their own. The majority of them were unaware of Gigi's lineage name. Some knew where her stage nom de plume originated. Others wrongly assumed. This proved successful to allay more unanswerable questions being lobbed in her direction by well-wishers. An undetectable smile broached the fringes of her lips. Once again Gigi had eluded the trap of disclosure. Bjorn and Kirsten smiled. Their age-old secret was safe, for the moment.

On the cessation of dinner, all adjourned to partake of liquid refreshments in the lounge room. To talk in privacy there was unthinkable, so the couple excused themselves. With everyone's blessing, they escaped the throng and took a short drive down to the river.

The crisp night air and the gentle flow of water lapping sedges nearby, lent peace to their surroundings. The cool atmosphere created a serene ambience, ideal to converse in passive tones.

They daren't sit on the dew-drenched grass, so Pierre retrieved a waterproof rug from her car's boot. Spreading the crocheted rug over a dry mound of earth in a secluded spot he sat down beside Gigi. With moonlight filtering through the tree canopy, he cradled her in blissful solitude.

Befitting his shyness, Pierre quietly broached the subject of his previous marriage. To relieve some of the tension, he decided to confide in Gigi, whom he sensed had suspected how he'd suffered for eons over his wife, Marion.

In the soft evening light, the gloomy shadows of yesteryear were put to rest. Pierre now wished to voice his anguish. 'I want you to understand, I love you like I've loved no other, Gigi. Still there's something I have neglected to tell you. You may choose now is not the right time to speak of it. Please don't hesitate to tell me if you object?' Underlying his moderate tone, there smouldered a huge amount of pain and a lot of anguish.

Without saying a word, Gigi carefully listened to him speak as to the reason for his building frustration.

'If it eases your mind Pierre, I'm eager to hear what has hurt you so dramatically over the years.'

'Gigi, in all good faith I cannot step into our new life together by hiding the truth from you, my treasured partner. My marriage to Marion was, to a great extent, farcical.'

'Darling, you don't have to explain your life with her to me. You had to know of mine. But whatever is making you miserable let it remain dead and buried.'

'Sadly, I cannot. Nor do I wish to withhold the truth from you any longer. Somewhere down the years in our journey together, my secret is bound to surface. Then my deceit will destroy our love, as it has done my entire married life.'

Difficulty in selecting the right way to break this news, Pierre realised it required a clear mind. Unburdening one's soul wasn't easy to relate without tears.

Gigi sensed his dilemma and gave her fiancé the space he needed to think. Pain manifested itself in Pierre's pitiful expression. She could see the pain mirrored in his eyes. His face lay directly in the ray of moonlight filtering through the broken canopy above.

Leaning her head on his shoulder, peace seemed so near, yet a millennium away. Bathed in silent thought, her womanly intuition begged her not to mention his wife of some years standing, unless the conversation warranted. By letting Pierre explain in his own way and time, only then would his agony be confronted.

'You remember Marion was my first prima donna. Equal in grace, we performed at all supreme ballet venues, the world in its wisdom offered.'

'I do remember that, Pierre. Your dancing and choreography were superb. I followed your performances and I ardently learned from your graceful movements.'

'What you do not know chérie, is that unless Marion married and produced an heir within six months of our marriage, she would not benefit from her promised inheritance. That clause was a condition of her granddame's will. In our poor circumstances money was vital to keep my company and us viable.' For comfort, he pressed Gigi's fingers to his lips. 'The old dear knew this and put her demand to both of us. I loved Marion then,' confessed Pierre as a sole tear blemished his pale and drawn left cheek.

Gigi noticed the drop glistening in the moonlight, but preferred not to comment. Instead she said, 'I was under the impression you were an ideal couple. Leave it at that. Please don't go on Pierre, your pain is too deep.'

'I had no choice then and I have none now. Please, let me confide in you. It is very hard.' Still holding her hand, he blessed her fingertips again with kisses. Poised as if forever, he held her hand in the tender clasp of everlasting love.

'Tell me what is worrying you now. I won't impose on your thoughts.'

'We were blessed in our union, and within twelve months Marion was delivered of a baby boy.'

Puzzled Gigi interrupted him. 'How could that be, Pierre? Mercedes is your eldest child.'

'No chérie, she was not. I loved Jacquan our son. On his second birthday, the lad was sucked out by a monstrous wave while building sandcastles on the beach at St Tropez. Jacquan drowned. Later the police retrieved his tiny fragmented body from those rocks. I felt it was my fault. I was on tour when I should have stayed with Marion and my boy, as I thought of him then.'

Perplexed, Gigi again intruded. 'Sorry Pierre, I'm at a loss to understand what you mean.'

'Probably,' he decreed. 'After the child's funeral Marion, in a fit of anger, denied Jacquan was mine. She told me about the moment of his conception. I knew the wretch who'd seduced my wife.' Embroiled in horror Pierre refrained from speaking for a second or two.

'At the time, I didn't know the full extent of their sexual liaisons. It appears, even before our marriage, Marion and Alexis were lovers.

Because she had never fallen pregnant, they assumed he was incapable of fathering a child.'

'Perhaps they thought that she couldn't conceive ...'

'No, Gigi. Traumatised as we were, Marion then fell pregnant with Mercedes. She decided to attend the same clinic, so she could have her original gynaecologist. Prior to her giving birth to our daughter, her doctor confided in me. "This could be a risky birth, especially after Madam Bouvier's last abortion". Of this I … I was totally ignorant. I can honestly say, I sensed who had fathered that child, but not why she aborted it. l do know that I could not have been its father.' Distraught, Pierre rendered a slight sigh and after a short duration he continued. 'This is my interpretation of what I assume may have occurred. I am sure Marion believed her lover was impotent, until she discovered she was pregnant with Jacquan.' Pierre hesitated until Gigi interjected and then Pierre motioned for her to remain silent.

'What hurt me the most was her deceit. I found out about that pregnancy by accident. Marion was my entire life. We'd grieved together. Made love in the same bed, as she and her lover had done in my absence. We cried over Jacquan's death with our hearts and arms entwined in sadness. Yet she still betrayed me.'

'In that wonderfully big heart of yours, couldn't you find one tiny corner of forgiveness for her?' Depressed, Gigi stood to pluck a sprig of native frangipani from the tree above their heads. Patiently she awaited Pierre's continuance.

'How could I forgive her disloyal conduct to me?' Pierre tried to stifle his emotions, which were building to a crescendo. 'I only stayed with her to perform my fatherly duties. Gigi, I alone was responsible for my family's welfare.'

'You've always loved and supported your family.'

'*Oui* chérie, I did. More so in later years, until I discovered Marion was dying of cancer.' Accepting the sprig of native frangipani from Gigi's hand, he sniffed its fragrant aroma. By bending his head to taste its sweet nectar, the spray concealed a tear. A drawn breath allowed Pierre the freedom to control his all-consuming thoughts.

'I wanted a companion for Mercedes. And I adore my sweet daughter. In hindsight, I'm not sorry Marion gave birth to our son, Raoul. A more lovable rogue I could never wish to see or meet. My children idolise me as I do them.'

'Rightfully they should do, Pierre. Your two kind and lovable young adults I could never fault. Their manners are impeccable. I love and respect them both, in a motherly sort of way.'

Gigi lent credence to the passionate tone in Pierre's voice. In an effort to understand his predicament she responded softly, 'I can feel your pain, your sorrow. In expressing your anguish Pierre, you have lightened mine. I hope by letting you expel this terrible nightmare and listening to you, I might, in some way have helped to alleviate your stressful anxiety.' Gigi cradled Pierre to her bosom. This was her way of consoling him.

'You have chérie, more than you will ever know. Deeply hurt by Marion's infidelities, I refused to believe that love still existed. I also refused to have intimate relations with her. And my life without you became inconsolable, really desolate.'

With downcast eyes, Pierre could feel tears building in his eyes. He sensed the same of this pensive woman who still cradled him and whom he idolised.

Reflecting on their recent ecstatic moments together on Abergeldie, Gigi said in a demure tone, 'Anger like thunder, is the rumbling of one's soul. Release your fury to the heavens, my darling Pierre. Or its power will forever hold your misguided love in bondage.'

'A wise Welsh philosopher, are you? I would be lost without your comforting words, Gigi.' One sniff and the crushed flower nestling in his palm blew free. Lifted by the rising wind, it soared to the heavens. This act of deliverance created the calmness which his aching heart desired.

'Burdens are only eased by a listener who cares for the one in turmoil. And you do care, my dear. I ventured in deep territory by confiding in you, on this our special night. In your wisdom you have removed the stigma of my failed marriage. I cannot thank you enough for your concern and genuine love, Gigi.'

Supporting her arm, together they climbed to their feet. Pierre collected the rug off the ground and smiled, holding her close. 'Now, my chérie, I think it is time we returned to your home. Everyone will surmise I have eloped with you.'

'What a delicious thought! We might just do it.' Gigi smiled up into his deep-set tear-filled eyes. Illuminated by the brightness of a midsummer moon she caught a glint of relief in Pierre's rich brown eyes. 'Can't you imagine our children's faces, if we did a 'flit' by night, Pierre? I think Kendall would have a fit or a temper tantrum.' Holding hands, they walked up the river bank.

'Not mine. I should say my children will soon be ours.' In a gentlemanly fashion, Pierre helped Gigi across to the dew-cloaked car, pausing for a moment to endow her forehead with a kiss.

Nigh on midnight the car's headlights shone on Abergeldie's front gates. The party still underway in the homestead was in full swing. A few neighbours and their guests had gate-crashed Abergeldie's Christmas festivities.

On their return Gigi was heard to say in Brianna's hearing, 'Pierre, the more the merrier to celebrate our terrific engagement. Also, my son's yuletide blessings in snow scenes somewhere in Austria or Switzerland. I envy Kendall's freedom and I do miss him tonight. It seems a millennium since his departure a couple of days ago.'

Alone in his room, Gigi spoke in a whispering tone to Pierre, 'Everyone hereabouts thinks Ian died while on an holiday overseas. I'm not enlightening them. Peter suggested I make his death known to prevent gossip spreading throughout our town of Yass. Tongues do wag in small towns.'

'That I can believe *ma* chérie,' Pierre agreed. 'What did Mr Bucknell say about your married name being changed?'

'Peter advised both Kendall and I not to reveal our new surname nor any of our plans for the future to our neighbours, friends or drifters. All except our close, reliable friends like those present, I mean.'

'A sensible move I think ma petite,' concurred Pierre, who understood their family predicament. 'Nor should we tell your uninvited guests of our betrothal. It is our families who should be told. Not strangers, Gigi.'

Unbeknown to their hostess, Bjorn had organised this late barbecue with a few select and trustworthy friends. Jalna and Abergeldie's female staff surveyed their workmen, who insisted on basting a whole pig and a lamb on both spits.

Some of their guests huddled around the barbecue. This protected area behind the homestead sheltered everyone from a crisp night breeze that had descended with vengeance. This unseasonable wind sweeping in across the plains carried a chilly breeze from the Snowy Mountains. These bleak winds froze everything in their path.

In the privacy of Pierre's room, they toasted in absentia, Mercedes and Raoul in Paris. They also raised a glass in honour of Kendall, who with the innkeeper, his wife and their guests were enjoying the penultimate eve morn of Christmas Eve in Ebensee, Austria.

'Well, my dear I now wish you goodnight. I leave overwhelmed with love in my heart for you,' Pierre confessed embracing her again. He looked down into her eyes and his lips roamed over her fingers. Engrossed in their fantasy he felt loathed to release their tips or her.

'Goodnight my love, I'll dream of this day until I can dream no more,' she whispered. Watching him leave, she replaced her engagement ring in its box.

By closing the lid on both the jewel and their gem of a day, it sealed her and Pierre's bond of love forever.

23

Disappointed over his failure to make contact with his mother in Australia by phone, Kendall decided to try a third time within the hour.

Utterly despondent, he chose to bury his miseries by choosing a delectable breakfast in the tavern's eatery. His nourishing meal consisted of oatmeal porridge doused in goat's milk. An array of nutmeat grilled on a skillet followed with fresh fruit and a pot of green tea. Something to fortify his strength wasn't on this morning's menu. Selecting a table by the bay window, he looked forward to enjoying this belated breakfast. On settling in a comfortable chair his order was instantly taken by Mein Host.

Half-awake and a touch maudlin, not able to speak to his fiancée, Kendall all but missed the magnificent panorama unfolding before his misty eyes. His view from the window was spectacular. Sleighs full of gaily-clothed children with their parents wrapped in warm woollen coats and fur muffs glided by under a cloudless sky on their way to church. Choristers singing Christmas carols with sleigh bells ringing in tune created a magical festival, all of which kept him and the guests of this quaint hundred-year-old building amused. Sighing, Kendall rechecked his wristwatch for the umpteenth time.

I better ask mien host if he can get a line now to Australia. The local receptionist may oblige this time. She has my phone number at home. Putting a hold on his dismal thoughts, he gave a hearty laugh at the children snow fighting several metres from the Appel-ski Tavern. Their excitement defrayed the high price of his meal and put paid to his miseries.

This small, welcoming inn where he'd chosen to stay was located on the eastern flank of Ebensee, a small village. Its grandly decorated cafeteria with yuletide trimmings looked spectacularly beautiful. Its magnificent views outranked the chef's miserable scowl and harsh mood while preparing their Christmas Eve food.

A misguided snowball collided with the leadlight's outer pane just above Kendall's head. The force of its impact cracked the glass, although it didn't shatter. He considered himself fortunate, for the double-glazing had protected him.

From where he was sitting, his attention was drawn to the frustrated innkeeper. The elderly yodeller cursed while shovelling fresh slush off the pathway up over a high bank of impacted snow. He reached for his yard broom. Using it as a weapon, he whacked the bottoms of three young culprits. Clutching their buttocks and hollering loudly, the trio headed for their brightly decorated sleigh.

The stern directive given to the children was in German, which Kendall's table-neighbour interpreted as, "That will teach you boys not to make mischief. Yuletide or not, your parents will pay for my broken window".

Scribbling their names in the snow, to keep them in mind, the innkeeper re-entered his domain to seek a pen.

Kendall laughed on noticing one of the boys return, followed by a pair of toddlers. Six small feet in snug snowboots rubbed out the shallow indentations. With one almighty swoop, their feet kicked a pile of loose slurry high in the air. Huge chunks of snow landed on the path that Mein host had painstakingly swept clean.

In his haste to chase them, he slipped in the wet slush and flew through space with the grace of a bull in flight. Landing face down in powdered snow several paces from the path he spluttered. This impromptu act brought a screaming chorus of small voices, the owners of which rollicked excitedly while building a snowman away from harm and danger. They used a carrot for its nose, two blue balls for its eyes, and a row of cherries formed the mouth. Woollen gloves stuffed with hay acted as hands. Caked mud moulded into shape were the snowman's boots.

Until his hysterics settled, Kendall couldn't eat breakfast. Eventually, he recovered enough to approach the desk where he asked the sore and sorry concierge to try again for a line to Australia.

Not having detected his quest's mirth, he politely handed him the receiver.

'You get through to your homeland nuw, *Mein Herr*. The line, she is cleared Doktor Gawalam.'

Kendall smiled over the man's mispronunciation of his surname. Careful in dialling, he mentally noted the exchange number. Speaking in precise syllables to the telephonist, he repeated the homestead's number in Yass and patiently waited to be connected.

Brianna came across loud and clear to which Kendall replied, 'Happy Christmas darling. Enjoy your day. We celebrate it here tomorrow. Yes, I also miss you dreadfully.' The time delay made him pause. 'Have one for me, Bridy. It's morning here, far too early for me to imbibe on champagne. I feel miserable without you. Yes, I'll call *your* time tomorrow night. My time on the phone is up, Brianna. Bye. Please darls, put Mum on.'

He hesitated until his mother spoke. 'Yes Mum, I will enjoy Christmas. The same to you and Pierre. All the best for today, and don't talk the ear off him either. Yes, I'm eating well and the choice of festive food here in at the Appel-ski Tavern is to die for. Trappings I never knew existed.' He heard the line crackle then it dropped out. 'Blast,' Kendall whined, 'that means I'll have to phone again later. Never mind, they're bound to be still nibbling until noon.'

Astride a bar stool, Kendall looked across the lower snowfields to the high alps. Caught in the sun's coral glow were silvery-grey reflections of a distant church spire. Nestled in among Ebensee's conifers the belltower glistened and looked resplendent. Sunlight reflecting off the snow shone on the church's leadlight windows in a myriad of colours. Looking out on this spectacular scenery, it reminded him of a recreated fantasy of his childhood days.

Festively groomed horses nodded their plumed heads as hooves plodded to the jingle of bells. The sleighs laden with warmly clad occupants who were decked in an array of harlequin colours. Joyful skiers on their way to morning mass sang hymns. The interwoven hues of their winter costumes contrasted with the vivid blue of God's summit, crowned in white.

Kendall noticed as the sleighs glided towards their destinations how their colours muted in with earth's pure mantle. This picturesque scene should have captured his imagination. It didn't. All this beauty was lost in a listless and worried mind.

Daunted in spirit and feeling dejected, he vacated the stool. Reticent of mood and with head bowed, he climbed a flight of stairs leading to the first floor, then his room.

In a courteous gesture he paused midway to allow a newlywed couple to descend the stairs. Arms intertwined, she giggled at the remarks passed by her newly acquired spouse. A brief nod and they moved on, mirthful in their wedded bliss.

Just managing to control his emotions, Kendall ascended the remaining stairs. 'What the devil did I come up here for?' he uttered. Then he recalled 'Oh yes, my ski pass.'

With a backwards glance, he blindly approached the closest room. The door was ajar. Not thinking, he walked in to collect his ski pass, goggles and gloves off the bed. Silently he cursed himself for forgetting them before he went down to breakfast.

'Put my brunch tray on the table. It's about time. I'm famished,' demanded the occupant of room nine over the whirring blades of an electric razor.

Kendall recognised the voice, even though its accent held an Irish overtone. At a loss to place where he'd heard it before, the gruff response made him realise this wasn't his room.

In the effort to creep out backwards, he glanced at the unmade bed. The item confronting him defied belief. He blinked, deflected his gaze and looked again. This time his eyes remained transfixed on the tan satchel with a bronze eagle under its handle.

Disguising an already breaking and strained voice, he stammered in a tone totally unlike his own. 'Sorry, wr … wrong room. Sir, I'll go down and tell the maid to hurry.'

Fraught with fear, his feet wanted to move, yet his brain resisted its commands. Regaining his courage, Kendall carefully backed out of the room and quietly pulled its door to a close.

The stairs seemed a thousand metres away. Still, he blindly plodded down each one. Safe on the ground floor, he bumped headlong into the concierge.

In a flustered voice, Kendall apologised. 'Excuse my red face, I've just done an awful thing. I walked in on the man in room nine. He bellowed for his brunch. Please don't tell him I entered his room, Mein Herr. I feel quite embarrassed.'

'You secret, she is safe vid me, *Herr Doktor. Herr* Frankie Byrne, he is rude man *und* angry all de time.'

'I hate to worry you again. I've forgotten to wish my fiancée a happy birthday. And if I don't phone Australia now, I will miss her.'

Looking over his bifocals, the man behind the desk understood. Nodding his peppery-grey head in answer to Kendall's request, he pointed to a room down the hall.

'I try, *Herr Doktor.* I vill connect the call to *die* library. *Die* key!' His stumped and aged fingers passed it to Kendall. 'Bring key back to desk ven you finish, *danké?* I vill ring *die* exchange, you vait.' His beaming smile was directed at Kendall, who by then had begun walking down the hall.

'I vas young and in luv once,' recalled the aged Jew. Smiling, he held the equally ancient telephone receiver close to his ear.

Kendall empathised with his Austrian host, trying to battle with English. *My German is far worse than his broken English, I know the agony he is going through, trying to communicate with me, although his heart is in the right place. It'll be early morning at home now.*

A considerable time elapsed before the innkeeper managed to secure a free line. The exchange receptionist connected him to Australia.

Enduring thirty minutes of misery, Kendall felt as if his mind dangled in suspended animation. Self-torment ceased instantly a buzzer sounded in the silent and musty library. Relieved, his eager fingers grasped the receiver. An Aussie twang from "Downunder" serenaded his ears.

'Bjorn, this is Kendall. Don't speak, just listen and let me do the talking. This is extremely urgent. Where will I find Peter Bucknell at this hour?' Surmising the diplomat would be at Jalna instead of home, Kendall smiled. 'Good, put him on and make sure nobody is within cooee of that hall phone. Better still, tell him to take this call in mum's study. I'll wait.' Time again stretched eternally. 'The escaped rat, I have it trapped. What should I do?' The instructions came over in precise terms as prearranged. Brief, and to the point.

'Yes, I'm certain it *is* him. What's more important is, I recognised his distinctive briefcase, likewise his voice. Although at first the Irish brogue threw me, then I twigged to whom it belonged.' Kendall spelled out the fictitious name as he understood it to be. He then repeated his father's room number.

Withdrawing a pad from his ski-jacket, he supported the receiver between his neck and shoulder while noting down the directions Bucknell gave him over an intermittent line.

'Go ahead! Salzburg Airport you mean? Yes, I heard you. Who? Neils Arnaldsen, spell his name.' A short respite confirmed Kendall's interpretations were correct. 'Right, a local police car should be here at the tavern in Ebensee within ten minutes. I won't have to identify him will I, at the police station? The Feds have sent his photograph through. Good, the one they procured from their agent in Berlin and in his Nazi uniform. What a relief. Oh, I have to collect everything from my room later in the day.'

The intermittent line cleared and Kendall sighed. 'Mum will tell you the phone number here. I understand that. They better not let the rat escape this trap either, not after I've signed my deposition.' Kendall listened intently to what Peter asked of him.

'Except from what I'm wearing, everything's in my room. Leave them? You have, thanks. My passport and personal papers are here, in a zipped pouch under my shirt. Yes, I'll go there now,' confirmed Kendall, assessing all the data he'd assimilated in a few short minutes.

Trying to make a discrete departure from the tavern he followed Peter's advice. Instead of going to meet his friends at the ski lift terminal, he went straight to town. Wearing his snow boots and carrying his shoes, stocks and skiing gear he casually made for a local transporter to the village. With the ski omnibus running on time he reached his destination in twenty minutes.

Unfazed by not being familiar with the village layout, he found its local constabulary. There he reported the situation to the Ebensee police. Everything Kendall had declared to the detective sergeant, Peter had already confirmed by phone. A carload of police officers followed by a contingent of photographers left the station almost immediately.

Having received AFP notification, the Austrian Chief Inspector was conversant with Kendall's plight. He ordered a car and driver to be placed at his disposal. This eased Kendall's mind, although he did have some difficulty coming to terms with spelling his new surname to the driver. A police van escorted Kendall's chauffeur-driven vehicle across to Salzburg.

With the data supplied and verified by ASIO, the Mossad and other sources, ex-Nazi Kapitan, Rolf von Breusch was arrested immediately all paperwork was completed and a warrant issued.

Constantly protesting his innocence, the ex-war criminal, posing as an Irishman, was transported in a police wagon from Ebensee to Salzburg for further questioning. From there the following day they listed him to be transported to Linz, the Austrian capital.

After an extensive and thorough interrogation by the senior police inspector in this city and under guard, his next stop would be to the courthouse cells.

This foreign scoundrel, known only under the name of Frankie Byrne in Austria, had eluded the hunters and their nets long enough. The police received information from the Jewish Mossad, that their captive was known as Rolf Max Breusch in Scotland. This data tied in with the report from his ex-university tutor, both of whom had lived in Edinburgh; until Rolf left university in 1930 at the age of twenty-three.

Immediately this data could be validated, a folder containing all his student and his entire war history travelled by courier from Salzburg to Linz. In the Austrian capital a panel of judges would eventually conduct the initial hearing.

Within two days of receiving this news, an agent representing the Jewish Mossad produced a glutenous dossier on von Breusch. Delivering it in person to the authorities in Linz, the agent observed two Austrian Police Inspectors intensively interview their captive.

Accompanied by an ASIO undercover agent flown over from Australia, the group of six men extracted snippets of truthful information from the lies falsely presented by this war criminal.

Trapped and confused, von Breusch's current story contradicted with his initial statements given in Ebensee. Each time his verbal deposition was recorded it countermanded the previous one. Most details he proffered conflicted with the data in his dossier, the document the Director of Police now held in his hand.

Challenged repeatedly about his identity, von Breusch adamantly denied ever being in Australia. He also stated that he'd never lived in Germany or Scotland.

The Austrian interrogator played his ace card, after reviewing the results handed to him from a forensic colleague.

'I'm duty bound to ask did you enjoy your drink *Herr* von Breusch? We can supply you with more liquid refreshments. Although I think that will be unnecessary, because we have a complete set of your fingerprints.' He held up the glass.

'What are you talking about,' Rolf snarled. 'Neither you nor any of your officers have taken my fingerprints. Therefore, you have none to compare mine with.' A snigger emitted from his curled top lip.

'No!' the interrogator exclaimed sarcastically. 'All fingerprints lifted from both the handle and lid of your distinctive briefcase proves otherwise. I'm not just talking of its outer casing. I'm referring to its concealed pockets. The fingerprints on the glass you drank from on your arrival compare favourably with those in our possession.' The interrogator paused as Rolf gulped. 'All your fingerprints have proved identical, this communiqué I'm holding from our forensic experts says.'

'My valise was locked. What have you done with it and the key?'

'Do not roar at me, von Breusch. Unfortunately, we found it necessary to force the lock that enabled our investigators to sift through all your important documents. The notes written on Reich letterheads prove that while you worked at the Chancellery you signed thousands of death warrants and sent more than five hundred innocent people to their deaths.'

A sly smile sat easily on the interrogator's robust features as he accosted his quarry with a challenging glower. Mordant in tone, he sardonically acclaimed, 'It pleases me to inform you that the Austrian authorities have examined your satchel's contents in detail.'

This affirmation stunned von Breusch. Pounding one fist against his open palm Rolf angrily demanded of his accusers, 'How dare you snoop into my private business.'

Withstanding the depth of anger shown by this man sitting opposite, the police interrogator countermanded his rudeness with an arrogant grunt.

'Spare yourself and us in this inquiry the embarrassment of foolish denials, eliminate all lies. These miniature photographs in my possession still resemble you even now after so many years, von Breusch. Two of these three miniatures taken of you in your Wehrmacht uniform in forty-four are self-explanatory,' retorted his interrogator, passing the snaps to a colleague.

'What photos? Let me see them.' Rolf made a swipe at the three miniatures an officer standing beside him was holding, only to have them savagely snatched out of his reach.

'Remain seated!' ordered the Police Inspector. 'Why even looking at you now, devoid of facial hair and bald headed, these images are of your likeness. Shaving your head was a futile exercise, von Breusch.' Bending, he appraised all the miniatures then compared them with the face not a metre away and smiled. 'Mm…mm! Yes, their likeness of you in person is remarkable.'

Looking up at the officer now towering over his head, Rolf decreed, 'How many times must I tell you, I am Frankie Byrne. My passport and official papers will verify …'

'Verify nothing,' scoffed the inspector. 'Utter rubbish! It is now a proven fact as to who you were, and are. Fingerprints don't lie, like their owner.'

Incensed by not receiving a response regarding his briefcase, Rolf incredulously accepted the truth.

'You … you didn't wreck my valise? The papers inside it were of a private nature, nothing to do with you or anyone here in Austria …'

'Oh, a million pardons *Mein Herr*,' the inspector jeered. 'I think you sir, have misunderstood what I said. Did I say we *NEVER* dismantled your satchel?'

This ruse was to force him into making a full confession of his past. In truth, the external leather and centre pockets of his briefcase remained intact. Only its inner-linings were prised apart, and they were not dissected or removed.

'By ransacking my valise and tampering with my documents you lot have trespassed on my privacy. Surely I have the right to object …'

'Will you allow me with my professional expertise to decide that von Breusch. To the contrary, our investigators found your private documents extremely informative. I must admit, it took quite a while for my officers to discover the valise's cleverly concealed compartments. I might add that your satchel and its contents are sufficient evidence to hang you,' concluded the chief inspector and standing back, held his hand out gloatingly to Rolf.

After this sardonic inference Rolf refused to answer. He scowled at those within reach. Restricted by handcuffs and with little movement, he felt entombed in misery. The four men crowding his territory in this small confined room hampered his ability to think. Feeling nauseated, he decided to bluff it out.

Solemnly and belying his fear, Rolf calmly looked at his accuser and demanded, 'To what were you referring?' Instead of a verbal response, he received bemused looks.

His pretence appeared convincing and with crossed fingers Rolf continued, 'Oh, of course, now I know how this terrible mistake occurred. I lent my valise to a friend who'd lost his briefcase in transit from Salzburg to Germany. He must've put the incriminating papers in my valise while on a business trip. There can be no other explanation.'

Reviewing the documents, the Chief of Police remained silent. *This ruse has more holes than a mosquito net. I immediately saw through his charade. Although it is possible, this man could still have affiliations with his Nazi homeland.*

Bemused by the prisoner's implausible and pitiful excuses, all those present wondered what lies he would conjure up next.

After a good laugh the interrogator rendered a sangfroid snort. 'And of course, fairies who sleep on your pillow whispered those lies.' Enraged, he then roared, 'Furthermore, NO other fingerprints were lifted from the satchel than yours.'

Quick in response, Rolf unequivocally retorted, 'If your men touched the valise, besides my friend, there must be fingerprints on its lining. I saw Hans pick up my valise and walk towards the aircraft on his way …'

'Would Hans be the same man whom you cunningly plotted against? An innocent countryman you befriended as a substitute in your phoney suicide and then murdered?'

Shit, this idiot has done his homework. Rolf's denial came as a predominate glower of his interrogator's insipid blue, bespectacled eyes.

Wrenching them asunder from von Breusch's impenetrable gaze, the interrogating inspector vociferously bellowed, 'Liar! These officers are not your Chancellery superiors listening to lies. Today you've confronted intelligentsia in this room, not Nazi rabble. By that, I mean the idiots you worked with under Hitler's regime.'

These insensitive remarks heralded resentment in von Breusch, unable to reply to his inquisitors who, to his warped way of thinking were the idiots, not him. *They accepted the false information I've voluntarily fed them until now. They considered it plausible. So why all this humbug and endless questioning?*

'I've heard enough of your insults regarding my ex-colleagues. Like myself, they were honourable men. As of now, I refuse to answer one more of your damn questions. You're trying to concoct fantasies to entrap me.'

It appeared to the interrogator that this latest act of defiance portrayed von Breusch's belief in his country's Nazi revival. Even though he'd deserted the Fatherland in its dying hours, Rolf harboured and nurtured the hope that the Nazi Party would one day be revived. The party would rise to govern, similar to the pre-war years. In his distorted reasoning, Germany prospered when the Wehrmacht governed its unruly and angry populous. Riots were common on the streets of Berlin and all throughout Germany due to lack of work, poverty and starvation.

Unknown to the authorities, with his in-depth knowledge of Reich policies, Rolf resorted to keeping notations of their intricate foreign dealings. He had hopes that the Hitler Youth's revival would stir to life the strength of his country's power, lost after being defeated in the Second World War. Over a period, he had stashed a huge dossier of incriminating documents and over one hundred Reichsmarks in a bank vault in Switzerland.

Now under pressure and forced to endure their bullying, he couldn't recall in which bank they were housed. *These fools must never discover either the funds whereabouts or my Reich diary. Undoubtedly the vital data in my notes could bring the hangman's noose closer to my neck. If I continue to defy their demands it might work to my advantage. The majority of our Reich records no longer exist. Most of them were obliterated in Allied bombing raids over Berlin in 1945.*

24

During his journey across to Salzburg, the Australian physician remained silent. Confused and with a pounding heart his rambling thoughts sped faster than the whirring motor of the vehicle transporting him to safety.

In a stressful daze, Kendall Gwlynne, before setting a foot inside the airport, stood to quietly calm his erratic thoughts. Composed, he walked to the departure desk and approached a female attendant. 'Excuse me Fraulein, I'm Doctor Gwlynne. Could you please tell me where I might find *Herr* Arnaldsen?'

The woman nodded to a gentleman walking towards her counter. Introducing himself as the Assistant Manager, he directed in English, 'Follow me please *Doktor. Herr* Arnaldsen is expecting you in his office.'

Kendall's long legs allowed him to keep stride with the blue-suited official. A short distance along the corridor they stopped outside a silver nameplated door.

Tired and feeling ill at ease, Kendall entered the room on Herr Arnaldsen's polite command.

Immaculately attired, the gentleman who'd just arisen from the far side chair shook his hand. Holding out his other palm, he gestured for Kendall to be seated.

'*Doktor* Gwlynne, I prefer you sit in my chair on this side of the desk. It is far more comfortable than that low chair.'

Kendall surveyed his surroundings. The small room, tastefully yet scantly furnished seemed insecure. Instinctively the manager sensed his uneasiness.

'The walls are soundproofed, *Doktor.* Never fear, no one can hear us talking. *Herr* Buckneel will call you at ten. Until then, please make yourself comfortable.'

Kendall smiled. *How quaintly this man of either Scandinavian or Norwegian origin had expressed Peter's surname.*

Arnaldsen tapped his finger on the underside of his desk, directly below a phone. 'Feel this button, Doktor.' Kendall did as he advised and then nodded. 'Should you require my assistance, do not hesitate to press it. Otherwise do not touch any buttons on the phone. It is the company's direct line. No one can listen to our conversation. That is the reason Peeter asked for my office.'

'Am I to understand sir, that you know Senator Bucknell personally?'

'I do *yer*; Peeter and I are old friends. In my position I come across many diplomats. Peeter has travelled extensively throughout this country down the years.' Collecting his hat from the coat stand he queried, 'Would you care for coffee and something to eat, perhaps a meal, *Doktor?*'

'Coffee and sandwiches will be fine, thank you sir.'

A nod and Arnaldsen departed the office. Letting the door close naturally it self-locked.

No sooner had his solicitous thoughts settled when the phone buzzed. Picking up the receiver Kendall spoke to the caller.

'Doctor Gwlynne speaking,' he said in a professional tone. 'Hi Peter, I've been expecting your call. I'm listening. Arnaldsen can be trusted and I'm to do as he suggests. Yes, I understand that perfectly.' His pen wrote down on a page what Bucknell implied. 'Why there? It's safe and private, okay. I expect him back shortly. While I'm talking to you, I've noted down everything.'

Static distorting the line forced Kendall to bide time until it cleared. 'Sorry, I missed what you said. Mentally note it, pen nothing. Fine. Yes, of course I do. How's my family and Mum's pup? Amber is very protective of all women in our household. I couldn't agree more. She's a terrific guard dog if danger looms.' While listening to the diplomat, Kendall ingested his written data. Destroying the scrap of paper, he returned both it and his fountain pen to the top pocket of his ski-jacket. 'Well, that's terrific news. I'm pleased Mum's settled down. I suppose it's some consolation. Goodnight Peter.'

Kendall replaced the receiver and feeling despondent he looked up as the manager re-entered his office. The man's candour appealed to Kendall. He liked his direct approach. He was not a liver-lilied individual who ran off at the mouth and got nowhere.

'I need the men's room urgently,' he said stripping off his fleecy-lined jacket because the room was stuffy and hot. 'Could you tell me where is it please, sir?' He began to feel nauseous from lack of sleep after ingesting unpalatable food. He needed a hot meal. Not the stuff he'd hurriedly eaten en route to the airport.

'The men's room is two doors down this corridor. Before you go Doktor, here are your flight tickets, your security clearance pass and gate boarding passes.' The manager handed an envelope to Kendall. 'First sign this receipt please *Doktor* Gladwyne.'

Accepting them, Kendall thanked the manager for his trouble. Pen in hand he signed. Passing it back, the inner corners of his eyes creased over how this airport manager had misquoted his surname.

On Kendall's return, Arnaldsen confirmed, 'The instant you leave my office Doktor, speak to no one. You will be escorted to a twin engine aircraft, which is being refuelled on the tarmac. For your protection, two UN plain-clothed guards will fly with you to Linz. From there an official government vehicle will deliver you to the Australian Embassy. Their staff will attend to your needs. Do you have any luggage?' Arnaldsen queried, a little concerned for his welfare.

'None at all, thank you sir.' Kendall didn't want to confess that he'd foolishly left his ski stocks and boots at the police station in Ebensee. 'Sir, do you have any idea, how long I will be detained in Linz?'

'Not really.' The manager's hand rested on the red phone he was standing alongside. 'Peeter informed me on this phone not an hour ago, that Linz was your destination. Until I read this communiqué, I cannot be certain.'

'I can't afford to be grounded in Salzburg, or in the Austrian city of Linz for an indefinite period,' Kendall confided to his host.

If the answer did not conform to his plans, it could expose him to danger. Considering his predicament, Kendall realised he would be treading on perilous ground if he remained in this city. He feared bumping into either his father or one of his contacts who may still live in this region. It wasn't impossible, because Salzburg lay in close proximity to the Austrian-German border. *Should anyone suspect me of being in Austria, in all probability that person, or persons would never think of looking for me in Linz. I'll be a man lost among a forest of faces.*

Herr Arnaldsen looked up from the paper he'd glossed over to ascertain if he had quoted the facts correctly. 'Doktor, how long you will be in the capital I cannot confirm. Under the circumstances, it could be a week or longer. All I can say, Peeter will contact you at the embassy in Linz tonight at seven.'

Passing Kendall his fleecy-lined jacket, Arnaldsen apologised for the delay of his federal passes not arriving on time. With a farewell gesture he nodded to his deputy. Accompanied by an elite squad of "heavies" they left his office. Each guard, massive in structure, formed a quadrant. Keeping at an unobtrusive distance, two men walked ahead of Kendall, two closing at the rear to escort him through the building. On the tarmac several more officials joined their entourage. This group of six uniformed officers accompanied him to their revving aircraft.

Kendall frowned. *The Austrian authorities aren't taking any unnecessary risks where my safety is concerned. Perhaps they suspect some form of retribution, should I be identified.*

Arnaldsen had insinuated that they expected some difficulties to arise. Fortunately for all concerned trouble never eventuated. Smooth in taxiing, their eight-seater sublimely lifted into the blue. Levelling, it settled on a new flightpath, with the instruments set due-east.

Observing it change course, Kendall felt emotionally drained and disappointed. Its flightpath lay directly over the Ebensee snowfields. This quaint little village shimmering in the noonday sun, looked totally different from an eagle's aspect.

Quite unexpectedly, Kendall and his bodyguards felt the aircraft tilt and swing north. It headed in a different direction towards the Austrian capital.

On finalisation of official embassy formalities an exhausted Australian physician was shown to his private suite. It surprised Kendall to find his luggage in the bedroom. Somehow all his cases, ski boots and stocks had beaten him to Linz.

'More of Bucknell's undeniable magic, I suppose,' Kendall uttered while unpacking his clothes. 'What a man to have as a guardian angel. Peter's a friend in a million. He's been terrific to Mum and me, for longer than thirty years.'

Kendall's main priority was to have a long hot shower. Feeling refreshed, he decided to unwind and access an unknown future, while dossing on the comfortable double bed. Relaxed, he tried to untangle the myriad of thoughts in an utterly confused mind. Unable to concentrate due to a massive headache, he tried to connect the fragmented visuals and recurring images of the day's bewildering incidents which destroyed any chance of a peaceful sleep.

For the first time in his life, he understood how his patients must feel when sleep evaded their waking hours. Kendall's body clock, totally out of kilter, failed to send his brain any switch-off signals. Fighting tiredness, his legs twitched in spasmodic fits of agonising cramps and restlessness.

Right on six the phone in his room rang. It couldn't be Peter. Instantly summonsed to attend the embassy office, he donned his suit coat. Carrying a tie, Doctor Gwlynne ambled down a flight of stairs. In the elegantly decorated room he observed its plush furnishings. Ushered to a seat by an attaché, Kendall observed a rather staid-looking guard standing by the closed door. The elderly attaché introduced himself as Robert Flagstaff, a retired Australian Colonel, who doubled as the embassy interpreter.

'The Linz Police have requested you to put in a personal appearance at their headquarters, Doctor Gwlynne. I have been instructed, your appointment will be at nine tomorrow morning,' the attaché confirmed observing his secretary taking notes. 'This procedure the police wish you to undertake will clinch the case against Herr von Breusch.'

'Am I to identify the Nazi in a line-up?' Kendall looked puzzled, and as the phone rang, he noticed the colonel nod. Something drew his attention to a nearby window. Looking up Kendall spied fresh icicles creating jagged necklaces under the eaves. A rather nasty wind had sprung to life, it brought in a flurry of sleet. Cascading ice particles drifted down to meet the bleak and frozen wonderland.

Kendall shivered at the thought of anyone venturing out of doors on this freezing winter's night. *Bed will be welcome for my tired body to embrace. Snug and curled up beneath warm sheets the world beyond my room will cease to exist.*

A lit match brought him back to reality. They were alone: the secretary had gone.

'You don't mind if I smoke, Doctor?' the attaché asked as Kendall swayed his head in reply. 'I'm dying for a puff. I usually light up in the men's room. Mind you, I'm not supposed to smoke on duty, or in front of guests. An old courtesy rule of politeness, you understand.'

'Well yes, but I don't mind Colonel, you go right ahead. I rarely smoke cigarettes. Occasionally I'll sneak a puff on a cigarillo. I have an Irish fiancée who detests me smoking.'

'Don't most women! Or the ones I know firmly object,' claimed the officer, taking a puff. 'Rancid breath, you know.' Colonel Flagstaff was aware Kendall detested his father. Therefore, as an embassy official he couldn't afford to insult his intelligence by referring to the Nazi under discussion as his father.

'Now Doctor, getting back to your interview tomorrow. There will not be a line-up like the movies dramatise. Far too dangerous in case some external source suspects the police have von Breusch in custody. If the *paparazzo* gets wind of the situation, they won't keep their noses out of this ugly business.'

'I'm aware that the knowledge I have of his service history at the Berlin Chancellery is too important to allow anyone to interfere with the prospects of him being brought to trial. Sir, I'm grateful for your assurance he won't be able to recognise me. Going by his false Irish name and what I've been told, I assume the man they want me to identify will be ex-Nazi Captain von Breusch.'

Stretching his legs Kendall sat back down. 'What I understand of the Wehrmacht and Reich regime, my ...' he almost said father, 'his abominable cruelty was indicative of their torture. My mother, apart from myself can testify to that. Having suffered at his hands, I won't be sorry to see von Breusch behind bars for all eternity.'

'Not behind bars, nor in the tiny cell he's housed in at the moment,' attested the colonel. 'Be assured Doctor, when his case does come to trial, as I've no doubt it will soon, that ex-Nazi will dangle from the end of a rope.'

Unable to disclose anything more on this topic, Colonel Flagstaff hesitated. In his role of attaché, he wondered if his guest's courage would remain stoic after viewing the absconder. The little he'd spoken to him, within the bounds of their law, he observed and admired Kendall's positive attitude. Honesty showed in his face and was expressed in his speech. It came from the heart. The colonel felt sorry for this fatigued and fazed physician in his company.

'Oh, before you go Doctor Gwlynne, a moment please. The ambassador has requested for you to dine with he and his wife this evening at eight, in their dining salon. Semi-formal, you understand. I'll come up around seven and escort you to their private apartment. It's on the top floor of this renovated, ancient building. Let me know if there's anything specific you need.'

'I do have a dinner jacket, although it'll need pressing and this navy tie will be most unsuitable.'

'A new white bow tie will be sent to your room. Please, leave your dinner jacket, shirt and trousers hanging on the doorknob, and shoes outside for the butler to collect at five. If you require anything, call me on this number.' The colonel passed Kendall a card. 'The blue telephone in your suite is a direct line to this office. You'll see which button to press. I'll be working here until just before seven.'

'Thank you, sir,' Kendall nodded accepting the card. 'I am expecting a phone call around that time.'

'Yes, I know that Kendall, if you don't mind me using your Christian name? I'll leave instructions for my secretary to have the call transferred to your room.'

Taking leave of the colonel, Kendall quietly retreated to his suite of four rooms, there to deliberate on the day's events. With no regrets pressuring a fatigued mind, he knew that he'd done the right thing by agreeing to identify his father as the wanted Nazi war criminal. Relieved of stress, his tiredness began to wane and he settled on the bed to rest.

6.30 pm: Bucknell's call disturbed him from a deep slumber. Precisely at seven Kendall showered and dressed for dinner with the attaché, his boss and his family.

While partaking of after-dinner drinks with the ambassador, his wife and their guests, Kendall spoke to Colonel Flagstaff who advised him there was a reprieve of several days, because of a delay in going through current data. He did however, in a professional capacity, state he would accompany Kendall to the Austrian Police Headquarters in Linz.

In pleasant surroundings, Kendall hadn't realised how quickly the night was drawing to a close. This unscheduled break enhanced his disposition. It allowed him to relax, and weather permitting, to stroll through the embassy's palatial grounds.

After a delicious supper, he found it challenging to exchange wits in a game of chess with a guest in their private lounge.

11 pm: Kendall chose to relax and read in front of a roaring fire in his suite. Reading his favourite novel, *East of Eden*, filled in the lonely hours until dawn that ushered in a wintery blast off the roaring North Sea.

He drifted off to sleep and as the book slipped off his knees, it hit the floor with an almighty thump. It didn't awaken him.

25

6.15 am. Wednesday morning: An ambassadorial chauffeur-driven Mercedes ferried Kendall and Colonel Flagstaff to Linz Police Headquarters. Accompanied by two Austrian detectives, they apprehensively observed the acclaimed war prisoner enter an adjacent room.

Under the pretext of his interrogator having to leave early, Herr von Breusch, whose hands and feet were shackled, believed this lie. Ordered to sit opposite the interviewer he quietly complied.

Rolf didn't realise the mirror he was facing held a secret. Nor did he suspect the son of his second marriage stood not three metres away ready to identify him.

Kendall felt safe behind the mirrored-window. If anything, the sound coming through the one way intercom fortified his courage to name his father.

'Herr Doktor Gwlynne,' said the Inspector, his voice clear enough for the tape recorder to pick up. 'I ask you to name the man sitting on the left of my colleague in that interview room. Do you recognise the prisoner?'

'I'd know him anywhere. Even without facial hair and though he's now bald, he is the person in the photographs you have in your possession. His voice, though disguised, is that of my …' On the cusp of cracking, Kendall requested, 'Sir, may I have a moment to compose myself?'

'Take your time Doktor. It is imperative we, through you, identify him. We have one more item for you to identify. Then where we're concerned, your part in this interview will be over. After which you will be free to leave these premises with the English Attaché.'

Feeling composed and relaxed, Kendall continued to identify his father. 'The prisoner, now known as Rolf von Breusch, who is talking to the officer in uniform is definitely … my father.' Kendall shuddered as he tasted the bitterness of recall.

'In Australia he went under the name of Ian David Ross. About England I can't be sure. Until recently I never knew his real name was von Breusch. I verify he is definitely the criminal you, ASIO and the Mossad have been hunting.'

'Now Doktor, please turn your back and face me,' the Inspector politely requested, and patiently waited. 'Describe the leather valise you say was in von Breusch's possession when you mistakenly entered his room on the twenty-fourth of December last?'

In clear view to everyone in the room, excluding Kendall, was his father's tan valise with the eagle emblem under its handle. The same valise Hans Selig had delivered to Rolf von Breusch's home on the night he murdered him.

Without flinching, the Australian physician described in detail the valise's colour and the bronze eagle situated under its handle. Distinctive in design, there could be no mistaking the item belonged to the prisoner. The briefcase was marked and listed as an initial piece of conclusive evidence.

The Chief Inspector then handed Kendall an envelope with a distinct sample of the German criminal's handwriting. He verified it as his father's neat script.

Without a word, the interrogator showed a photo of a Weimar officer to Colonel Flagstaff. He in turn placed it in Kendall's hand.

This photo of an elderly man in uniform astounded Kendall. The Reich Captain, with a similar build to his father, worked at the Chancellery in 1930. Although their uniform tunics had similar services patches, there was no disputing the men's likeness. They were father and son, both with the surname of von Breusch.

Kendall now knew his paternal grandfather had also been a Nazi officer which disgusted him. *This knowledge I will carry to my grave. My mother must never discover the truth, it will kill her.*

After this long, harrowing experience, Kendall discreetly walked from the soundproofed room with his head lowered. With the day staff not yet in, he was escorted to a private room.

The colonel waited behind to speak to the Linz Police Inspector. Escorted by another officer, Colonel Flagstaff then re-joined Doctor Gwlynne in a small annex until their car arrived at ten. Then they both left the building by its rear exit as their vehicle pulled to a halt kerbside in front of them.

Warned not to say a word in the car, Kendall remained silent until their vehicle cruised to a stop outside the ambassadorial private residence. There the colonel quietly suggested they speak well out of his chauffeur's hearing, perhaps on the front verge.

This recent incident at police headquarters not long before was the best kept secret the world over, and so must it remain.

'Colonel, will von Breusch discover who identified him? Whatever happens, my identity must never be revealed. It would ruin not only my life, but also my mother's. And then it would be impossible for me to continue in my profession.'

The attaché assured Kendall Gwlynne that neither his family name nor country would be disclosed before or during the trial. Colonel Flagstaff also confirmed the police had intentionally withheld showing von Breusch the miniature of his deceased wife, until his pre-trial. A chateau's name and date of a Reich Ball were written on the reverse side of this tiny photo, indicating it was taken at a function in Hamburg, Germany.

'I can tell you this much, Kendall, your positive identification of von Breusch was enough for the magistrate to serve him with an indictment. It must be cited before proceedings can begin here in Linz. No evidence will be released until the inaugural trial sitting. Be assured it will take time to organise. The cogs of law turn slowly here in Austria.'

'I'm interested to find out where the photo of his father originated. I know Rolf von Breusch's Godfather, a dying General in the Wehrmacht, deserted to pass vital information on to the Allies in the latter part of the war. Will the miniature photo be used as evidence, do you think, Sir?'

'No, I very much doubt if that photo will be produced at all. It was only used as a comparison, the officer in charge just told me.'

'Now I've a favour to ask of you. If possible, I want my sister's original miniature returned. The photo of Erika, her birth mother means a lot to Arneka.'

Flagstaff pulled a small packet from his vest pocket. 'I've been requisitioned to hand this to you, in person. You'll find it's a copy of her miniature photo.' He paused while Kendall ascertained if the copy was correct.

'In answer to part of your first question, when asked, the prisoner refused to verify where that sepia print of his father was taken. However, historical experts have established both photos of the von Breusch men in uniform were taken by an official Reich photographer in Potsdam. The photos were snapped a decade apart of course; yet in almost an identical spot, and at different Reich conferences.'

'Well, at least I now know and won't forever be wondering who they were.'

'Have a safe trip home to Australia, Doctor Gwlynne. This is where we part company, at least for the time being. I'll enjoy a puff of my pipe sitting on this bus seat, until my chauffeur returns after he delivers you to your embassy quarters. I have an important appointment to attend in town.' The colonel tapped Kendall's shoulder, as he approached his idling vehicle.

He did, however, reiterate to Kendall that he would be called as a witness at his father's trial. Flagstaff also disclosed, in all probability his testimony would be held 'under glass', meaning behind an opaque panel. And as promised, his voice would definitely be disguised.

Flagstaff acknowledged his wave and recalled Kendall mentioning his wedding plans in Australia. Purposely the colonel refrained from saying who his appointment in town would be with. He glanced at his wristwatch and knew if his chauffeur hadn't returned in ten minutes, his two friends would be marking time in the Linz Airport Executive Lounge.

Flagstaff had advised Kendall to be ready when summoned to testify in this Austrian capital. He promised to personally arrange through embassy channels for his return flight from Australia. With this thought in mind, Flagstaff re-entered his vehicle and departed to meet his professional friends in town.

2 pm: Kendall placed his heavy luggage outside his room, ready for a butler to carry it down to a chauffeured vehicle. Meanwhile he searched through his grey vested suit pocket to make sure his flight ticket and gate pass were in their folder. Minutes later he nodded to the chauffeur as their car headed to the Linz International Terminal and his flight home.

26

With his connecting flight from London to Australia running behind schedule, Kendall settled in the Executive Lounge, where he was given royal treatment. In the interim Heathrow authorities had examined, stamped his passport, checked and cleared all his travel documents.

7.30 pm: A flight attendant ushered Doctor Gwlynne to his first-class seat aboard their international carrier. Kendall was looking forward to his long-awaited journey home.

Loaded with its full complement of passengers and their luggage, the commercial jet taxied along an icy runway. Taking off effortlessly, the aircraft speared through a clouded ceiling. Levelling, it peacefully cruised above a quilt of stars. The moon, hidden beyond earth's dark corona, had not yet risen.

Comfortable and relaxed, Kendall lolled back in his seat to meditate on his traumatic experience in Linz. Relieved his part in this indescribable episode had concluded, his tired eyes closed as he pondered on future ventures in the secrecy of a passive mind.

Striking turbulence from Singapore through to Darwin on this forty-hour flight from Heathrow, touchdown in Sydney was welcomed by all.

Kendall Gwlynne was the first passenger to descend the aerobridge steps. The instant his feet touched Kingsford Smith tarmac, a federal officer ushered him through to customs. From Mascot's International Terminal their vehicle taxied across to the domestic terminal.

With a special passenger and his two bodyguards on board, the pilot of a Piper Aztec received clearance from the control tower to roll. Lifting off a wet runway, the small aircraft gracefully purred skywards and set its course for the Australian capital. It landed around four in Canberra where Peter Bucknell and his entourage instantly converged on the Piper.

'Hi there, Kendall. I bet you're glad to be home. Come, my chauffeur is waiting to drive us to Federal Police Headquarters. It's merely a formality. Their officers require a brief deposition of what occurred in Austria. Our top echelon, including the Prime Minister, is fully conversed with that situation. The federal boys need to compare your written affidavit with all of ASIO's long list of data.'

'Can't this interview wait until tomorrow, Peter? I'm exhausted. Literally out on my feet.'

'Sorry Kendall, I have my orders. By the way, I've booked you into my hotel overnight. Your rooms are spacious and quite comfortable.'

Upon stepping inside the building, a federal officer ushered both Senator Bucknell and Doctor Gwlynne to a private room. After a solid hour of strenuous debriefing, the Feds allowed Kendall to leave.

Peter Bucknell's chauffeur drove him to their upmarket hotel for his overnight stay. Afforded time to shower and change in his executive suite before dinner, the prospect of a snooze was soon thwarted. Pinned to his pillows was a note. It explained why the huge basket of fresh Australian fruit sat on his bureau.

"I'll meet you in the lobby at eight. Don't feast first, PB."

'What has Bucknell lined up for me now?' Kendall laughed. Smartly he changed his tune on realising rest was now a forbidden dream.

'I need some sleep. Oh, what's another hour or two,' he grumbled. Then a brighter side to his dilemma emerged. 'Well, I suppose a superb meal will banish these blues of mine to hell.'

Precisely on the dot of eight Bucknell, escorting his secretary Deidre Sanderson, met Kendall in their hotel lobby. They walked together through to its small, though charmingly decorated restaurant. The enchanting atmosphere in this private and intimate eatery accompanied by soft music created an ambience for relaxing in congenial company.

Freshly cooked lobster topped with Russian caviar were worlds apart from the indigestible meals served on his recent flight home. Kendall enjoyed this delicious feast. The cuisine was superb. Luscious desserts were to dream over. To boost the evening's enjoyment all three toasted his successful venture to Austria as they imbibed on a nightcap in Senator Bucknell's suite of five lavishly decorated rooms.

As morning crowned the dawn, sad farewells filled the air. Dampness imbued the diplomat's eyes while observing Kendall depart for home and Abergeldie. Bucknell knew his new tour of duty would commence shortly in a foreign country. This final break with Doctor Gwlynne, his family and their friends at home in Yass wrenched at his heartstrings, particularly as a latent thought intruded on Peter's intensely occupied and investigative mind. 'Oh blast,' he growled, gently thrashing one fist against his left hip, 'I forgot to tell Kendall my terrific news. A letter to Abergeldie will have to suffice, once I'm settled. My main concern is for his mother's health. Gigi deserves a better life now she's free of that traitor Ian David Ross, or as we now know him, Rolf von Breusch.'

This small town of Yass was shrouded in darkness and in the dead of night thunder growled, accompanied by lightning as a dark purplish-green halo encircled the distant mountain range. Dry earth beneath the baked topsoil on Abergeldie and its fellow properties cried out for steady rain to drench their crevassed mantle of brown.

Until late January fine rain was inconsistent. It caused the harsh crust to remain unblemished. With moderate falls sweeping over the Snowy Mountains their autumn crops might survive. Every pastoralist and landholder prayed for rain to keep their lucerne, sorghum and wheat alive. Abundant crops of corn were sown and were also at nature's unyielding fury.

However, if heavy storms flood Yass, the backwater will swamp and destroy everything in its path. Now the wet has drifted in with a fresh smell of cleanliness, things have returned to normal on Abergeldie. With the prospect of bumper crops on the way and with fences mended, no sheep or any wild beasts can trample our crops. For the first two months in seventy-three, Abergeldie has remained peaceful. Kendall mulled over the note he received from their manager while unpacking his cases in Jalna's guest bedroom with an ensuite. He selected a perfumed rose from the vase on his fiancée's bedside table and sniffing it, put the bud on her pillow. Setting the mood for love he dropped a bundle of lint towelling with a tube of lubricant in the top draw of his bedside table as he thought, *Every job has progressed without hassles, hitches or glitches, Kirsten told me on our way home from Canberra's airport. On both our properties harvesters have gleaned all the lower paddocks. This is the best crop ever of lucerne and hay we've baled by far this season.*

Sheep crutching and dipping had gone according to plan. The milkers were in full udder. The drifting shearers had done a successful clip and gone their own ways, with only the regulars staying on Jalna and Abergeldie. No longer could entrails of dust be seen careering across the horizon, or willy-willies dousing the homesteads with grit. With both bores in and pumps working the top dams stored their overflow. The run-off had also bolstered their water supply.

Now most properties around the Southern Highlands of New South Wales could climb out from under a quagmire of debts, which were mainly debts accrued by banks who threatened to foreclose on their mortgages. Small landowners and home farms relied on their financial suppliers to keep their properties viable and productive.

Abergeldie and Jalna were safe. They were self-sustainable up to and including a point. Now they also were climbing out of the red that had threatened to engulf them.

During the first week in March, Brianna with Harriet, her daughter, sister Helen and niece Jasmine's help were packing everything in sight, ready to move out of Abergeldie.

Gigi had returned to Paris with Pierre and Amber, her adorable red setter pup. Kendall was working in Jarvis's surgery. He and Brianna had accepted the Svensson's offer to stay on Jalna, until they moved to Queensland, to organise his own medical practice. Brianna's gratuity and the $900 000 down payment on Abergeldie and all its landholdings would keep this couple viable until the equal amount was paid in full. This gratuity Gigi had given to her son and Brianna in lieu of a wedding present. Another trip home to see her parents in Ireland would be reorganised once they were settled. With good fortune this journey may be averted. Everything balanced on the winds of time.

27

1974: A blustery wind ascending across the continent brought with it the spring melt. The severe winter weather and bleak snow flurries had dissipated. As each day lengthened the icy blows were forecast to moderate. Cold nights spun over into warm mornings. In Paris heavy drifts of snow no longer affected its populace, although the sting of its passing lingered.

With their lives back on track after twelve months of utter hell, Gigi and Pierre's wedding was announced. The ecstatic couple had chosen her birthday in March to say their vows.

At Pierre's new apartment in Foch, an exclusive suburb of Paris, excitement reigned. The four women of three different nationalities congregated with Pierre's daughter, Mercedes, to toast the bride and to assist her in dressing for the most important day of all their lives.

7.39 am: Before a glorious sun arose over the Seine, these ladies were ready to attend a hair salon on *Rue Fabourg, St Honore*. On completion of their coiffures and facials, Arneka, with Brianna driving, stopped at a florist on their way home to collect the men's buttonhole sprays, forgotten by a boy, who at seven, had delivered all their bouquets.

Meanwhile Mercedes, in her Citron had taken Gigi and her Matron of Honour, Kirsten back to Pierre's penthouse, where they were staying. A small adjoining apartment, reserved for guests, was let au gratis to the younger two women. Pierre was due to arrive at this apartment block within the hour.

When home, Mercedes resided on the second floor, one below her brother's apartment. For convenience she'd now chosen to stay in the penthouse with her future stepmother.

Kendall and Bjorn shared the small flatette next to her usual flat.

Pierre would temporarily bunk in with Raoul in apartment 'Nine C', nestled at the rear of this third floor. Father George Brady and Kurt Baumer were situated in a quieter suite alongside Raoul's apartment. These flatettes allowed his family the freedom to pursue their interests undisturbed. With businesses to run, the Bouvier family's individual privacy was paramount. Pierre allowed his company executives to occupy other apartments in this tower of seven floors, on a permanent basis.

Arneka had just finished arranging boxes of cut flowers when a familiar buzz rang through the intercom, which she was first to reach.

'*Oui,* who is it?' She seemed quite perturbed over having to answer the buzzer again within minutes.

'*Fleurs pour Madame* Gwlynne,' the florist's boy called back. On releasing the button, he heard a familiar voice telling him to enter the lift. Passing three boxes to Arneka his hand fluctuated twice. He expected a sizeable tip for the third time to this apartment with an hour.

Scanning through her change purse, Arneka dug deep to find the required French francs. Its bottomless pit of tipping funds would soon be expended.

'Gigi, more flowers for you. I'm blowed if I know where we're going to put this lot.' Handing the prospective bride two boxes, one of Cattleya orchids and the other long-stemmed roses, the sixth in as many days, Arneka remarked to Brianna, 'Guess who this lot are from?' Concealing the third box behind her back, she winked to Gigi.

'These magenta orchids are beautiful, and the lemon rosebuds smell divine. The card says they're for my birthday. How sweet of Pierre to remember. He really is thoughtful. It'll break his pocket if he keeps sending me beautiful bouquets.' Intuition made Gigi walk behind Arneka where she spied the third flower box hidden under her skirt.

Brianna also twigged her partner in devilment was up to mischief. 'What are you hiding, Miss? You haven't fooled me, Arneka.'

'Here, take the box Bridy,' she sniped, grabbing the florist's card. 'Ha, look what I have.' She jiggled the gift tag in front of her and giggled.

'You little minx, hand both it and the box to me now please Arneka. Don't keep taunting me. You should finish dressing. Kendall or your father might come up and find you wearing only Pierre's robe and half-naked.'

Furious, yet trying not to show it, Arneka waved the card in her face. 'Not before I read what it says,' she laughed. 'It says here my brother does love you, Bridy.'

In pretence not to care, Brianna walked off to fill the vase of rustic roses delivered earlier. Aware who this latest bouquet came from, she pretended to ignore Arneka, who followed her through to the solarium.

'Can't keep a secret from you, can I? Men never send me flowers,' she pouted.

'No. I saw Dorian pinch a gardenia from the vase on our table last night at dinner. You smiled with delight as he handed it to you.'

'That's different. Dorian's a good friend. He didn't mean anything by it.'

A frown from Brianna indicated Arneka should keep quiet. One more word and she'd spoil the surprise for Gigi.

Still to be completed were the final touches to their nails. Last minute jobs of putting on gloves or adjusting jewellery could wait. First, they wished to celebrate this happy moment with an extremely nervous bride.

Before accepting her goblet, Gigi retied the cord of her fuchsia gown. The silk robe was another expensive gift from Pierre for her birthday today.

'Here's to the happy couple.' All four women charged their glasses.

'May your lives be long and you have a dozen screaming kids,' Arneka touted in gest as Gigi's bridesmaid, Mercedes joined them just in time to toast her Godmother.

'At my age?' Gigi frowned. 'I hardly think that likely. If I produce another brood, I would be a statistic in "Ripley's". The qwerky thing is, no woman can conceive if she's neutered. De-sexed, Arneka, my friend, that smacked your little joke right into the impossible basket.'

'Well, bigger marvels have been known to occur. I had to wish you something. And it was the thing first that hit the grey slate of my witty noggin.'

'Rightly so! And you my dear have imbibed long enough. In other words, Arneka lay off Pierre's expensive whisky,' Kirsten decreed in earnest.

Unable to keep a straight face, Mercedes thought their argument hilarious. 'Arneka, you have a funny way of saying things.' She expressed these words in slanted English while retouching her cheeks with a translucent face powder.

Casting a glance at the wall clock Brianna sighed. 'Gee, it's getting late and I must fly.' She kissed Gigi's cheek then stepped towards the door. 'I have my suit to press before Kendall comes up here. You promised to fix my hair once I'm dressed, Arneka.' A smug look passed between the girls. 'I can't possibly let my darling see me in these rags. He'd have a fit.'

'You've forgotten to show me your new suit, dear. Fetch it while Kirsten helps me to slip on my bridal gown,' Gigi called from the hall.

'Sorry, you'll have to wait. That's if you want me to be at the church on time.' One glance in Arneka's direction and a finger embraced her lips. 'My blue suit you've seen, Gigi. All I bought in Paris yesterday were new accessories.'

'Oh, never mind.' Gigi sounded disappointed as she struggled into her cream silk petti. 'You should hurry, Brianna. I don't want you to miss Kendall escorting me down the aisle. Neither would he. Get a move on, you two gigglers or we *will* be late. The wedding cars will be downstairs soon.'

Incessant chatter in the hallway indicated that the spritely pair had returned to their flatette.

In Raoul's apartment the men, including Pierre and Bjorn his groomsman, congregated to fortify their courage. Kendall's nervousness almost exceeded Pierre's anxiety.

Sneaking in to join their friends, Father George Brady, Kurt and Christian Baumer put in a late appearance. All those present, excepting the priest, took a swig of vodka to fortify their nerves prior to leaving for St Madeleine Catholic Church, on the Avenue des Champs Elysées.

The latter two men were not included in the bridal party. Knowing he must refrain from imbibing, Father George refused to take even a sip of whisky. Permitted by the church hierarchy to say Mass, he felt obligated to remain sober.

All morning it seemed a dauntless task to get himself and the Baumer men motivated. Finally, things were beginning to go right for Kendall who sighed. Trouble loomed. He cursed the detestable bow tie. No matter how he persevered it refused to sit right. 'This thing has a mind of its own. I give up.' Removing the bow tie, he sought Pierre's son and his best man's help.

'There, your tie is perfect now,' Raoul confirmed.

Father George had misplaced his clerical stole. Without it he couldn't perform his part in the wedding nuptials. After a fruitless search, he discovered it under the lounge. The final rehearsal for Pierre's part in the ceremony could now proceed while the others, including Kendall and his best friend, Dorian, observed the procedure.

Pierre, the stable one of this group, fumbled his lines while crookedly buttoning his grey silk-brocaded vest. Nothing would go right. Still, he seemed relieved after having a man-to-man talk with Kendall about how his mother might accept him as a husband.

Dressed ready to leave, Kendall pointed to the cars lining up by the sidewalk below. 'Hurry you men; time's ripping on. All my blessings, Pierre,' he embraced the bridegroom's hand. 'See you in church. I'm off upstairs to make sure my mother isn't having last minute jitters. With your shaky hands Pierre, the last thing you need to be is left standing at the altar.'

'Thank you, Kendall,' Pierre clasped his hand with the firm embrace of trust. 'Whisper of my love to my chérie. I know she will look magnificent. Today, I think I am the wealthiest man in all Paris. Rich in love I mean! Give mine to Brianna …'

The priest intruded. 'Away with ya Kendall, or ya mother will be late. Then I'll be way up the spout with me notations fa the ceremony.'

Arneka greeted Kendall at the door of her and Brianna's apartment. She planted a kiss on his left cheek. He reciprocated with a hug.

'Sis, give Brianna a hug for me. Mum's ready. I'll make sure she doesn't trip down those steps like she almost did this morning. She can be bumble-footed at times, my mother.'

'Dressed in that tight-fitting gown Gigi could fall. I offered to unpick and lengthen the side split three centimetres, to give her enough space to walk. She wouldn't hear of it,' Arneka laughed. "It'll spoil the design of my sleek-skirt and leave holes in this magnolia delusted-satin," she informed me.'

Arneka flaunted her near-naked shoulders in dismay. 'Take special care of her. Your mum is very precious, Kendall. I couldn't wish for a better mother than yours.'

'No darling, she's *our* mother now. Gigi's no longer just mine; she's as much yours as Brianna's. You're a treasure for saying so. From today on we'll be one contented family, and that makes me a proud man. Most of all, I'm proud of you for being here with us on her special day.' A peck was imparted on her powdered nose as he said, 'You look beautiful in orange.'

'Yuk! This ain't no orange, brother dear. This is apricot chiffon, none-on-over-ninon they say. Well, not quite,' responded Arneka cheekily and as she twisted her body, her tone-on-tone satin petticoat swished. A hitched skirt showed her apricot-tinted shoes and trim ankles. Squinting she added, 'Kendall, you look pretty much of a spunk yourself in that penguin suit.'

'You think so? More like a straightjacket. I hate being dressed like a prized prig.' He smiled and saw her smirk at his outmoded slang. 'I'm anxious to see Bridy. Hey, I must push on, otherwise we'll need a crowbar to move my mother. I popped in next door and found her almost buried up to her neck in wet tissues. I can't have a teary-dearie walking arm-in-arm with me down the aisle.'

'Your slang is tops. It's fab and sounds surreal coming from you. More of my company and you'll be converted.'

Kendall laughed. 'No thanks; I prefer my diction to be correct. Precise English with an Aussie twang, aye Arneka.' They laughed.

Meeting Kirsten in the penthouse doorway Kendall called, 'Come on Mother, or Pierre will think another Parisian has eloped with you. He's very nervous, so don't keep him waiting.'

'In case you haven't heard, it's a bride's prerogative to be a little late on her wedding day. And I've waited twelve months for mine. Don't hassle me, son. Just walk me to the lift.' Gigi clutched her missal with a cream Cattleya clipped to its satin cover. She preferred not to mark the empire bodice of her tight-fitting frock by pinning the corsage to its delicate, cream fabric.

'I'll do better than that, Mum. When we reach the main lobby, I'll carry you down those curved stairs. Then we *will* avert an accident. Only don't knock my buttonhole spray from the lapel of this monkey suit.'

'You're not carrying me anywhere. This gown will crease if you do.' Alighting from the lift, Gigi took a moment to arrange the pin-spotted veil of her cream pillbox in place. 'Kendall, a grey morning suit and matching tie does not constitute a "Monkey suit", as you uncouthly put it,' she said, while being escorted down the seven stairs.

'No, it's more like a straightjacket. Don't distract me Mum, or I'll have an accident trying to manoeuvre my shoes on these stone steps.'

Large feet in the dim light struggled to find a comfortable foothold upon their narrow treads. Usually Kendall jumped over the seven steps in one Herculean leap.

Kirsten and Mercedes had long departed for the church in a limousine, her namesake and one of their company's fleet. Apricot ribbons atoned to their gowns fluttered from the Mercedes silver emblem. Likewise, satin streamers decorated its door handles.

The duco on Pierre's highly-polished limousine glistened in the late noonday sun. This sleek Rolls-Royce, with the smell of newness still impregnated in its cream leather, was his wedding present to Gigi. It doubled as the bridal car. Right on time his personal chauffeur pulled the vehicle kerbside adjacent to St Madeleine's porch steps.

Kendall assisted his mother from the car. Adjusting her pillbox, Gigi flipped the spotted veil down to protect her privacy and to hide a radiant smile.

'Well Mother, within the hour you'll be Madame Pierre Jean Paul Bouvier,' he proudly said, handing her the prayer book with a kiss on the cheek. 'The best thing is, I'll no longer have to worry about those naughty escapades of yours from dawn until dusk. From today on you'll be Pierre's responsibility, thank goodness. Although I must admit, you do look beautiful.'

She never directly answered. It took fifty-seconds of solicited anxiousness before she responded, even then it came in a brief reply. 'Kendall, I'm worried. I've not seen that mischievous pair, Brianna and Arneka for well over an hour. What can they be doing? Surely their car wouldn't have arrived here before ours.'

'Listen Mum, they're probably in the church. Who wants to arrive after the bride? Nobody I know. So quit panicking, or you'll lose your footing on these irregular stone steps.'

The instant Gigi entered the Galilee-alcove leading to the nave the Bridal March began. Kirsten walked in slow graceful steps, five paces ahead of Mercedes, who preceded the bride in her final walk of freedom. Holding her son's arm, Gigi's heart gave an excitable leap when, halfway down the long aisle, Pierre turned to watch her walking towards him.

'Keep looking straight ahead, Mum and don't falter or you might trip on this carpet. My arm will guide you until we reach the low altar.'

Her son's smile lingered until she stepped in beside Pierre, who'd observed every nervous and graceful move Gigi made. He tried to contain his pride, as every footfall brought her one step closer to their new life together.

In a euphoric daze, Gigi focused her eyes only on Pierre. She was oblivious of the guests standing either side of the blue carpet she walked on. Nor did she notice vases filled with huge sprays of exquisite blooms adorning the nave and aisle seats.

Long-stemmed apricot and lemon rosebuds and white baby's breath interspersed with mauve carnations peeped from among the greenery. Every second pew was adorned with a bouquet and a silver horseshoe. They were decorated with lilac bows and sprays. The small posies of briar-roses Mercedes, now thirty-eight and Arneka, her similarly aged guest had arranged. Shafts of sunlight filtering in through the stained-glass windows highlighted all these floral arrangements.

As the bride took her final steps, the soft swish of her gown inspired the congregation to cherish this moment. A hushed moment of silence blended with a serene ambience of the church, which looked magnificent. The serenity enhanced its aged and mythical beauty from time immortal.

Caught in the moment of bliss and embracing Pierre's hand, Gigi almost dropped her missal while passing it back to Kirsten, her Matron of Honour. Her bridesmaid, Mercedes demurely lowered her eyes, to focus on the bouquet in her own hand. Then equally as shy, her gaze lifted to the high altar prepared for mass.

As the conventual priest began their nuptials he asked, 'Who giveth this woman's hand in marriage?'

Kendall stepped forward and holding his mother's hand, he presented it to the priest. 'I do,' he said proudly and then retreated to be seated on the front pew.

Solemn in mood the service continued. When the time came to announce Gigi's name everyone twittered with excitement. Only a handful of people were conversant with her real Christian names. Everyone smiled and looked at their neighbours. No longer could she conceal the signatures on her birth certificate, so jealously coveted down the years.

'Do you Gwladys Idelle take Pierre Jean Paul as your wedded husband?'

'I do,' she shyly declared. As requested, Gigi repeated the oath of marriage.

Pierre recited his vows in a quiet, sombre voice. The priest then nodded for an exchange of rings to proceed. The instant Pierre lifted her veil to kiss his bride; a sigh, if expelled, could be heard echoing though the hushed nave.

Their nuptials finalised, Father Brady nodded for the married couple to follow him to the High Altar where they received Holy Communion.

On completion of this ceremony, the wedding party moved into the vestry. While Gigi and Pierre signed the register one of his Parisian friends sang Gounod's *Ava Maria.* Her beautifully clear lyrics soared beyond the church dome and ascended to a wide cloudless blue abyss.

With all official documents signed, Father George Brady introduced the newly wedded couple to the congregation as Madame and Monsieur Bouvier.

From where Gigi was standing a ray of sunlight infiltrating through the stained-glass dome struck her eyes. Temporary blinded, she shielded her eyes from the glare with a prayer book. Accepting her husband's arm, they moved out of range of this strong beam of sunlight. A little confused, she failed to hear a soft strum of a familiar tune pervading the nave.

'Darling, aren't we supposed to walk down the aisle …'

'Shush chérie, come and sit here with me for a moment,' his voice cadenced to a whisper. 'We cannot do so now. The Monsignor has not finished prayers. Let's move over to those chairs.' Pierre nodded sideways towards the vacant pair of burgundy velvet carvers. The third chair Father Brady had not long vacated.

In full view of all, excepting his mother, Kendall stepped across to the altar rail and then stood to one side. The soft music increased a crescendo as the organ chords pulsated again to the Bridal March.

Escorted by her father, Brianna stepped in behind her bridesmaid. She paused to allow Arneka, who'd been warned not to gallop, elegantly precede her down the aisle.

Tears cornering his eyes, Kendall watched his bride serenely step towards the altar. Fascinated by her gracefulness, he longed to touch Brianna's hand.

Pierre encouraged Gigi to sit in the empty carver nearest the aisle, so she could witness her son's wedding ceremony in comfort. Utterly amazed, she looked up as he winked.

'Darling, why is Kendall standing at the altar?' Gigi gasped in disbelief. 'Look, Brianna is about to stand beside my son.' With pools building in her eyes she softly enunciated, 'Pierre, she looks divine, so elegant and majestic in her guipure lace wedding gown. This is unbelievable.'

'Hush chérie, or you'll miss them saying their nuptials.' Looking down Pierre lovingly kissed his bride's forehead while clasping her left hand.

Gigi refrained from speaking to listen to her son take his vows. Peering up into Pierre's dewy eyes, she gently squeezed his clammy fingers.

Brianna looked resplendent in ivory delustered satin. The sleekness of her embossed gown displayed her curvaceous figure to an advantage. The fabric beneath her guipure lace bodice glinted in the light and her long train and beaded veil spread fanwise behind her lace-covered high heels. Her bouquet shook as she elegantly approached Kendall.

Accepting her lily-of-the-valley posy, Arneka held the spray in one hand. Juggling her own apricot and lemon bouquet of carnations in the other, she stepped back to observe their ceremony.

The nuptials between Doctor Kendall Gwlynne and Miss Brianna Skye O'Shea began in earnest. As promised, the priest whom she'd known all her life in County Clare, officiated at the final part of their ceremony.

The proud expression on Father Brady's face was no less engaging than the smile on her mother-in-law's face. Madame Bouvier's day was doubly blessed.

Signing and witnessing of their marital names in the church register took minutes. Proudly holding his bride's hand, Kendall whispered, 'Darling, today you've made me the proudest man in this historical city of Paris. Your beauty outshines its glorious pavilions and the majestic gardens designed by King Louis. They called him the Sun King.'

'Yes. I studied Paris and its historical past at school, in County Clare. Your mother wants to go home to Australia tomorrow night with the Svenssons.'

'Why? I know Pierre has a conference to attend in London on Friday.'

'She's scared someone will put both your horses down. And she's fretting for Amber. As you know Kendall, her pup's in a kennel here in Paris.'

'Amber will be quarantined for another two weeks. Potchkin and Wallaby Downs were put to pasture with Bjorn's ageing horses on Jalna. He organised everything. All our antique furniture and her priceless jewellery arrived at Orly yesterday as air cargo. Mum should have told me about the pup. She didn't mention it to me last night. I'll phone them before they leave for Switzerland in the morning.' Kendall ceased speaking as the couple approached him. Brianna was ecstatic because Gigi, with Pierre's sanction, offered to let her and Kendall precede them down the aisle.

Followed by their dual entourages, they moved out into the fading dusk as carillon bells stirred the night-owled Parisians into wakefulness. A flock of white doves, released at Pierre's request spiralled high. To the throb of chimes, they soared to paradise.

A procession of guests filed past the christening font and down a short flight of steps. They mingled with a crowd of onlookers on the church purlieu beneath its aged and fearsome gargoyles. Strangers gasped in delight to see two brides and their exquisitely gowned female attendants. The men chattered until an official photographer moved in to take wedding photos of both couples and their friends.

Promptly ushered into their respective cars, they travelled a short distance to a parking area adjacent to the Eiffel Tower. A twitter of excitement surrounded each bride as she exited her vehicle. Seeing two brides exquisitely gowned in one day amazed the children and visitors. A clear sky reflected in the eyes of these disbelievers and youngsters who twittered and giggled with delight.

28

On alighting from their cars both wedding groups separated. All official photographs were taken against a magnificent backdrop of the Eiffel Tower. The couples then regrouped. Forbidden to walk on the grassy verge, family photographs were hurriedly snapped before a full moon crept above the horizon. Kendall feared if they dallied there too long, they may receive a reprimand from the gendarmes who patrolled these parklands from twilight until dawn.

Little did he know that Pierre had prearranged with one of the officers on duty to grant them a fifteen-minute respite. Their privacy was a courtesy gesture of the Parisienne gendarmery. On sighting their dual wedding parties, the four armed gendarmes nodded to Monsieur Bouvier, then kindly extricated themselves to a distant location. Batons were waved at some outrageous louts who tried to infiltrate their privacy.

Kendall smiled as his groomsman, Dorian Payne, placed his left arm around Arneka's slender waist. Together they stepped between he and Brianna for a final photo. Smitten with his partner, Doctor Payne kissed her cheek.

Arneka didn't attempt to object to this manoeuvre. Mesmerised by Dorian's quiet unassuming manner, she smiled to think this young physician considered her a close friend, rather than a stranger. There were no comparisons between him and Kendall in looks. Dorian wasn't tall or handsome like Kendall. His charismatic charm she found alluring; so much so she couldn't take her eyes off his athletic physique. The young physician's keen mind astounded her and she admired his majestic stature.

During the final photo sessions under the Eiffel Tower, all their guests were driven to *Quai Des Tuileries*, their boat's departure point. Moored at a wharf, the luxurious *Bateaux Mouche* tossed to a mystical sway, while awaiting to ferry her passengers on a scenic tour along the River Seine. With everyone on board, the *Queen of Paradise* slipped her moorings. Her skipper manoeuvred his floating *palais* well out to the mainstream traffic.

Important and personal photographs consumed the first half-hour on board. Although most of the guests preferred to enjoy light refreshments on her upper deck, some snapped the spectacular views of boats traversing this majestic river. Others took in the sights, through binoculars, of this ancient Parisienne city with its cobbled capillaries winding through lengthy veins and busy lanes which wended their way from the boundary to its countrified lungs. Grasslands and forests where the occupants breathed fresh mountain air of its provinces stretched further than its majestic horizon.

On board, time dictated whether to allow more guests taking photos to pursue their interest, before the bewitching hour called the bridal couples to return to their respective hotels.

Brianna unclipped her gown's train and the veil. Folded neatly in their box it allowed her the freedom to dance. Instead, she and Kendall chose to mingle with their guests on the top deck.

Gigi's pillbox found a home in the same box as her daughter-in-law's veil. Comfort for all on this unseasonably warm evening was paramount.

Guests coped reasonably well, until a sea breeze took hold to filter the smoke-imbued air on its lower deck. This allowed the wedding breakfast and speeches to proceed, while every guest appreciated the coolness without sweltering in this unseasonable heat.

Raising his glass, Pierre's best man, Bjorn Svensson announced, 'I take the opportunity of toasting Gigi and Pierre. Not forgetting Brianna and Kendall, of course. Their marriage certainly came as a shock to you, Madame Bouvier. Kendall got his own back on you Gigi. You had not the foggiest idea he plotted this double wedding with Monsieur Bouvier. Please be upstanding ladies and gentlemen. To the brides and grooms; let those goblets ring with your chants of wishing them long lives in harmony and peace.'

A harmonious refrain shook the timbers. Ironically at that instant the cruiser shuddered. All laughed, as the skipper announced they'd struck the wake of their sister ship, *The Fleur de'ley,* descending portside.

Few guests bothered to look overboard. The majority were intrigued to hear Gigi's response. To their dismay, they were disappointed. Instead, Pierre began his speech.

'On behalf of all our guests and myself, I offer my apologies to my darling wife, for the deceit of uniting both our wedding ceremonies. Kendall will later explain his side of the proceedings.' Holding her hand, Pierre bestowed Gigi's fingertips with a loving kiss. Winking, he then addressed his wife. 'We certainly fooled you Gigi. My faux pas to tell you was thwarted by your son.' Finding it hard not to double over with mirth, Pierre quietly added, 'I have no other apologies to make. I hope in time you will forgive me, ma chérie.'

'I'll say my piece to you tomorrow and in private, my darling man,' sprouted Madame Bouvier who remained seated. 'Pierre, don't you and Kendall think you'll get away with your deceit. Be assured, neither of you will. The same goes for Bjorn Svensson and you lot sitting at this official table.' Appraising everyone over her spectacled eyes, Gigi affirmed, 'I will endeavour to keep my reprimand short, to the point and sweet. That's if I don't laugh.' Standing, she held up her goblet. 'To my conspirators, may God bless you one and all.'

Having listened to long and tiresome speeches, Kendall capped the dawn hours with a finale. Holding his glass aloft in salutation, he proposed several toasts. 'I speak for Pierre and myself in saying this. Thank you all for travelling long distances to Paris just to be with us on this, our special day. On behalf of my wife, I propose a toast to her beautifully gowned attendants. Not the men, although they are a handsome bunch.' His response brought those on the lower deck to mirth.

'The men, in their own way, have carried Pierre and myself through harsh ordeals, emotional and otherwise. Likewise, the ladies have assisted my wife Brianna, with donning her gown for our wedding, the church decorations and through all manner of difficult moments. Weddings aren't easy to arrange. Believe me, they're not Arneka, so watch out.'

Unembarrassed, she grinned and pressed Dorian's hand under the table. Distracted by his charisma, she didn't respond. He whispered something that excited her. To those treading the boards, it was obvious that these two young people were smitten with one another.

A guest asked Kendall where he and Brianna intended to spend their honeymoon. Until now he'd warded off answering. Time demanded a reply.

'We'll be touring all through the British Isles, France and some of the continental countries for a month. On our way home, we'll be stopping in Singapore for several days to take in the sights and enjoy its mystical nightlife.'

'Where to then, Kendall?' queried one of his skiing friends.

'Australia! Where else? I'm not like you lot of nomads who can roam forever.' Proudly he then enunciated in his Elysian mood, 'I'm a married man now and can't stray. Bridy won't let me out of her sight. Not that I mind, quite the reverse actually,' Kendall proudly confirmed. Endowing his wife's hand with a kiss, he waited for the twittering to abate.

Smiling, he then confirmed, 'I must say, we are not the only newlyweds here tonight. Please charge your glasses to our friends sitting on my right at the next table, Mr and Mrs Bucknell. Deirdre and Peter were married a week ago at the Australian Embassy here in Paris.' The couple stood and nodded to a roar of three cheers.

When silence reigned Kendall endeavoured to explain how, to get their Australian legs back to a relaxed mode, the Bucknell's had lent them their holiday cottage at Kendall's Beach. Casually poised, he awaited a response. Receiving a lot of flak over his namesake's beach, he refused to disclose its location, in fear of their privacy being invaded by hordes of friends.

'Now, if I may finish? It's time I danced with my wife before the night ends.'

'What future plans do have you for a medical practice, Kendall?' queried another of his university and skiing friends.

'Well, as most of you know, I've just completed working with Doctor Jarvis in Yass.' Stephen nodded. 'My wedding present to my wife is I'm moving on to a bigger, more lucrative practice in Queensland. I've secured a place in a Brisbane university to study paediatrics. Being contented in a specialised field and with a lovely wife to come home to, what more could a man want? That's all I wish to say. Topic closed.' As his hand wavered Brianna touched his sleeve.

'I knew you were plotting some mischief, thank you darling for my gift. You've made me proud of you, Kendall.'

Arneka must have her say. She stood and bowed, much to her father's consternation. Embarrassed, he and Christian lowered their eyes, although neither man commented.

Dorian smiled. *This girl shows courage and I admire a spirited woman.* He stopped thinking when he saw Kirsten's arm gently nudge Arneka's elbow. The prod failed to deter her.

'I can vouch Queensland's beautiful beaches and the hinterlands are calling everyone to visit us in Yeppoon. The Capricornia Coast is the icing on the cake as a tourist destination.' Pausing to sip her drink she then announced, 'Thinking of cakes, come on you married couples, how about cutting yours? I'm dying from the lack of a sugar fix.'

All eyes at the official table focused on Arneka's enthralled expression. Blissfully gazing into Dorian's eyes, she laughed as he again said something endearing to her.

Kendall arose and smiled at Arneka. 'Should you not be aware, this delightful lady is my long lost sister. Dark horse, aren't I? Arneka has stolen a special spot in all our hearts, especially mine.' Charging their glasses to her, Kendall then added, 'After twelve months,' he purposely faulted, 'Brianna and I will enjoy having,' another brief pause ensued before he said, 'visitors.'

Sitting down he whispered to his wife, 'Mark my words, darling. Arneka will owe me that ten dollar bet because she'll be the next to breast the altar. I think she's met her match with my medical colleague. Dorian will make her toe the line, or tame her. Time will tell if I'm right.'

'I think you are right darling. His eyes have not strayed far from her excitable smile all evening. We can't discuss this now Kendall, or perhaps later.'

Bjorn announced, 'It's time for the cake cutting ceremonies to begin.' A hush invaded the gathered throng. Tuned to the gentle sway of the *Bateaux Mouche's* hull and serene ambience of its lower deck, each groom held his bride's hand and in accord, their serrated knives speared each individual cake's icing. The floral arrangement of green sprigs of holly wreathed in Australian native wattle, a huge crimson waratah and wild pink boronia encircled a Celtic harp which caught everyone's attention as together Kendall and his bride cut their two-tiered chocolate cake's pale lemon icing.

With silence prevailing and while still upstanding, he quietly orated, 'Seriously, I've left the best for last. My wife and I are sincerely pleased that my mother has found genuine love in Pierre. His family and mine go back decades. Mercedes and Raoul, along with the Baumer family have this day joined with us in loving unity. My mother and I welcome Brianna's parents into our conjoined family.' Mopping his sweaty brow Kendall, red-faced, decried his own speech. 'I know I botched that. Don't expect a doctor to make speeches. Attend to patients, yes. Lectures 'a-must-do' exercise. Make speeches, never. What I'm striving to say is from today on our life is blessed by all those I've mentioned. I must not forget to include the Svenssons and Father Brady. He's been our guide and stay all these past months.'

'In response, I think I can speak for all concerned,' Bjorn, the MC piped up, a smidge tiddly. 'Kirsty and I are honoured to be numbered among your family. A final toast to this superb day and the weddings we're celebrating here in Paris.' A glass brimmed with wine was raised by an unsteady hand. Beer steadied his nerves; wine accomplished the opposite. 'We've come together as friends from all over the globe, so let's celebrate. Three cheers for all.'

Kirsten tugged his coat sleeve. Gigi and Pierre smiled, and the girls giggled.

Pulling a face Kendall topped the lot by saying, 'Excuse me, Bjorn. I've been granted a brief say, and I couldn't agree with you more. I'd like to thank on behalf of my wife and I, my mother's friends who couldn't be here today. She'll know who they are. Now I'll read their Christian names from this cablegram, which I received a moment ago from the skipper. "Gigi and Pierre. Stop. All our love and best for your future happiness together. Have a wonderful life. Don't stop. Special love, Henry and Margo". This gram is also signed "Homer and Sharon". Both those couples are living at present in Switzerland. Oh, there's a PS. "Letters and parcels in transit to Paris".'

Kendall passed the gram along to his extremely delighted mother, who was excitedly conferring with Pierre. 'Thanks darling,' she blew Kendall a kiss. 'Oh, isn't it sweet of them to honour our day. I was disappointed none of them could be here with us,' she said over the noise of musicians tuning their instruments. As the music pulsed to the Bridal Waltz, Bjorn challenged the newlyweds to dance.

Having played her 'finale', Gigi enjoyed Pierre holding her close while they patiently waited for Kendall and Brianna to step on the dance floor. Deliriously happy, all four moved in unison to the gentle sway of the River Seine.

Looking down into Gigi's misty eyes Pierre whispered, 'I can't wait to have you to myself chérie. It is divine that at long last I can call you my wife. Oh, how I have longed for this moment.'

Pounced on by intruders, he eased her closer. 'Be careful dear, we have company.'

Kendall overheard his comment and jocularly responded, 'Don't hog the floor, you two lovers. And don't believe half of what she tells you, Pierre. My mother tells whoppers.'

'Kendall, your mother does not lie. It's as much their night as ours. You're supposed to be dancing with me. Not chatting to those two darlings,' chimed in Brianna, smiling. Nestled in his arms, they danced to the lilting tune of *Paris in the Spring*. The melody lingered.

Gigi turned and discretely spoke to Brianna. 'Take him away darling, before I … Well, you know where to. My son is disrupting our pleasure.'

The music reached a crescendo and they swayed to its harmonious rhythm. Having heard their friendly debacle Pierre, who'd said nothing, laughed. 'Chérie, to witness your, no our son's happiness is sheer delight to me. I think it is not before time.'

Feeling secure in her husband's arms, Gigi felt her heart skip a beat. Every graceful step these two professional ballet dancers took was poetry in motion. As the tempo quickened, they escaped to the observation deck, where they could be alone to talk in peace.

Holding her close they danced cheek-to-cheek to the intimate tune *Love Me Truly*, with waves gently lapping the cruiser's hull. On a down surge, Pierre lent over until his lips gracefully touched her forehead.

'Gigi, I do not know how I have existed without you. All through the desolate years of my marriage, I have pined to be with you. From the deepest depth of my heart, I'll cherish and adore you. And I will always love you, ma chérie.'

'I appreciate you not pressuring me to consummate our love. I understood it must've been hard for you, dearest. Suffering my tantrums and abuse wasn't easy for you to tolerate. When I thought I'd lost you forever, I wanted to whither and die.' She brushed a wisp of hair from her eyelashes. 'I've been so lonely without you, Pierre. Phone calls every other day aren't the same as personal contact with the one you adore.'

'Well, I tried to talk you into coming to Paris many a time. You refused to believe my feelings for you were genuine. Somehow, I sensed you needed time to adjust to your stress.'

In a euphoric daze, Gigi flicked a tear of happiness from her left cheek. 'I asked you something on our engagement night, yet you didn't reply. I knew you had more pressing things on your mind. My letter, wasn't it?'

'Oui Gigi. Your letter contained nothing I did not already know. You fooled yourself. A man in love can tell if his darling loves him as deeply as he does her. I smiled when Kendall gave me your sincere billet-doux penned from the heart.'

Resting the heal of his palm against her breast, Pierre could feel the rhythm of her heartbeats as they swayed to the river's cadence. It mingled with the gentle breeze weaving a flurry of autumn leaves through her hair. By moonlight he noticed how her rosy cheeks glistened with soft spray accosting their beauty. Relinquishing the irresistible urge to gently crush her cheeks between both hands, his lips brushed their florid blush.

Arneka, with Dorian's arm around her waist intruded upon the couple's privacy. Their desire to be alone now lay in tatters. Seeing their elders in an endearing tête-à-tête they turned and strolled arm-in-arm towards the bow, there to continue their endearing courtship.

Pierre patiently watched the couple disappear and then quietly decreed, 'I wanted to leave on the first plane to bring you home to Paris. Unbeknown to me, your dear son, knowing how miserable I was, advised me not to approach you. Kendall arranged my flight before he returned to Austria. Without saying a word, he left my ticket and a brief note attached to your letter on my pillow. I thought at first it may have been a note from Raoul, or Mercedes. Before I opened the blue envelope, I sniffed its delicate aroma. That instant I knew it was your billet-doux.'

Pausing to kiss her on the tip of her nose Pierre whispered, 'A sentimental man, I still have it, my chérie and I have no desire to destroy your letter. To the contrary, I found your billet-doux delightfully overwhelming. It brought tears to my eyes and I'll treasure both it and you.'

'You sweet sentimentalist. I'll cherish those words always.' Gigi stopped to think and then responded, 'There's something I can't quite fathom. What I'm endeavouring to say is, how you an astute businessman, found the time to confide in Kendall and if you would be free to travel?'

'We talked over coffee in my office about me leaving for Australia. Aware how deeply I loved and wanted to marry you, Kendall could feel my anguish. I think it stunned him not being home with you and Brianna for Christmas. Now we're all together and you are mine. And my chérie I will be yours, forever.'

'I know darling,' she replied vaguely. 'Is that when you both arranged our dual wedding?' He nodded. 'I thought so and I'm truly grateful, Pierre. I'll love you until all the seas run dry. Brianna says that often. We'll make every moment count in our married life together, if we do what you proposed long ago.'

'Now, what might that be, chérie?'

'We should leave the past, not only in the shadows, but dismiss it altogether and begin life afresh. Because we idolise our three children, our marriage will grow stronger every hour. We've earned our divine love, you and I. Now we are free to live our dreams, which we created ages ago for our life together in your apartment building in Montmartre, Paris.'

During the first week of their honeymoon on the continent, Kendall was recalled to Austria. Before flying out from Heathrow for Linz, he took Brianna home to Ireland to be with her parents to await his return.

The Baumer family and Father Brady were scheduled to accompany the Australian diplomat, Peter Bucknell to Linz in several days. Required to give their initial evidence "in camera", the five Australians would be housed at the Australian Ambassador's residence, until such time they were called to attend court, as witnesses against the ex-Nazi captain and wife murderer.

Kendall feared his father standing trial over his evil misdoings in Germany, as it would impinge on his private and professional life. Laymen might dispute his genuine motive for testifying. They might label him with the same hateful tag as this Nazi whom he intended to expose. Whereas Arneka, with her feisty temperament, swore to bring him into disrepute with his own kind of bombastic troublemakers.

'Kendall, I *will* confront that mongrel in court and I *will* accuse him of murdering my mother whom I absolutely adored. As a child of three,

Erika favoured Chris more than me. I've never known such a pompous prick as that Nazi. He makes my skin crawl and I shudder whenever I think how he murdered masses of people in our homeland. I wish one of his junior officers had fired a bullet through his head in Hamburg.'

'Arneka, I promise you that von Breusch will dangle at the end of a rope before long. Evil people like him don't deserve to live. Don't get upset or cry again. He doesn't deserve your tears, or anyone's pity. My mother detests the cruel way he mistreated her and me as a child. Once the rope tightens around his neck, he will never hurt or brutalise her, or any innocent and unfortunate person again.'

'Not where he'll end up. I hope ole Nick makes him stoke the fires of Hell until he's incinerated in its flames.'

Kendall thought the same although he didn't respond. 'Come on, it's time to enjoy our morning tea with your father and Chris. I have good news for Kurt. The report on his last blood test shows a remarkable decrease in his cholesterol. His heart should keep beating, God willing, for another twenty years, give or take a few months, free of infarctions and angina.'

'Who sent you the report? More of Jarvis or his locum's work, I guess. We're in Linz, Austria now, remember. Not home in Aussieland.'

'The technician who sent me your father's results did a thorough job. That's for me to know and for you to keep guessing, sis. A luscious morning tea awaits us. You love munching on delicious cream cakes and lemon meringues topped with gooey caramel.'

'I wish we could go home. I'm dreading the pending trial. Aren't you? God knows if they will hang that damn Nazi. Some idiot might suggest otherwise. Kendall, I miss your mother and how she corrects me because of my abruptness.'

'Mum and Pierre are due home soon. Touring through the ruins of Poland's extermination camps would've been horrific. Pierre's maternal grandparents died in Treblinka. Their brutal deaths occurred late in forty-four. They were innocent people caught in the hustle of war.'

'Your mother says she mistrusts most Germans. Gigi refuses to discuss the rough treatment she suffered on Abergeldie. She can't abide the despicable and inhuman way the Nazis treated most Jews and less fortunate people in Europe in wartime. And I agree with her, Kendall.'

'She'll be keen to collect Amber from quarantine in town and anxious to see us all here in Montmartre, Arneka. They'll have lots of nerve-wracking tales to tell of the time they spent in Poland and touring though Switzerland and quaint Austrian alpine villages.'

'What will happen to your horses and where are they now Abergeldie's sold?'

'They're both at pasture on Jalna. Pierre bought Mum a horse and its stabled on his business manager's farm. I'll miss Potchkin, he's ageing fast. Brianna and I will buy two young geldings from the Magic Millions sale on our way north to wherever we decide to settle in South East Queensland. I'll register my thoroughbred horse as Son of Potchkin. He's been a true and faithful steed and I will never forget how he saved Mum's life after her so-called accident, which occurred in my youth at a place we called Wave Rock on Abergeldie.'

'Kendall, will you teach me to ride your horse at home in Aussieland?' With a mouthful of cake, he just nodded.

29

Forerunner to the tribunal to be held at the Linz Courthouse in Austria. Supported by conclusive evidence that proved the prisoner in custody was the escaped war criminal every agency had chased for longer than a year, the UN Security Council's order was for the pre-trial date to be arranged in the very near future. This move was sanctioned in conjunction with the other agencies, with the prospects of producing witnesses to testify against the ex-Nazi captain. As the pre-trial approached, arrangements were in place for the prisoner's immediate transfer from Strasburg to Linz, on demand of the tribunal judges.

After endless months of confinement in his cell, Herr von Breusch was served with an indictment for his war crimes. Having represented his client in Salzburg, the appointed barrister instructed Rolf to read the six-page document and waited while he did so. The pre-trial had finally been sanctioned to go ahead in this beautiful and historical city.

As the time and day drew closer for the initial hearing, von Breusch was extradited from Salzburg by vehicle to the Austrian capital. From the airport Rolf was expedited under guard to jail within the courthouse confines where he was sentenced to solitary confinement in a two by four metre cell. He felt discontented with being confined, as it severely inhibited his freedom.

Rolf had the option of selecting a self-appointed solicitor, or one would be assigned from the list provided. The choice was his alone to make, not counsel's. Instead, he demanded to have as his defence counsel, a friend and lawyer whose reputation was of good standing in Linz and Salzburg.

The ruling stood as in previous hearings. Rolf's lawyer, Terrill Scharles insisted on being present at every interview, whether in court or in his client's cell. This suited both men, because Scharles understood Rolf von Breusch's militarised and authoritative mind. Their trust was mutual.

The selected judicial panel for this trial were appointed and approved by Austria's legal counsel, a close fraternity. The prosecutor and his privately designated defence counsel's undertakings were underway in chambers. Facing their legal teams was the mammoth task of collating reams of data and contacting witnesses for the forthcoming tribunal.

A clinical psychologist had previously interviewed Herr von Breusch and classified him legally sane to stand trial. Prior to the pre-trial's commencement, his signed report read: "My professional assessment of von Breusch is that he is of sound mind, and well enough to stand trial for all his indicted war crimes".

The extensive pre-trial of war criminal Rolf von Breusch concluded on Friday the 29th of March 1972. Three weeks to the day later, the actual trial date was listed to commence on the 19th of April. This allowed twenty consecutive days for the remaining witnesses to be transported from their hometowns or flown in from overseas countries. The majority, registered under nom de plumes, were already in residence at their hotels in Linz or its surrounding towns.

During this brief respite, the prosecuting and defence counsels had ample time to be personally acquainted with their witnesses. It also allowed them time to go through their testimonies before the appointed judicial panel for this tribunal arrived in chambers, and well before the spokesman delivered his aperture. This speech he reviewed to correct the mistakes. Selected to sit on the panel were three puisne judges and two judicial members to pass judgement on the prisoner, should it be proven he had violated human rights and was a criminal.

Warned of in-depth inquisitions leading up to and including the trial, all concerned were aware they would be under a great strain, but none more so than von Breusch.

The prosecutor would undoubtedly delve into all facets of his private life. Mr Michael Reagent would also peruse the lives of the witnesses who volunteered to testify against von Breusch, prior to and during this tribunal.

Doctor Gwlynne and Senator Bucknell's testimonies were delivered "in camera" much to their relief. The origin and names of both men were withheld, in fear of a rising Nazi splinter group's reprisal in Austria. Death threats were always an imminent risk in trials where interviews were conducted under strict security.

To compound an already difficult situation, news had filtered through that the criminal in custody would possibly be charged with only six of an endless string of indictable war crimes. Personal depositions of sworn witnesses, apart from signed affidavits accompanied by private letters, would be produced as evidence. Lacking were some official Reich records. They were unattainable due to the annihilation of the Chancellery building in Berlin towards the latter part of World War Two. Other methods had proved successful in obtaining this ex-Nazi's war records and his personal war history. Data relating to both issues came from several valid sources plus substantiated evidence. This included his personal and Reich diaries and official Reich documents, confiscated prior to his escape from Germany in 1944. An array of passports and other documents were also salvaged from his valise. This tan briefcase was von Breusch's constant companion until his capture at Christmas in the winter of 1972. Its data revealed secrets that surmounted the prosecution's highest expectations. All items proved damaging enough to bring this criminal to trial.

Legal justices chosen by the UN Security Council to represent this tribunal were: Mr Justice Michael Reagent, a British judge appointed as the court prosecutor.

President: Presiding Justice of the Court, the Honourable Mr Justice Edward Strezlewski, an American. His selection as Senior Justice of the panel was as spokesman for the puisne judges.

Accompanying him on the panel was a Queen's Counsel, a High Court Judge, Mr Justice Robert Spiller his right-hand man, who took his role of deputy prosecutor seriously.

A third judge of French origin and duly qualified was Monsieur Justice Hérbert Pautenel.

The fourth judge, requested to sit as an unbiased observer was a Russian. Equally qualified as his colleagues, Leonid Pyotr was appointed to the judiciary panel for his fair mindedness and forthright attitude.

The fifth man was Justice Brage Ingivar, Swedish by birth. This independent adjudicator, specially chosen for his unbiased thinking, was delegated to the panel for a specific reason. His Honour stood as an independent adjudicator, should one be required to break the bar in the event of a locked, or equally divided verdict being delivered in Linz, Austria.

Due to the delicate and harsh nature of this trial, no public jury was selected to either attend court or to pass judgement on the prisoner, Rolf von Breusch.

The defence lawyer Terrill Scharles, an Austrian-German, was known to Rolf. This man with whom he had previous dealings in Vienna offered to defend him for a nominal fee.

The prosecution's team were in the process of tackling budgets of legal data their overseas agents had procured. They constantly burrowed through all the information procured, to challenge the nonsensical arguments put forward by the defence counsel at the previous hearing. Once all the prosecution's witnesses had given their testimonies, they would shoot his deliberations down in flames. Mr Scharles assumed his three witnesses would arrive long before the trial was due to begin. Only two had confirmed the arrangements. The third witness was unobtainable and must still respond to his cablegram, if she intended to appear in court.

Commencement of the tribunal at Linz: Court in Session. Friday 19th of April:

How ironic was the date? The eve of this exact date and month von Breusch had found so predominant in his past life, including the eve of Hitler's birthday. An indicted criminal, now it was von Breusch's time to face retribution for crimes he had committed against humanity, in the name of war.

Bitter winds, both of history and of nature's wrath, inflicted their powers upon the immense crowds cloaking the footpath outside and adjacent to the court building. Intense interest was drawing the public's attention to the pending trial of this indicted ex-Nazi officer.

Media hounds wedged themselves in every foothold they could force their equipment and bodies into outside the courthouse. Their crews hung from balustrades, fences, perched on rooftops or sheltered in doorways out of the icy blast descending on the city. Yet each photographer retained a prominent view of external proceedings. Some members of the television fraternity were stationed with their cameramen on stationary vehicles. Others settled on street corners, balconies or atop steps in nearby buildings. Some gawked out windows or other positions of advantage to get the best scoop of each session.

News posters and banners fluttered in the breeze. Street placards littered every vacant spot on cobbled pavements. They brought to notice the trial of this once prominent German officer, who at this time was perceived to be a war criminal. The judicial panel's main job in court was to prove the accused man guilty or innocent.

The panel of selective judges comprised four main world powers, plus a neutral adjudicator to preside over this unusual case. At present the justices sat in their respective chambers discussing their responsibilities during the impending trial. They were endeavouring to clarify some nondescript and illegible items of evidence with their teams. The panel must be impartial to all the unproven facts of von Breusch's affiliation with the Gestapo's criminal elements in Berlin. If not handled correctly, the recorded history of him signing death warrants at the Chancellery could still incite hatred towards his countrymen.

Once all preliminaries were finalised, a cortege of wigged gentlemen entered the courtroom. In an orderly file they proceeded to their seats at the bar.

'Oyez,' the clerk of the court called three times. 'All rise for their Honourable Justices.'

The Senior Justice and spokesman, Mr Strezlewski acknowledged the prosecutor and defence lawyers who also nodded courteously. The legal fraternity in turn reciprocated. Then, with the panel seated, Strezlewski took his position at the mahogany bench. His judicial colleagues followed suit, two sitting either side of him. Immediately the inaugural proceedings were over when a disturbance in the public galley occurred. This forced the spokesman to call the court to order.

Addressing the prisoner, who'd just been sworn in, Justice Strezlewski spoke. 'Herr von Breusch, you have been served with a copy of the indictment. As you voluntarily rejected the right of appeal at the pre-trial hearing, you are now charged to enter your plea. How do you plead? Guilty or not guilty?'

'Not guilty of all charges, your Honour,' decreed von Breusch, whose acrimonious snigger astounded his counsel. Scharles scowled at his client, as a warning for him to remain silent and not to antagonise the judicial panel.

'Your plea of not guilty is entered in the court records.' The spokesman fumed over the prisoner's uncourteous snigger, as did his colleagues.

Strezlewski then addressed the prosecutor. 'Deliver your opening address, Mr Reagent.'

Solemn and staunch in carriage Michael Reagent, the prestigious British High Court judge, commenced his initial speech. 'May it please Your Honours, I will endeavour to produce a variety of issues concerning this case. It is never easy to put forward evidence, weighted with responsibility of the trust of free peoples of the world. Perhaps it might be prudent to allow the facts to speak for themselves. History of Germany's part in the Second World War is well documented in the annals of time. Let it be noted here we do not seek to crucify the populace; only those who perpetrated crimes against innocent victims during Hitler's terrible regime, and of the atrocities his officers and their men conducted under the Reich banner. The plundering, murders, rapes and criminal acts were instigated by Wehrmacht soldiers in uniform. One ex-army officer my witnesses have identified, by photographs, will be called to testify in due course.'

The constant barrage of muffled twittering made it impossible for Michael to continue. He paused until the court was again called to order. 'It has been proven the terrible acts of inhumanity were, in truth, the degradation of every individual that evil Nazi regime touched. Now it is the duty of this court to hear this trial in fairness. Only by listening to every item of evidence and weighing the inaccuracies as they present themselves, will we arrive at the truth. Decisions that we make must be honest and just to allow only the truth to surface.'

Reagent hesitated to draw breath. While turning over the page of his notes a growling disquiet ascended from the rear of the press gallery.

An echoed sob, half choked by an early to middle-aged reporter in the throes of mental agony rented at his tie in an effort to gasp for air. Sweating profusely, water cascaded down his distorted features in torrents.

'What's wrong? What has upset you to this extent, pal?' his American sidekick whispered, assisting his stressed companion to release the restrictive neckpiece.

Beneath his hand Andros Stroud mouthed, 'My God, the prisoner in the dock ordered my mother and sister's murder at Ravensbrück. My father worked as a printer, a forger. That swine instigated his murder and signed his death warrant during the war,' claimed Stroud who ripped off the offending necktie. 'There, that's better. At least now I can breathe.'

'How do you know it was von Breusch who sanctioned their deaths?' his colleague queried softly, utterly astounded by this revelation.

Out the side of his mouth, the reporter responded low to his senior, 'Shush! You'll get us thrown out of here. I'll tell you everything during a break; once the prosecutor's address speech has finished.'

'All quiet in the court,' the Presiding Justice warned, his voice echoing through the room. 'Silence, or I'll have the press gallery cleared.'

Having said his piece, the spokesman's angular features glowed red and his mouth twisted. The savage glint in his eyes expressed anger. Justice Strezlewski's power was exercised to its limits. His patience had worn thin. This was evident, and both the public and press galleries ceased their low-key rumbling.

'Continue please, Mr Reagent.'

Taking up the address, the public prosecutor proceeded. 'We of this international tribunal acknowledge the presumption of innocence. We also declare guilt will override that innocence and truth will eventuate. Nevertheless, during the course of this trial and in coming weeks we will endeavour to seek to deliver the truth of this prisoner's years in the Reich Chancellery.'

After a sip of water, Reagent resumed speaking. 'We have irreputable evidence pointing to the prisoner's part in the annihilation of not only Jews, but also of innocent people, many of whom were not of the Jewish faith and whom he unjustly accused, to satisfy his own wicked desire to annihilate them. Also, we will endeavour to establish to show how von Breusch set up a colleague to promote his own status in the Weimar. Everyone who crossed the accused man's path, he had charged as alleged criminals against the Reich, that all previous documentation has been firmly established will verify. They were read and recorded in the original court hearings, so shall these declarations be entered and recorded.'

The peak of emotions again reached high expectations. The hushed voices ranged from chosen representatives of the countries concerned, down to invited dignitaries from all walks of life. Several pronounced and stifled cries accompanied by moans came from those moved to tears in the courtroom.

For the second time within minutes its patrons were warned by the voice of authority to desist with their noise. The rumpus then dropped to a deathly hush.

Permitted to resume, Reagent, who was running true to form, never faltered in his address. It related to how all judgements must conform to Geneva Convention guidelines. The court's arduous and harrowing task was to clarify all evidence thus presented, and how the judicial system must see justice was accomplished, even at this late stage.

The prosecutor then continued. 'Therefore, the prosecution will cleave to all documented evidence. Given time we will prove without a doubt, this ex-Reich officer's guilt. It is our solemn duty to bring forth many witnesses and signed depositions including sworn affidavits to enable everyone present to justify this court's dismal task, in order to set the world straight, as to the purpose of this tribunal.'

Reagent again paused to study his notes. 'If we rebuild the confidence of free peoples, their lives should not be ruled by dictators. The masses should never again have to tolerate criminal codes, as they have previously witnessed in past years. Nor will they suffer infamous wrongs instigated by evil and perpetrated by officers under the umbrella of duty. It is within our power to determine that no criminal elements will again set precedence for future regimes to copycat or recommence, either in war or in peacetime. Perhaps then, such outrageous and dastardly dictatorships that bring tyranny to their fellow men will never again be created. Under our statute laws only we members of the judicial system are empowered to prevent cruelty from ever rearing its serpent's head again. Only by honest governments coming down hard on individual terrorists or splinter groups will this ever be accomplished.'

Another short respite allowed the prosecutor time to review his remarks, prior to completing his inaugural speech.

'I conclude my address by stating we have the enormity of making this a fair and just trial in the view of the world, the press and notwithstanding, this courtroom. Should we fail to bring justice in this case, we fail our fellow humans and leave them open to pain and ridicule which can be unjustly and cruelly metered out by anyone in authority. I thank you, Your Honours.' Reagent bowed to the bench and returned to his seat alongside his legal secretary.

At the conclusion of the chief prosecutor's address, the defence counsel took the floor-well. Mr Scharles put forward his statement, which to all accounts lacked substance and sounded spineless. It lacked finesse and expunged the truth in favour of his client. He then stood down. Until the chief judge declared proceedings over, nobody in the room dared move, bar the press contingent.

Under strain and while gathering his notes, Andros Stroud remained silent. Reticent of mood, he waited until the courtroom bustle subsided before speaking to his colleague. 'You wanted to know how I found out about my family. A Reich officer I'd befriended in Austria informed me of my father's plight. Upon my return to Berlin, I discovered Dad hadn't slept in his bed, he was missing.' Stroud hesitated to think.

'After months of delving through masses of papers in his workroom, I found out my father had been arrested by the SS. From memory it must've been around the time my mother and sister were trucked off to a concentration camp. After I escaped from the Gestapo in Berlin, I searched every known place for them. And I never heard from my family again. A neighbour, who witnessed my home being ransacked, told me what happened to my parents.'

'Shush, keep your voice low, in case someone hears you. One never knows what big ears are lurking in this gallery. You should relate what you've told me to the prosecuting chief wig.' With one ear glued to the fading rumpus and the other tuned to his sidekick, the senior newsman nodded for them to leave the press gallery.

'It appears the officer who spoke to me about my father's death, once worked in the Chancellery. Over a few beers he opened up about finding something which related to a certain captain's supposed death. Much to his detriment, I discovered later who he meant. That swine in the dock was the officer he had referred to. I'm positive von Breusch was the instigator of my entire family's demise in two separate concentration camps.'

'How?' queried his colleague.

Keeping their voices to a minimum, Stroud clarified what he'd said. 'Some time after my father disappeared, I disposed of the clutter in his workroom. It astounded me to find a snippet in a book, which the Gestapo thugs had missed. The scrap of paper with a Reich letterhead listed several items required by a Reich captain, or kapitan as the Germans say. The strange handwriting was impressive and clear. The name written beneath my father's was in a precise script. For an illiterate man his handwriting was distinctive, so I couldn't mistake it. An interrogator from the Chancellery wanted documents forged. But I never knew why, or who had signed it, but I'm in no doubt now. I still carry that scrap as a memento. It's in my camera bag here beside my knee.'

'We may find out the reason during the trial. Hey buddy,' the Yank pulled on Stroud's shirt sleeve, 'as I just said, you should make those facts known to the prosecutor. You do know that our chief editor will release you from your contract, especially if it's due to extenuating circumstances.'

'About the facts and those forged documents, I suppose it'd be advisable to tell the prosecutor. I'll catch him now, before he leaves the building,' confirmed Stroud in a low voice and he clammed up as the spokesman nodded towards the dock. 'Forget me pulling out though. Not on your life,' Andros Stroud scowled. 'For me to turn tail and run would be tantamount to being a coward. I've never shown a yellow streak, and I don't intend to start at this late stage.'

This Associated Press delegate's emotions tinkered on the boil. Anger stirring through his veins caused his heart to gallop. How he managed to remain mute and not scream he couldn't imagine. It wasn't easy to gag a rebellious tongue, especially with vital news to relay.

Others whimpered with emotion after casting a glance at the dock. In the course of the prosecutor's recent address and while discussing his father, Stroud noticed an acrimonious smell of evil pervading the courtroom. Influenced by a rancid smell of sweating bodies, he felt nauseated. Mental stress was rife and to the extreme.

Upon the panel completing their brief assessments, Justice Strezlewski arose from his chair. 'Court is now adjourned,' he declared nodding to his colleagues.

Nobody stirred until the panel deserted their posts. Then polite mayhem took hold to swell the noise of anguished departing observers.

The judicial panel collected their briefs, discussed something in low tones for a minute then retreated back to chambers.

So ended the first day.

A grave quietness descended on the pathetic faces in the process of exiting this courtroom. Every person was cocooned in misery; their expressions astounded Stroud. His eyes settled on the handcuffed prisoner, now in the process of being led away.

'Did you see the smug expression on that Nazi swine's face?' Stroud felt like smashing it, as a heart-wrenching array of feelings stirred bitter memories to life. 'Hot butter wouldn't melt on his moosh, or blemish that bastard's self-righteous smugness. The snotty look he gave our press contingent was undeserved, I think.'

'It's just as well I'm writing this one up. You'd be a biased journo,' affirmed the AP's senior reporter. 'One of us better get out there, or the stringers will get their scoop in first. See the prosecutor while I'm talking to base. We must be hot off the press with this scoop.' Not waiting for a reply, the Yank rushed off in pursuit of a row of press phones.

Stroud's short discussion with Reagent was extremely enlightening. The reporter was requested to follow Michael to his allotted chamber. With members of his counsel present, Justice Reagent's stenographer took down Stroud's entire statement. In detail it fitted in with data already on file regarding von Breusch's past. It also tied the jigsaw together with scraps of written material secured from his army satchel.

Satisfied and elated, Mr Michael Reagent embraced this witness's declaration with gratitude. Accompanied by his legal secretary, he departed for their hotel. On the way down the courthouse step he stood for a second and smiled at the press delegates. Michael's eyes focused only on Stroud's contingent.

After reviewing the reporter's signed testimony, the prosecutor held it over until a more favourable time. Other information received came from a valuable foreign security source. It backed Stroud's report of his father's work and portrayed a huge amount of data not on the procured Germanic official listed records.

30

After a brief recess, the prosecutor recalled Rolf von Breusch to the stand. Reagent immediately moved over the floor-well to cross-question him.

'Remember you're under oath, therefore I charge you Herr von Breusch to answer honestly. In 1943 you worked as an interpreter-interrogator in the Reich Chancellery, did you not?'

'That's true. I have nothing to hide on that score. From the early 1930s until 1944 I served the Führer and my Fatherland honestly and dutifully as an officer, firstly in the Weimar and later in the Wehrmacht.'

'I believe part of your duties during the mid-1940s was to question Reich prisoners then pass sentence on those unfortunates,' demanded Reagent.

'Yes, that was only a small part of my responsibilities. My job entailed dealing with saboteurs; traitors who tried to destroy my country's facilities. All manner of situations arose after the traitors were captured. I was responsible for ridding our populace of scoundrels whose ideals conflicted with the good of our German people. It wasn't an easy task. Most were liars, thieves or murderers that belonged in cesspools. Those gutter-rats had nothing to lose by misleading the authorities and by destroying their trust invested in my judgement.'

'So, on a whim these said traitors and other prisoners were all Jews that you persecuted?'

'*Nein*, not entirely Mr Prosecutor,' answered Rolf whose eyes remained transfixed on his defence lawyer. 'If Jewish scum or other rebels damaged Reich property they were rounded up by the Gestapo, who occasionally referred them to me in my office. The same applied if they murdered German soldiers. It didn't matter what religion they were. They did wrong, and they were punished by whatever means our superior officers deemed necessary. All officers were entrusted to make hard decisions and to stand firm in their decisions.'

'You admit you held high-ranking authority.' The answer Reagent sweated on would most probably be another lie.

'Only inasmuch as to weed out criminal elements and to declare them guilty, if need be. If proven guilty that person was delivered to the SS or Gestapo for further assessment. They dealt with that prisoner from then on. NOT me or the officers under my control. The men under my command were loyal officers. They, like myself, only followed orders. As soldiers we were all responsible to a higher echelon.'

'You talk of orders,' Reagent roared, 'orders that you delighted in delivering. Orders to kill! Orders to maim people and leave them half dead strung up on wire structures in the Chancellery courtyard. A great number of prisoners were shot by your firing squad, while you gloated over their bodies.'

Rolf's counsel objected, nevertheless the panel quashed it.

'Never! I never gloated over anyone's misfortune. I had a job to do which I did to the best of my ability. You must remember, "me Lud",' Rolf's rancorous tone stank of bitterness, 'many things occurred in war time that my men had to cope with under duress. As to your ridiculous assumption that I prized myself by gratifying the prisoners in our care misery, I find an atrocious notion,' he objected in a rambunctious tone. 'It's totally absurd. That is nonsense, hearsay on your part. I never knew what happened to the prisoners after they left my office.'

'Oh no, Herr von Breusch. These documents in my hand prove that was precisely your intention on more than one account.'

'Mr Reagent, I signed documents suggesting the prisoner in question should be shot once sabotage was proven. You tell me, would the Allies not have the same rulings on traitors and saboteurs? Of course, your military would,' Rolf scoffed, not flinching an eye.

'Never would an Allied officer condemn a prisoner, not without a fair and just trial,' countermanded the prosecutor. 'No Allied personnel would sentence an innocent person to death without proven evidence shown. It is you whom we wish to set the record straight on. Not the Allied army or other forces. Situations place you, von Breusch, at the centre of those atrocities. That I can prove. Just for interest sake sir, what floor was your office on at the Chancellery in 1944?'

'Second, why? I've told you under oath that I was not guilty of any atrocities. My duty was signing death warrants for prisoners to be taken into charge by the Gestapo. They had the right to see criminals were dealt with, in the way their superiors knew well. Not me.'

'Oh, come now!' Reagent scoffed. 'Do you expect this court to believe you were a completely innocent party to all those murders? I will submit in evidence photographs of mutilated bodies and murdered victims hanging in the Chancellery compound. Photographs of innocent victims you, between the mid-1940s until just prior to the war ending, sent to concentration camps.'

'I'll tell for the last time. I had nothing to do with any killings. And, I certainly never sent one criminal I interrogated to a concentration camp. Whoever supplied that rubbish, wouldn't know what happened to any prisoner who left my office, restricted and under guard.'

'I contend you sir, are a born liar.' With this stormy accusation from the prosecutor, court whispering was silenced. Mutterings ceased amid anguished cries for peace to reign.

Defence Counsel Scharles, desirous of rebuking prosecution's defamatory remarks, declined to comment. One snide glance from the Panel cautioned him to remain silent.

Justice Reagent rebuked von Breusch's last comment by declaring in a reserved tone, 'The evidence in my hand indicates that you personally were responsible for an unaccountable number of captives being put to death either by hanging, or being shot by an SD firing squad. There's a third group I can name, the Schutzstaffel or SS firing squad.'

The chief prosecutor's gaze defected towards the judicial panel; then his eyes surveyed the item in his hand. 'I declare as evidence this distinct black and white photograph, taken in 1944 from an office window overlooking a courtyard. It is witness to eleven lifeless bodies strung up by piano wire on bowers. The other two snaps show victims dying, after being tied to posts by your guards and then shot. They all support the theory of your cruelty in Berlin's Chancellery.' Reagent's nodding gesture angled at the prisoner in the dock.

Voices rippled in disgust. They mingled with pitiful cries coming from the public gallery as well as the upstairs press gallery. Even the thought of human beings strung up, as described was repulsive. Women gasped in horror and it bought official visitors to tears.

'What photos?' barked von Breusch. 'I don't recall any photos taken from my office. I would never have condoned that idiotic idea.'

Reagent handed the three small snaps to a court clerk, who passed them to the spokesman for his colleagues' perusal.

'There were no photographs taken of the Chancellery courtyard, nor its compounds. Unless the Führer requisitioned photographs, none were taken. Of that I can be certain. Nor did anyone take snaps of me, in or out of my uniform.'

'What photographs indeed, you might well ask von Breusch?' Reagent smirked. 'Strange, I could swear I never mentioned where those exhibits were actually taken in Berlin.'

An enlarged ten-by-twelve black and white print was held aloft in front of the prisoner. 'The photo I just showed you clearly depicts a blond Reich officer in uniform standing by an open window. A note scribbled on the back of the original states this photograph was taken from an office opposite your window in the Chancellery building. Must I paint you a picture depicting who that officer without a cap on was? This enlarged photo is the clone of a younger you, von Breusch.'

This indeed came as a shock to Rolf, who silently mused: *I never dreamed there was ever a likeness of myself in existence. It must've been taken from the opposite side of that Reich quadrangle. Why wasn't I alerted at the time that someone had a camera? I would've confiscated it and had the traitor shot.*

Justice Strezlewski demanded the constant low mumbling in the courtroom to instantly cease, or he'd have the public arena cleared.

'Your Honours, I object. The chief prosecutor is trying to badger my client into submission. This is not the correct procedure to arrive at the truth, in my learned estimation.'

'Request denied. Be seated Mr Scharles.'

Justice Strezlewski then addressed the prosecutor. 'Mr Reagent, we Justices of the Panel wish to re-evaluate that last photograph you've withheld, so it may be verified. If you will, pass it to the court clerk.'

The spokesman ordered him to collect the photo plus all those in the prosecutor's hands. 'This enlargement the panel have just viewed indeed verifies the truth. Photographs do not lie.' Reaffirming their decision, he passed the print back. 'You may continue, Mr Reagent.'

'To every free-thinking man, it is unbelievable to imagine men and women being shot for small misdemeanours. It's abhorrent to picture their life's blood ebbing into fresh snow. The photographs I now hold are listed thus.' Reagent named the numbered items listed in evidence. 'They came to my hands courtesy of a member of the Urban Underground; a man who is unavailable at this time to attend court, although he is now living in France, Your Honours.'

Striking the dock with his fist, Rolf angrily retorted, 'Huh! Any idiot can rig photos. I suggest they're all fakes. Doctored negatives are easy to do. I've done a few myself over time.'

'We of the prosecution have a signed affidavit and a deposition stating the date, place and the photographer's surname, a man who was a guest of a major. An officer I believe you detested, von Breusch. It even indicates the office where these photographs were taken. There's no mistaking this evidence is genuine and authentic.'

'You lot are setting me up,' von Breusch bellowed, again pounding the wooden rail with both fists. 'I confess to nothing. You lot are ganging up against me. Lies expelled from liars' lips. I'll have none of it, you hear, none of it.' Rolf clammed shut when a spokesman glowered from the bench. It cautioned him to be silent.

'I suggest you modify your voice and keep your temper under control, or I will place you in contempt of court and have you thus charged, Herr von Breusch. Then you will be forcefully removed from this courtroom.'

Squinting, Rolf appraised his accuser before attempting to answer the prosecutor. 'I took no part in the atrocities you have accused me of doing. I am innocent of all indictable charges against me. You're trying to crucify me, an innocent Reich officer who …'

Justice Strezlewski again interjected. 'Herr von Breusch, curb your tongue. Or we, the Judiciary will have you re-shackled and led from this room. The panel will not warn you again.' The spokesman had stomached enough insults levelled at his colleagues during this tribunal from this belligerent criminal. His anger, though controlled, deepened the scarlet flush of his cheeks. It complimented the red-lined cowl of his collared robe. 'One more outburst and you'll be charged with inciting a disturbance and taken down to your cell. There you will be barred from hearing all further evidence. Prisoner, answer with civility.'

Given verbal permission to proceed, Justice Reagent continued. 'In other words, you never sent an Irish priest to the SD firing squad, indirectly under your control in 1944. Nor were you responsible for numerous railway gangers being whipped, tortured and then shot dead, under your orders.' Michael Reagent awaited a civil reply while strumming one finger on his notepad.

'*Nein*! You're trying to trap me into making a confession. What is it you want of me? I've told you all I know of those terrible times,' retorted Rolf whose frowning glare warned his counsel not to interject or speak.

'So, you do admit they were terrible years! I suggest they were not for you, with your lifestyle and freedom to enjoy lavish meals. You and your Reich colleagues were thugs who dined on Russian caviar and swilled on wine, while your countrymen, their womenfolk and children were starving and beaten to death in their homes or on the street. Innocent people of many nations were maimed, starved then shot and left hanging on wires in the Chancellery confines.'

'Objection, Your Honours,' Scharles again interjected. 'The prosecutor is trying to trap my client into confessing to lies. Everything Reagent has related is mere superstition on his part.'

'Mr Scharles, I have had cause to warn you twice today. Rephrase your question, or move on.'

'Let my accusations and questions speak for themselves,' the defence counsel confirmed and sat down to review his notes.

This left Reagent free to speak. 'At a more convenient time, Your Honours, I will resume my inquisitions of this prisoner in the dock.' After exhaustively championing his argument, blustery at times, the prosecutor reserved his judgement. With no more pressing questions to put to von Breusch, Reagent asked to be excused. As the hour was late, both he and the panel closed their portfolios, nodded to their colleagues in court and retired for the day.

Back in chambers, one of the tribunal judges feeling uncomfortably hot, removed his alabaster goat-haired peruke. Placing the wig on its stand he dropped his briefcase on the table and ruffled his hair to let it breathe. Then he decreed in a low monotone, 'It looks like a long stint this time.' While stripping off his outer garment its red cowl caught on something. Justice Spiller freed the garment and dropped it on a chair. Awaiting a reply, he unclipped his white cravat and removed the black robe worn underneath. When comfortably attired, he added, 'Won't I be glad to be back home in England.'

'*Oui,* I agree. Still, justice must in due course be served however long it takes. I need to phone my wife in Paris. It is forbidden until this trial is over.' Hérbert Pautenel shrugged one shoulder. This prestigious case was far too important for him to break an oath given under tribunal ruling.

'Discussing the case already, are we?' the Senior Puisne queried removing his cloaks of justice. 'Proves one thing, an honest decision will be the right decision. And we, the judiciary must access all the facts before reaching any conclusion,' Strezlewski sniggered, 'and in my opinion, gentlemen, it won't be cut and dried.'

'*Mon die*,' decreed his French colleague. 'I do not think it will be pleasant listening. We all understood that would be so, before we were delegated and became committed to this trial.'

Not wishing to breech his oath, the French judge changed the subject. 'I am going for a drink at our judicial Club Bar uptown. Do you wish to come, Edward?' queried Hérbert, while hanging up his robes in the locker of their chamber. His stilted English sounded a little too exact for a "Frenchy".

'No, I'm off to review my briefs.' Collecting his legal data, Edward Strezlewski added, 'I must get some rest; or I'll be nonplussed in the morning. We'll be bracing the bench first thing, so I'd rather retire early. Suppose you guys won't want a late night either. Can't have the media spoofs glaring at one, should one yawn. That would fire their greedy, rabble rags. Besides, it'll be a drag tomorrow,' drawled the Yank, while endeavouring to cover his mouth. This feat was almost impossible to accomplish with an armful of briefs ready to stash into his briefcase. Overstuffed it fell on the floor. 'Blast the damn thing. Now I'll have to re-collate all the pages of those documents. They'll need browsing through again tonight to make sure they're in the correct order. What a job,' he growled.

'How about you Robert,' the quiet man asked, removing his judicial robes. 'Are you coming with Edward and me for a quiet drink? Our Russian counterpart, Pyotr has gone on ahead with his Swedish companion ...'

'Hardly Pautenel.' Justice Spiller feeling overtired and crotchety interjected his French counterpart. 'Shopping beckons. I have accidentally left most of my underwear back home in Norfolk. A man can't run around with a bare arse,' he grimaced at the Frenchman, 'unlike a certain acquaintance of ours. That would cause a sensation and create an embarrassing headline. "Leading Judge bared all before the Bar, thus he gave the world an eyeful of his precious jewels." Can't you see media foxes scavenging for that spectacular mainliner? I can.'

'Leaving oneself open to litigation, is not my idea of a joke. Personally, I did not enjoy your *bon mot* Lord Spiller,' growled the French advocate. Quietly excusing himself, he left his legal companions to battle on regardless in this, their only free strike at liberty for some while.

'Huh! He's got the huffs. Let him go, Ed. I was only making light of a difficult situation. Some Frenchies can't enjoy a joke. I was serious; I did leave them at home.'

'What, your precious jewels or your undies?' The spokesman gave a mannish chuckle. Then in a harsh tone Strezlewski stated, 'Might I suggest Spiller, during this case we must all work together, or we could stuff everything up and make the wrong decisions. Settle your arguments before we breast the bar tomorrow morning is my advice, Robert. I don't care to know what caused them.'

Strezlewski knew their confrontation was over his nomination as spokesman, including that of senior and presiding judge. Not wishing to be further embroiled in their touchy little spats, he left his remaining colleague to mull over his advice.

Settled in his hotel room Michael Reagent, the senior prosecutor, stood passively by the double windows. Taking in the early night scene of what appeared to be ants bustling far below, he lit a cigarette, took a few puffs and stubbed the butt in an ashtray. He then perched his bottom on the cold brick windowsill.

People hurrying home in the twilight hours gave neither this tribunal nor its delegates a thought. A large extent of daily travellers took the court proceedings in their stride.

Giving latent thought to his opening address, Reagent found himself floundering with his delivery tomorrow. Today had proceeded well and according to plan. It was often the situation in a rare and a most unusual case, rather than the norm. *Shit, I meant to phone my wife's solicitor earlier. Barbara might not be home now. She often visits my wife and sometimes they enjoy having supper together in London.*

Somehow his brief lacked decisive substance. Having constantly gone over his notes, Michael discovered the reason. 'Where the hell could my original draft be?' He queried his stupidity while stripping the room bare to find his folder.

Overtired and feeling frustrated, he cast his lot by rummaging through his briefcase for the umpteenth time. Sure enough, he found the document caught in between an English newspaper, *The Daily Mail,* he'd purchased from a local stand on his way back to the hotel and his notebook.

Unable to concentrate and constantly cursing his foolishness, Reagent cast the notes aside. Taking a hot shower followed by a cold one, a habit of old when things hit a blank wall, he could relax an overtaxed and exhausted mind. To make matters worse, he reflected on a miserable marriage, separation and pending divorce, still unknown to his legal colleagues.

His private life, such as it was progressing, would not impinge his right as a talented and fair-minded adjudicator. Reagent had been denied the privilege of winning an unproven case only once earlier in his career; no more or less than most legal beavers in the initial years of their chosen profession.

This important case meant life or death to his career. With fifteen years life under his judicial belt, this chief prosecutor needed to present the panache of an honest and clear-minded adjudicator to win this case and secure a viable future.

All the details of his professional duties in court during this trial were unknown to his wife, with the exception of what everyone now knew, through a bevy of nosey stingers from their International Press contingents.

To his way of thinking, his estranged wife Jenny had no legal justification for abducting their son. She knew Michael would never risk his professional integrity, where a special case was concerned, by fighting a custody battle. And to drag her through the judicial system at this time was unthinkable.

Instead, he sensibly impeded that hearing until the panel of judges had drawn their final conclusions on the impeached war criminal, and delivered their verdict.

Reagent was specifically chosen by the UN delegation to defend the victims of this ex-Nazi's atrocities. not only for his skill in solving awkward and difficult cases as they were presented; the main reason put forward was his integrity in delving into the hidden conflicts of von Breusch's military mind, including his personal history, which were

Reagent's prime targets. He vowed to seek the truth, even if it failed to eventuate, due to a lack of official documents and records, most of which were destroyed in the Allied bombing raids of the Berlin Chancellery.

Michael had sworn to uphold Austria's Criminal Law Act and the Tribunal's Code of Ethics. He and the judiciary were vested in keeping oaths of secrecy regarding top secret documentation. His mouth, similar to his ethics, were bound by official tape.

Sane logic didn't alleviate the personal trauma building in his frustrated mind. Lateral thinking on his part didn't help, or enhance the bleak mood tormenting his usually positive attitude and stressed brainpower.

'A hot meal's out of the question,' he retorted lighting another cigarillo. 'I've nibbled all afternoon, so I shan't starve from lack of one dinner,' Michael scowled, somewhat mesmerised by the erratic traffic ploughing through wet stormy weather. The roads clogged with stalled vehicles and half-submerged hay wagons up to their axles in muddy debris looked pathetic. Feeling isolated without anyone to console him, the chief prosecutor felt desolate and miserable.

In retrospect Michael wished he'd accepted his legal secretary's challenge for dinner. 'If I don't shelve my thoughts into their night closet, talking to myself could become a habit. Well, I suppose it dusts the cobwebs to hell.'

Shunting his personal frustrations into the lost pocket of his mind, Reagent finished a drag on his smoke and stubbed the butt in an ashtray. Because his skin was oozing buckets of sweat, he headed for another shower, then bed. Sleep usually came immediately his head hit the pillow. A sledgehammer might be needed to let him settle into his usual sleeping routine, that now seemed impossible. His over-anxious and alert mind refused to close down.

An hour of revising his drafts made him determined to procure enough conclusive evidence to gain his teams' goal to see justice would be done. After another cold shower, and a mug of warm milk, refreshed he endeavoured to cope with solving some of the problems facing him in court tomorrow.

Tonight however, sleep in its entirety proved him a liar. Lack of it made him subject to his own miseries. Ever-present were his matrimonial worries and the loss of his child to a mischievous and doting mother. Eventually he decided to dismiss them in favour of going through his briefs again.

Michael lit a cigarette, did the drawback and lolling against his pillow, he rescued a huge file of discarded drafts from the bin. They allowed him to reconstruct and alter a few minor improprieties that were inadvertently listed.

Re-penning the corrections he noticed what his drafts lacked. 'Ah, that should do it.' Michael again scanned the brief to ascertain if it sounded correct. 'What I needed was a positive outlook to ingest the true facts.' Reagent rambled on, 'The rest remains to be seen, by the response I receive from my scheduled witnesses. I can't rely on truthful responses from von Breusch, nor his defence counsel. Scharles is a witless idiot who will not confront me, even after I put a couple of irregular suppositions to him. I doubt if he is a qualified lawyer. They argue over the slightest provocation and end up disputing them. Hell to the lot of them and their pettiness and unsubstantiated quarrels that create unrest in court. I've drawn a list of my clients who will be the first to take the stand. That's apart from what defence council might throw my way.'

Desperate, Reagent sipped a dram of whisky and forced himself to think sanely. 'A man needs the peace of Morpheus, to be fresh enough to answer awkward inquisitions first thing,' he retorted, selecting a light blue shirt and a paisley tie to match his steel-grey worsted suit for tomorrow.

2 am: Sleep still eluded him. With both legs straddling a chair Michael rested his head on folded arms on its top rung. Peering through his window in the pitch of night he solemnly observed a set of traffic lights at the crossways change. The sudden screeching of brakes along its main thoroughfare anchored to a halt.

Finally, as a glow flickered to life across the fading blackness the sky opened and released a torrent. Having resolved every problem for tomorrow's opening, he smiled contentedly.

Peering through the wet window, Reagent watched an endless red-eyed snake far below slowing down to a diminishing crawl. Shadowed figures clinging to umbrellas slithered on greasy pavements, while trying to protect their heads and shoulders from tremendous winds. Then as the wind eased, people strived to keep upright on icy walkways. The gliding red-eyed monster took along with it his wretchedness. This scene of utter confusion allowed Michael to recoup some peace to accept the waning night's call. At three, his mind finally closed down.

31

Monday 5 am: A listless hand switched off the alarm clock, Michael cursed its loud bellow which had awakened him. Naked and shivering he grabbed a robe and scurried to the shower; then shaved and dressed in the smart attire he'd laid on a chair in the bleak dawn hours.

Packing his briefcase with the data required for today, he headed downstairs for breakfast, contented in the knowledge his mind was alert and receptive enough to face all challenges and proceedings which confronted him. How he managed such an amazing feat after a restless night, this judge of criminal law daren't think. But then that was part of being in this game. Nonetheless, he intended to harvest the truth with a righteous scythe doused in victory.

9 am: Running ten minutes late Reagent crossed the courthouse top step. The quintessence of this sanctuary allowed him peace to read through his prepared notes and to memorise or minimise them, if necessary. Sam, his clerk would be in a little later, well before the trial was called to session.

Dropping his briefcase on a chair in the anti-chamber provided for council, he selected his most important papers. Once they were correctly collated, he headed for courtroom four. Reagent was prepared to face the challenge of his life, with the trial beginning its second round. Mentally running through his notations, he pondered on what today's proceedings might bring. What fretting sacrilege and petty innuendos would be put forward pertaining to the life of this ex-Nazi captain? On principle he refused to acknowledge von Breusch's wartime rank in the Germanic term of kapitan.

However, he must positively reconstruct each situation to reach the real truth behind this officer's past in the Berlin Chancellery. Outside the courtroom he paused to reflect if he had forgotten something. Out of the shadowed hall a feminine hand rested on his shoulder. Quiet and alone she whispered, 'Guess who had a terrific dinner last evening? And without you, Michael dear.'

'Oh Sam,' his indolent answer sounded harsh, 'you startled me.' His chief secretary had intruded on his private thoughts and Reagent, rehearsing his brief, misread it. Michael laughed as Samantha tossed her head back, her dark mane flowing free in the breeze of an open window. She looked at him, frowned and grinned.

'Good morning to you, Milord,' she sprouted jocularly.

'I'm glad you enjoyed your meal and a good rest last night. My brain wouldn't relax or hit the sleep button. Listen and I'll read you what I'm going to open with today. "May it please Your Holinesses?" Frowning at the goof he'd remade Michael growled. 'Don't laugh, it's noted correctly, I just read it wrong,' he challenged her smirk with a broad grin.

His faux pas amused her. 'Oh Mike, you're a silly goose. Thinking what we'd discussed about churches, were you? What were you drinking last night? Vodka, or whisky?'

'No, not a dash thing. Well a nip of scotch,' he assured her. 'You know the old judicial idiom, "Sober as a closeted wig".' He exerted a sigh as Sam exuded a charismatic smile. Reagent then added, 'Just now I was in the process of re-evaluating my briefs when you barged into my thoughts. Not that I mind a darling girl impinging on them. Still, a man must shelve all licentious ideas, especially in court.'

'Shush, you'll get us … dare I say it?' Sam grinned and tenderly squeezing his arm, she welcomed him escorting her along the hall in this male dominated fraternity.

Pausing to review all the data in his possession, Mr Justice Reagent spoke softly to his clerk. 'How's this sound Sammy? "The depositions of today's witnesses will be living proof of listed atrocities perpetrated against their innocence. They speak of this Nazi interrogator-cum-interpreter's guilt while working in the defunct Reich power in Berlin". What's your verdict?'

'Well, I did have a hand in typing the first brief after you dictated it to me. That draft sounds reasonable. Personally, I disliked your initial speech. This one,' she waved a sheet of paper in his face, 'you have the original there. Mike, stick to the facts as planned.' Lowering her voice Sam spoke with an enormous grin, 'Take me out for dinner this evening and I won't tell. Then Milord, I'll see you are paid for your trouble in kindness.'

With the same sombre note and straight face, he nodded to a colleague who passed them, then replied, 'Promise, or threat? Sounds terrific, if that's a promise and how could I object Miss Huntley. Now, I suggest we settle down to business.'

She looked around the courtroom which was beginning to fill with patrons. Historians, dignitaries with passes provided were already seated, apart from the entire press contingent with their camera crews already to roll. Some pressmen began shuffling to their allotted enclave.

'We should take our positions Michael. Court will be in session soon, and I for one don't wish to be left standing here. "Lord" Judge, I need you to go over these notations I compiled of the briefs that you gave me to type yesterday.'

'Are those notes urgent and to do with today's session?'

'Not really Michael. You will need to revise them before tomorrow in court.'

'What would I do without you, my faithful beauty? Okay, put the file in my briefcase after this morning's challenge with von Breusch. He's a devious customer. Lies and untruths dribble through his undignified and ugly mouth. I doubt if his brain is ever in gear with his mind.'

Reagent needed to confer with the deputy prosecutor and nodded for his legal secretary to go ahead into court. He needed to compare briefs with his right-hand man, a young advocate whose talent outstripped the district attorney's caustic witticisms.

Miss Huntley moved with dignity through the courtroom and left her legal superiors to sort out their priorities. Then, they too moved in to be seated. This was accomplished just in time. The prisoner stood ready in the dock.

'All rise,' the chief clerk cried as the five judicial men filed in, bowed and were seated.

Once initial formalities were over Mr Justice Reagent acknowledged his colleagues and took the stand. When ready he presented his opening remarks.

'Your Honours, Ladies and Gentlemen of the court, today I have the obligation to bring forth witnesses. Each person will testify about how traumatised he or she had become in Berlin during the war. I am talking of the period between 1942 extending to late 1944. I also intend to revert to data describing the prisoner's acts of aggression prior to the Second World War. Then I'll lead up to the accused man's ghastly misdemeanours during that war. At times you will subjected to gruesome episodes referring to concentration camps. These include Auschwitz and Belsen, including the Ravensbrück camp for women, all of which I will touch on later.

'Before calling my initial witness, I will present professional details of the prisoner's army history. It stretches from his sojourn as a Weimar cadet in the mid to late 1930s as a second lieutenant while stationed at Hamburg, in 1938. At times my assessments will reflect on his private and home life during those years. Then I will move on to 1942, prior to him taking up a higher command of captain, in Berlin. I beg the indulgence and tolerance of my learned colleagues. To discredit Rolf von Breusch's testimony, I will then endeavour to produce enough solid evidence for the judiciary to evaluate the true facts of how that situation evolved in 1944 and how it now stands.'

'Explain to the Bench, Justice Reagent why you intend to delve so far back in time. And is it really necessary to set this panel straight on your delving into those past events?' inquired Justice Strezlewski, the spokesman.

'It most certainly is necessary, Your Honour, to disclose the power this prisoner wielded during those horrific years, while posing under the umbrella of duty. Even as a Reich cadet he caused havoc within the ranks; further to his final day stationed at Hamburg and the tragic circumstances stemming from that specific day. Those events will clearly be magnified in due course. Therefore, I need to lay the groundwork now.'

'Sobeit, continue.'

'I intend to refer to this prisoner under the English term of captain instead of the Germanic kapitan. Captain von Breusch commenced duty at the Chancellery around April in 1942. He held the position of Interrogator-cum-Interpreter until just prior to Christmas in 1944 as the records we have obtained verify. They were confiscated from sources other than from official Chancellery records. Nonetheless, all the data we legally obtained is authentic. The majority of official documents were burned either by Reich officers, or accidentally destroyed in AIF bombing raids over Berlin. The remainder were confiscated by the Russians before bombs annihilated the Chancellery building. All documented evidence is proven to be correct. Not the prosecution's assumptions.'

Rolf uttered, more in a whisper than aloud, 'That bastard and his crew have really delved into my Reich history. My only recourse now is to bluff it out and obstruct their ploy to discredit me.' He concealed these mutterings behind one hand and treated the prosecutor's sombre remarks with contempt. Under his breath Rolf cursed, '*Verdamnt noch-mal die schweinehunde.*'

Reagent, who walked across to the floor-well heard von Breusch's mutterings. And having understood them, he glared. Ignoring the swearing, he waited for the response he expected from Mr Justice Strezlewski. He wasn't left in the wilderness of wondering for an impenetrable time.

'As you stated, those facts are well documented. Where is this all leading, Mr Reagent? Are your deliberations constructive, and relevant to this hearing?'

'It is imperative to clarify the facts to consolidate future questions which I will put forward at a later time. Yes, they most certainly are important, on both accounts, Your Honour.'

'You may proceed,' confirmed Strezlewski, who sanctioned the panel's deliberation.

Reagent related how Rolf von Breusch, as a cadet, had entrapped one of his junior colleagues for self-gain, and how he perpetrated and indirectly caused that cadet's death. Then how he absconded down through Germany and on into Switzerland. Reagent proved this, by disclosing the documents in his possession were signed depositions from official mediums. One came from an ambulance nurse, another from a doctor, both of whom had accompanied the absconding deserter over the Swiss border. He explained how von Breusch then made his own way to safety, without going through "Odessa" like many Nazi officers had done late in 1944 to early 1945.

Another affidavit came from a general's batman who was a retired lieutenant at the time, stating how he helped both Herr von Breusch and a well-noted general escape in that ambulance. Their testimonies proved invaluable for the prosecution. Neither man of high integrity held a grievance against von Breusch. Their affidavits stated that they never approved of the methods he had used regarding his fake suicide, before escaping from Berlin.

The prosecutor concluded his speech by disclosing, 'With Your Honours' permission, I shall confirm what my research lawyer has just discovered in Schaffhausen. It is well documented while posing as a civilian the accused man had stayed there briefly. This document states von Breusch then journeyed to Sevelen, near Liechtenstein. He lived in that town until he remarried. He and his second wife, who must remain nameless for legal reasons, moved abroad. Her name and the country where they have lived for years are fully documented on her verified affidavit. They have one son whose signed deposition is in the judicial hands.'

Reagent paused to acknowledge von Breusch's look of unmistakable horror. This non-verbal response was indeed gratifying to Michael and he revelled in its meaning.

The spokesman's nodding gesture confirmed what the prosecutor was referring to. It related to Doctor Gwlynne's testimony which had previously been heard "in camera".

'Before I bring forth my witnesses to allow the situation to be fully evaluated, I ask the court's indulgence while I read this snippet of new evidence.'

Reagent narrated how he obtained a letter written by a woman from Dresden. It stated that she'd found a letter her husband had received in the latter war years. It came from her brother. Due to an oversight, the letter had gone missing until her family moved house. It disclosed the murdered man, whom everyone presumed to be von Breusch, because of their incredible likeness, was in fact Hans Selig. The frau stated that her brother Hans had resided in Potsdam until he went missing. (This coincided with the time von Breusch had deserted his Berlin post.) Her signed deposition contained a sepia photograph of two men in uniform.

An elderly waiter at a café apparently had taken the snap in Potsdam. It clearly showed von Breusch talking to a man, an actual clone of himself.

'This photograph relates to an innocent man, whom you, von Breusch lulled into believing you were his trusted friend. Subsequently, you premeditatedly murdered him in your home on a freezing winter's evening. You then rigged his murder by making it appear as if you had taken your own life by suiciding.'

'That is a blatant lie. I never murdered Hans. He was alive the last time I spoke to him on the phone.'

'Oh, so you do know to whom I was referring. Enough said,' the prosecutor smirked. His previous question exuded enough data to prove his point.

Defence counsel took this break to object. 'Can't you, Mr Reagent as prosecutor, see the fool you're making of my client and this court? Especially if you think your suppositions, for that's all they are in my estimation, will stand up against the truth. Any person can write and sign letters or produce photographs. They mean nothing. Therefore, I suggest you desist with your unfounded accusations against my client, or irrevocably withdraw your last question …'

'You are covering used ground with your objection of the prosecutor affirmations, Mr Scharles. The Judiciary suggests you be seated and we countermand your argument. It contradicts an earlier statement put forth by you, regarding your current client.'

After hearing four verbals for the prosecution, Justice Strezlewski, noting the time, declared an adjournment due to the lateness of the hour. He confirmed that court would resume precisely at 10 am the following morning.

11.30 am Friday: Several members of the Jewish community from Germany and the Netherlands were called. Similar to their predecessors, each independent testimony was examined in-depth. When cross-examined at length by defence counsel, all persons without question accused von Breusch of atrocities. Crimes pertaining to their next of kin flowed incessantly like the ripples of a river in flood.

Signed affidavits played a huge part of the inquisitions put forward by this prosecuting counsel. They confirmed a string of witnesses' accusations of how this ex-Nazi officer in the dock had mistreated their family members, while empowered to interrogate them.

Reports of atrocities listed on Reagent's agenda flowed rhythmically throughout the day. With a long weekend in sight everyone from the law fraternity down to laymen were grateful to see the closure of today's session. Up until the present time, intrusive inquisitions had culminated in documents related to his cruelty to prisoners in the Berlin Chancellery which had resulted in three oppressive weeks of intimidation and stressful hours of utter hell. Over the ensuing week, four more witnesses would take the stand to have their cases for the prosecution heard. All their acknowledgments led to von Breusch's guilt being established of the indicted war crimes.

32

After a relaxing three-day break in proceedings, Monday morning brought forth reams of confirmed data and selected witnesses all of which flowed in swift succession. Bound with grim facts, proven and documented, all data related to the prisoner's past.

Having dabbled in terrorist stratagems, under the guise of following orders, von Breusch's fear escalated as his victims emerged in rapidity. Their testimonies caused strong overtones which rippled through the courtroom and lacked substance until the lunch recess.

2 pm: Reagent took the prosecution stand and opened by saying, 'With an extreme number of affidavits reviewed by my colleagues on the bench, I think it fitting to produce my first witness. With a firm grounding of her case history, it will become self-explanatory. In the prosecution's interest, and within limitations, all relevant facts must be produced and be shown in their proper perspective.'

Reagent coughed. Excusing himself to his learned colleagues, he proceeded. 'I have brought to the Judicial Panel's notice affidavits from accused victims, others are already documented. The one I now put to you, relates to a prisoner who was entrusted to Captain von Breusch's care in 1943. This woman, like special Reich prisoners, was housed in the cells below the Chancellery in Berlin. This dossier in my possession and her personal statement will verify how people were murdered simply to appease the accused man's lust for revenge.'

'Objection, hearsay evidence is not admissible in court, unless substantiated by positive and proven evidence, my lords ...'

The prosecutor interjected, 'Mr Scharles, the letter from a witness I intend to produce is here in *my* hand. I believe every word she has written. Proof is in the reading as the panel have acknowledged.' Reagent impugned his opponent's comment as bunkum. To him this remark made by Scharles was merely to gain a point. Reagent was furious. He showed his meticulous force of a prosecuting judge, by presenting this well-orchestrated argument.

An inanimate court stirred. The Presiding Justice Mr Strezlewski declared his objection with a raised hand and in firm demand, behoved his authority, to which the panel agreed. This rebuff came as a shock to both Rolf von Breusch and his counsel, who disputed their comments. This limited his chance of appeal to deliver an honest opinion of the next witness whom he intended to cross-question, in due course.

Rolf's brain ascended into overdrive of silent monologue. *How could that be? If the next witness is who I think she might be, I was informed that she was executed along with the last lorry load the Gestapo trucked up to the Rüdersdorf Forest. I was a fool because I should've shot her in the compound below my office window. I'd already signed her death warrant on that day.* Observing the beligerent attitude being non-verbally delivered in his direction from von Breusch, Mr Reagent lunged straight into his next case for the prosecution.

'I now call Mademoiselle Longuine to the stand,' Michael stated noticing the accused man gulp. The court clerk repeated the call loudly. His voice rebounded through the room and corridors beyond.

A slight woman showing her years well beyond the age of forty, shuffled to the witness stand. Short of stature and disabled, she found it difficult to stay on her crippled feet for long periods. The court indulged Mieze by acknowledging her plea to be seated.

After being sworn in, the grey-haired German woman disclosed how she had suffered while in Reich custody. She declared the office where the initial incident took place was on an upper floor of the Chancellery, in August 1943.

'At the time I was a young girl of seventeen. The Gestapo arrested me on my way home from work. My first encounter with that man,' she pointed to the accused prisoner, 'happened the day after the Gestapo arrested me in Den Linden Platz. The following day they drove me to the Chancellery. Their officers threw me in a tiny cell, no bigger than an average toilet. Kapitan von Breusch tried to force me to sign a false confession in his office. When I refused to sign, two big guards dragged me back down to my cell.' Frightened the French woman's angry gaze again pinpointed Rolf von Breusch.

'Mademoiselle Longuine, I have two miniature photographs in my hand. Please name, if you will, who this one resembles?' Reagent held it within her reach.

'Sir, they are all of the man I have already identified. Kapitan von Breusch had me beaten by his men. Using pliers, they loosened my finger and toenails in a cell near mine. That's why I have difficulty in walking. A hot poker pierced my heels and both my feet were blistered. They oozed puss for two weeks, before a doctor dressed my festering wounds. He told the guards to carry me to my cell where I collapsed … because of the pain.'

Cautioned to state only the facts as described in her affidavit, Mieze Longuine was asked to explain what had occurred in von Breusch's office. Reagent knew he would be striking a raw nerve here. Only he had no other option. The response to his last question was vital and as such, it must be recorded as important verbal evidence.

Ignoring the twitter and faint murmurs in the courtroom, he assumed they came from several of her friends. Not from relatives. Mademoiselle Longuine lived alone. All her relatives were deceased, ashes of a crematorium, discarded by the guards in Belsen Concentration Camp. He also noticed how her eyes constantly appraised the rear of the courtroom.

Mieze, who'd heard the slight disturbances, remained silent until Mr Reagent clicked his fingers in front of her. 'In his office I refused to sign a false deposition accusing me of being a traitor, a saboteur. I was innocent of all charges laid against me. He, that man sitting there, physically tried to throttle me with his riding crop. I again refused to sign and said I wasn't guilty of sabotage, or anything he'd accused me of. He knew I was innocent, and he still threatened me with torture. Not by him personally, of course.'

Relaying her testimony wasn't easy and Mieze broke down and cried in the witness stand. Regaining some composure, she bravely battled on. 'In my cell two guards ripped my blouse to shreds. Half-naked they whipped my shoulders and back until I collapsed in a pool of my own blood. My shoulder blades were laid bare and bleeding as the leather whip cut into my flesh, but I still refused to sign. After von Breusch witnessed my torture, he laughed and said he wasn't satisfied with my replies. He then left my cell …'

Agonised by stress she procrastinated. In the process of giving her verbals she held up the proceedings in court. While she drained a glass of water the panel bided their time.

Reagent was so incensed by her verbals of von Breusch laughing over her pain, he almost commented on the situation. He realised this would make matters worse for the Frenchwoman. Instead, he glared at the prisoner, who caught his meaning.

'She's lying,' Rolf decreed of her testimony. 'Can't you lot see that woman's lying?'

'Prisoner desist your ranting. I won't warn you again.' The spokesman's eyes flashed fire and his caution came over loud and clear. 'Remain silent, or you will be removed from this courtroom forthwith. You will have your say under oath tomorrow, or at the court's pleasure.' Having given this demand, he then instructed the chief prosecutor to proceed.

'*Mademoiselle* Longuine, try not to be flustered. Quietly think and state what your mind dictates. We understand this is painful, it cannot be easy for you. Please take your time.'

'Thank you, sir. I feel much calmer now.' Mieze sniffed, wiped her nose and then resumed speaking. 'Later I cried with shame, because one of the officers who had molested me was not much older than myself. You do realise, I was just seventeen. I'd lost my dignity and my virginity in one, unspeakable day.' With her head bowed and in a genuine display of embarrassment, the French woman clutched her skirt tightly around a pair of shaking knees, the force of which whitened her knuckles. The court prosecutor allowed her ample time to become composed.

'Sometime, it may even have been days later, a Gestapo truck full of injured people drove us to the Rüdersdorf Forest. We were forced to strip and throw our clothes in a pile. Immediately guns began firing from the back of another truck. Someone pushed me back onto bodies of children, mothers with babies, old people, the lot. All that saved me were the corpses that had fallen on top of me. Wounded, I lay perfectly still. When able, I clawed my way free of those writhing bodies. My face was soaked in blood,' Mieze sobbed, 'partly theirs and partly mine. About then I must have fainted,' she gasped and swooning a little, Mieze clutched the stand with both hands to regain her balance.

Muffled comments from the gallery disorientated her. Feeling faint and confused she tried to recall what had occurred after escaping death. A male voice calling the court to order, she vaguely heard above a constant rumbling and groans of its elderly patrons.

Justice Strezlewski demanded, on conferring with his colleagues, for a clerk to provide the ailing woman with another glass of water. He then challenged Reagent whether his witness was too distressed to continue.

A slight delay ensued while Michael Reagent conferred with Mademoiselle Longuine who confirmed she was composed enough to finish testifying.

After a sip of refreshingly cool water she proceeded with dignity. 'Yes, I am ready to continue, sir. I will tell you how I survived their beatings on that horrible day.' Mieze raised her closed fist in an acrimonious gesture and glared at von Breusch. 'You thought you'd succeeded in having me murdered. You failed miserably. None of those brutal officers knew I was alive, so I lay perfectly still until the trucks loaded with machine guns pulled away.' Her soft voice oozed vengeance. 'I dodged the Nazi's bullets and I now walk with difficulty. Look how my fingernails have grown. I can hold this pen with some difficulty.' Two fingers reef her blouse free. 'See the scars on my shoulder, Herr von Breusch. I'm alive to testify against you. And as God is my judge, I will today in an honest way.' Tears channelling down her cheeks expressed not only her anguish, but also her triumph. She'd won the battle to survive, not for herself, for all those who had suffered through this ex-Nazi captain's men cruelly whipping her senseless.

Elderly and maimed observers in the courtroom twittered. Neighbour peered at neighbour with looks of anticipation. Hushed by the hand of authority, Mr Justice Strezlewski and his panel colleagues accepted her plea as being borne of agony and torment. They patiently awaited the prosecutor's reply.

Reagent pleaded for the court's indulgence until his witness again became composed.

Sanction given, her spirited drive encouraged Mieze to continue. 'Around dusk I discovered I could move my bleeding shoulders. My arms and hands were soaked in blood, and I then realised I was still alive. After my blood-shot eyes adjusted to the misty twilight, I could focus on my surroundings. In that pit my body had been supported by dying and dead people. Some were still writhing beneath me. From where I had fallen, I managed to fight free of clawing hands. Finally, as night fell, I peered over the rim of that long, smelly grave hiding a hundred despicable secrets.'

Mieze mopped her glistening cheeks free of tears, took a sip of water and then continued. 'As darkness fell, I could just make out by the fire's glow dark shapes of soldiers in mud-soaked boots standing a fair distance away from me. Some were smoking, others laughing. I thought they might be waiting for the next truckload of prisoners to arrive.' Mieze hesitated to use her handkerchief again. 'I remember small fingers touched my knee and I thought they belonged to an infant. Unable to help the whimpering child, I struggled over more bodies until I reached the top of that death pit. It seemed ages before I could move my legs and as I told you, my arms were covered in blood. Not mine …'

For the second time within minutes Rolf von Breusch interjected. 'You are a liar. None of what you've stated occurred. Some of my most trusted men were among the guards.' Although warned, he still lashed out with a rampant tongue. 'Every word she's spoken is gutter rubbish. Later that day, I received a report from the commandant in charge of the soldiers. It stated his men fired at a group of detainees who were escaping. They were not shot and tossed in trenches. The Gestapo guards drove your witness to that forest. Not my men. I suggest you make sure of the facts Mr Prosecutor, before accusing me or my officers of murder. In my opinion, the Frenchwoman who just testified, should be certified. Her abuse of me was unjustified. Her lies are baseless. Fabrications of a feeble mind. None of what she's told this court is factual. And she expects you gullible lot of lawyers to believe her lies.'

Mieze looked at the panel of judges then deflected her gaze to the rancorous scowl on von Breusch's face. Ignoring his recent libellous intrusion, she finished what she intended to say. 'When the soldiers turned their backs, I crawled on bleeding knees to a small mound of earth and lay perfectly still. I remember shuddering as I heard heavy footsteps approaching me. The guard relieved himself and some splashed on my face. It stung my eyes. In fear of him seeing me naked and alive I still didn't move. After he walked back to their truck, I covered my body with dry leaves and grass. My bruised and cut fingers were bleeding, and I stayed there until a dull cloudy day dawned. By then the soldiers had gone. I think I must've fainted again.' Mieze momentarily stopped speaking to pat dry her swollen eyes.

'Twin boys from a nearby farmhouse told me later that they found me next morning, battered, bruised and lying in a pool of blood. My head was almost submerged in water. The youngsters of ten saved me from drowning. They ran home to their parents who nursed me back to health.'

Both the boys, now men in their mid-thirties, corroborated her story. They added, "As boys, we thought the injured girl was dead until her hands flexed in pain. We ran ahead of our father who carried Mieze a hundred yards back to our house."

At the conclusion of his cross-questioning and today's session, defence counsel's negativity was obvious. Scharles had barely interjected until the weeping woman finished speaking. Why, was anyone's guess. And his direction, or lack of same, disgusted his client. An argument ensued between them, until the guards escorted von Breusch down to his previous cell.

Around seven as promised, Michael Reagent prepared to accompany his secretary to dinner. His mood had heightened and in a more positive frame of mind, his attitude was polite and affable. Amid congenial company, he and Samantha enjoyed wandering through the Linz beautiful gardens at twilight. Following a delightful meal, they toured the nightspots until dawn. Feeling the harshness of a cold, winter zephyr swirling around their legs, Michael suggested they should return to their hotel for a nightcap.

Tempted to make a night of it, he thought better of starting a full-on sexual liaison. There would be time for love once the trial concluded. Their passionate desires must be suppressed. It seemed different somehow. Now was not the time, with a divorce hanging over his head. Instead, he downed two more whiskies with Samantha, kissed the new partner in his miserable, though improving life and said goodnight.

The crux of his dilemma lay in having his trustworthy secretary being cited as a co-respondent in an ugly divorce. Michael feared Sam being labelled as "the other woman". What really devastated him was how the slightest hint of scandal would lean heavily on a promising career. A divorce would obliterate all prospects of him seeking prosperous clients in this honourable profession.

In a silent room, his mind regressed to their arrival in Austria, where he and Samantha had succumbed to temptation, enjoying an amorous petting session in Salzburg. The following morning, they travelled to Linz and booked into this hotel. Their enthusiasm for love was put on hold by the pending trial. Instead they joined a conducted tour through the magnificent art gallery Hitler often visited as a youth with its drab wartime façade and exquisite interior.

Drawing on his cigarillo, Michael recalled how Sam had remarked, 'I've never seen such magnificent paintings, some are ancient.' He consequently remembered strolling through and gazing at the exquisite artefacts the building housed. They triggered her imagination and she had inquired if the artists were buried in Linz. They both knew Hitler had grown up in this city, not Linz in Germany. Later as Chancellor, he visited there on numerous occasions.

Monday: Lunching in a secluded restaurant proved an unusual experience with yodellers singing local folk songs. Samantha spent time roaming through the city markets, while Michael purchased new skis, boots and stocks plus two woollen jumpers in a haberdashery shop. Tired and exuberant, the couple sauntered back to enjoy an evening meal in a café adjacent to their hotel.

Tonight, after spending a few hours chatting over their after-dinner cocktails, Michael escorted his secretary back to her room on the seventh floor. There he presented Samantha with a bunch of long-stemmed crimson rosebuds. She challenged him to have a whisky while she changed into a nightgown of cream satin. To appeal to his better graces, she donned a matching robe. He accepted her challenge, by removing his grey leisure coat, loosened his tie and flopped on a divan. With arms intertwined, they sipped the delectable nectar and whispered of their desires for a lucrative future together.

Disguising a yawn behind his left knuckles, he apologised to his hostess. As the clock chimed eleven, he blew her a kiss goodnight. 'I'll be here around seven in the morning with a wad of documents. We can browse through each page together in case I've missed or forgotten something vital for tomorrow's briefing.'

Samantha slammed the door on his heels. 'Oh, that man exasperates me at times.'

Within two minutes of climbing in between clean sheets Michael slept until the alarm clock beeped six. He cursed the constant buzz and knocked it flying with a sharp swing of his nearest arm. A hot shower soon enlivened this sleepy officer of the criminal law court system.

33

Tuesday 7 am: This magnificent day was ideal to pursue the truth from an evil man who had chosen to defy every word his opposing counsel put forward. Not a breath of wind nor cloud blotted a brilliant sun and its warm rays bathed the city pavements in rivers of gold. Its warmth enhanced his frivolous mood. Mr Reagent enjoyed his early morning constitutional down by the riverbank. Strolling in the brisk air stirred to life the man within.

Full of fire and bubbling with enthusiasm Michael hiked up several floors of their hotel to pay his respects to his secretary. Taking care not to crush the bunch of colourful blooms in his hand, he knocked on her door.

'Oh Michael,' Sam responded as a bleary set of eyed tried to focus on both him and the wall clock, 'what time is it? I must've overslept. Hold on a minute while I finish dressing.' She demurely retied the cord of the silk robe around her waist.

'Shush, it's just on seven. I've come to take a fair damsel to breakfast. But not before I've presented her with these.' From behind his back he produced the bunch of flowers.

'You dear, sweet man,' she sleepily whined and edged back towards the bed, 'these buds are beautiful, and their aroma is delicate. I can't really go out half-naked. Give me a second to shower then dress. Be a honey Mike,' she trilled from the bathroom, 'and put the pink peonies and crimson carnations in a vase. There's one on the small table.'

'Orders! Orders! And I should be giving them, not you. Well, what a turn up for the legal ledger. Don't be long.' Reagent turned while searching for the vase. 'Otherwise I might be tempted to come and wash more than your back.'

'I dare you Mike. Sounds terrific!'

'Don't tempt a man on an empty stomach. This is one man who's likely to take you up on that proposal.' Sticking his head around the bathroom door, the legal beaver caught sight of her slender shadow through the shower's opaque glass. 'Here, I've got the soap. Close your eyes, I'm coming in,' he jested jocularly. He'd stripped naked. The temptation proved too much for Michael on such a terrific morning. Caressing her bare flesh, he felt her body ripple with ecstasy.

Presentably clad and with a full stomach, plus a contented sex drive he strutted peacock-fashion in to the courtroom. Miss Huntley tagged close on his heels.

This morning he thought Sam looked ravishingly sexy, dressed in her alluring white linen blouse and navy, formfitting suit. The skirt was short and trim with a side split of nine centimetres. As she cut in front of him, it parted showing her thighs, which tantalised him. Sam chose her usual seat, while his attention was drawn to her physical attributes, more than to his notations.

After a gentle, though reluctant nudge from him, Sam uncrossed her legs and demurely sat with both knees together. By elegantly swinging her stiletto-heeled feet to one side, it allowed their schedules to be perused for the days hearing, without further sensual distractions.

In the impeached man's cell, a different frame of mind had developed, prior to the awaking of this session. About to partake of breakfast, Rolf von Breusch took one look at the burnt offerings. Disgusted by the smell of uninteresting food, he flung the well-done cinders in the guard's face.

The dreariness of his mood created a bleak ambience within his cell. *How can I face today or a meal, after the conflicting discussion I've just endured with my defence lawyer? It seems the compiled evidence is strongly building against me in that damn court. Negative thoughts and baseless suppositions don't enhance a bleak mood. What have I got to lose by refusing to answer more of that pompous snob, the prosecutor's questions? Nothing! At least the bastards could feed a man decent food. Not burnt scraps unfit for pigs to swill,* von Breusch mentally mused. Fed up to the ears of being disciplined like a naughty child, he continued to scowl and in the blackest mood of moroseness he thumbed through his notes on the bed. The partially-finished memoirs lay underneath his father's gold fob watch, which could be used as a bargaining tool, if his plea of innocence was rejected during this hearing. 'That prick on guard duty outside my cell seems gullible enough to be bribed. He's shown an interest in this valuable oil painting *A Nude in Love* and other things, especially my engraved cigarette lighter, even though it needs refilling and a new wick.'

Rolf von Breusch searched his conniving and intellectual mind to come up with an imaginary wall to block any undesirable questions put to him, during this final day. He surmised in this hearing he'd be subjected to the waffling of supposedly proficient imbeciles. The prosecution would be embroiled in fabricated ideas built of clay pipedreams, ready to crumble. Possibly some lies might be strung together by threads of fragile data to ensnare him or maybe to catch him unguarded with any means possible in officialdom's devious hands.

The desperate hands of time were set to wreak havoc on his future, ticking away like a time bomb, to distort factual events that surrounded his years working in the Reich. 'That idiot prosecutor will endeavour to discredit me in front of his colleagues. Well, he and his cronies won't succeed. I'll make a show of their miserable efforts to disgrace me. Eventually they'll tire of accusing me of false deliberations. If that idiot lawyer of mine doesn't present enough witnesses to clear me of all listed misdemeanours, I'll sack him tonight.'

Consistent with the accused man's devious ploys, he considered this farcical trial wasn't only directed against him personally. It also incited hatred towards the Reich. *The general public's antagonistic bitterness against the Fatherland is due, in my estimation, to the Allies having gained power in that damnable last war. Subterfuge and propaganda will always be the Allies backstop.* Thinking of his past life in Germany, Rolf surmised that in the traumatic weeks ahead the legal bigwigs, with their raked schemes now exposed, were plotting to crucify him.

Tight manacles linking the chains around his ankles avoided the risk of him escaping from either his cell or this courtroom. UN guards followed or escorted him everywhere he went or walked. Rolf was subjected to this indignity, under guard, to work out in a small exercise yard at the rear of his cell in this building. For an hour each morning he marched around the quadrangle, growling about his enforced incarceration. This privilege as such, made him feel like a caged lion denied its last meal.

Bound by silence, he considered rewriting his true memoirs. The revised manuscript would be camouflaged by an inflated satanic ego. Now strategically placed within his cell, the unfinished manuscript lay where the interested guard would find it.

Rolf assumed this particular officer had German lineage. After discussing recent topics with the corporal, he knew this man would accept a bribe. His fascination with the Iron Cross, presented to Rolf's father posthumously, always intrigued him. Each morning he greeted von Breusch with a smile and was outrageously courteous towards Rolf, whom he found exhilarating, even a touch bewildering. With little coercing and a twinge of exploitation concealed under the banner of generosity, he knew this officer could be tempted. If handled with decorum he might offer his documented memoirs to the press, but only when time was irrevocably in his favour. At times he sensed their association was reaching this climax.

By parting with the medal awarded to his father for services rendered to the Reich, he could possibly worm his way into this young man's good graces. Other than his private diary and the engraved gold lighter, a gift from his seductress in Berlin, Rolf only had the valuable oil painting to use as a bargaining tool.

In dramatic succession his ugly moods gathered momentum, until he began to feel the pressure of this trial building to a crescendo. It could in time crush his spirit. His irrationalism stemmed from dwelling on his inevitable demise by shooting. What bewildered him greatly was data the prosecuting team had gathered of his military service in the Chancellery.

'A few miserable scraps of unsubstantiated evidence prove nothing. Any idiot with a smattering of knowledge regarding Germany can fictionalise details of its horrific events. Those smug bastards at court, who call themselves judges with their antiquated and jackal judicial system, are ready to sanctify their glory at my expense. Just to prove to the world's dignitaries how proficient they've become.'

Rolf ceased verbally fantasising when the elderly guard challenged him about why he kept mumbling to himself. He ignored the guard and continued writing his memoirs. *Thank heavens this smug bastard can't read my thoughts. I intend to ignore every accusation levelled at me today by the prosecutor in court.*

Reviewing his notes, and feeling disgruntled, Rolf realised what Reagent had said yesterday didn't conform with his challenging address. *He made me feel uncomfortable in the dock with rows of idiots gawking at me when I tried to assimilate his ridiculous assumptions. My interest lay in the visual, rather than the verbal aspect of his witnesses' testimonies. A glare at the right moment could be enough to throw a feeble mind into turmoil. I'm a professionally trained tactician, and having studied mental seduction in the past, I've twice used this stratagem in court to confuse my opposition and his crony witnesses.*

34

8am. Wednesday: Penning his memoirs at the desk with the early morning light just pervading his cell, Rolf von Breusch, denuded of head hair, rubbed his facial fungi. His features distorted with frustration as he discarded all notations, crumpled the page and tossed it over his shoulder. The scrap landed inside his excrement pan.

After weeks in solitary and with the resumption of his trial a step away, he attended to his personal hygiene. This he accomplished over the small washbasin provided. When finished, he cleaned his teeth. Clad in a suit his counsel had provided, he sat on the bed to review his compiled notes. Rolf preferred to don his Reich uniform while waiting for his sentence to be announced. His Dress Greys should still be in pristine condition. He'd left his new uniform in Berlin prior to escaping down through Germany then to Switzerland. His regular uniform Hans Selig had worn on the night he'd visited von Breusch, who murdered him.

Distracted and with his mind sheathed in worry, Rolf shivered in his freezing cell. In a vain effort he tried to consolidate his thoughts for today's proceedings. 'No way will I let those self-opinionated pricks strip me of professional dignity asunder, which I nurtured long before I began work in the Chancellery and silently harboured all these years. It's the reason why I spent hours locked in my miserably damp den in Australia creating my memoirs. Valuable as this document is to me, I'm not prepared to let the bastards destroy it. Nor will they hang me because of the false charges filed against me. Like all officers in the Wehrmacht I followed orders and like my Chancellery colleagues I was obligated to carry them through without question.'

'Taking to yourself again aye von Breusch. Stand back against the wall and don't move.'

'I'm not ready for you two idiots to drag me upstairs in leg-irons again. I need a piss. Turn away now, or I'll refuse to comply with your blasted orders.'

The guard unlocking the cell door ignored his sarcasm. 'Piss in the pan and not on the floor, like you did this morning. The barber shaving you was disgusted. He reported you to the authorities. We haven't all day to waste while you relieve yourself. Get moving, or this young guard and I will delight in making you toe the mark.'

Rolf knew the guards meant business. He wasn't prepared to have another charge lobbed in his direction by the judicial panel. 'While I'm in court no one is to enter this cell. Those documents and the papers on my bed are private. I'll know if they're moved or read. After I've rezipped the fly on these uncomfortably tight trousers, you can cuff my ankles. Not before.'

At 9.38 am, Rolf von Breusch had scribed more historical notes while he'd waited under guard, for an orderly to shave him. Electric or strop razors were forbidden. Although he'd earlier enjoyed trying to wrestle the sheathed blade from the barber's hands, this had proved a useless exercise.

2 pm: Prior to court commencing, von Breusch dreaded the idea of being subjected to drivel or unsubstantiated crap put forward by his opposing team. Escorted by officers and with pinioned ankles he shuffled to the door leading to the courtroom. There, as in previous times, the manacles were removed. Those responsible for his welfare couldn't afford him going berserk a second time in one morning. After his latest attack on the officer who delivered his breakfast, Rolf was under threat of his cell being stripped bare, except for essential items. All writing implements, bedding, books and other small incidentals would be confiscated.

His power of inducement had drastically diminished where the junior officer, his "friend" was concerned. The corporal had taken his initial bait, but von Breusch puzzled over whether he might eventually betray him. Rolf realised that up until now, his prospects of a dismissal looked decidedly bleak. However, to express his fears would be tantamount to suicide.

Reagent had purposely withheld playing his wild card until the precise moment when it would deliver a real punch. Now he felt was the right time to produce an important witness. With the court reconvened and all formalities completed he immediately opened up with this address.

'Before I challenge the witness who is about to take the stand, I wish to ask the prisoner in the dock a vital question.'

'Very well, make it short Mr Reagent,' responded Justice Strezlewski.

'Rolf von Breusch, tell this court if you requested Herman Stroud, a master forger, to draft documents allowing you to use two, no pardon me, three aliases over the years. When you received the papers and passports and had done with his services, you then ordered Stroud to be shot. To compound your evil deed, you signed a deportation slip in late 1943 for Stroud's wife and daughter to be transported from Berlin to Ravensbrück concentration camp in Mecklenburg. At that time, Ravensbrück was located in northern Germany. We were all aware that the carriages of those trains were unhygienic cattle trucks crammed full of innocent people ...'

Defence Counsel Scharles interjected, 'I object, the prosecutor is out of order. Reagent is trying to subject my client to ridicule, your Honours.'

Mention of that concentration camp caused a bevy of emotional mutterings from the courtroom. The public and press galleries were warned and Strezlewski demanded the prosecutor's last comments be struck off. Promptly the female stenographer deleted Reagent's remarks.

Reagent then demanded in subdued tones, 'Herr von Breusch, this court is awaiting your answer.'

'I never signed a form for the Stroud women's deportation. The Gestapo were responsible for all deportations. Sir, I have never seen any cattle trains transporting prisoners to concentration camps. What the opposing counsel just stated is utter rot. Unadulterated garbage he must have concocted to satisfy his egotistical manner. How dare you accuse me of those atrocities. I am innocent. As I told you yesterday, I followed official orders in the Chancellery.' Rolf showed his anger with a threatening glare at a guard. Restrained in handcuffs, he then scowled at his lawyer for him to interject again. This ploy failed miserably. As a warning for him to not retaliate, Scharles nodded his approval.

This communicative gesture indicated for Rolf to continue. 'The truth is that Stroud died shortly before I escaped, as was every officer's duty in wartime. I decided to confiscate all the documents from his vehicle.'

'Would you prefer I ask Stroud's son to testify under oath what he has in his possession? And what happened to his family? I have no qualms in challenging him about both. Never fear on that account, Herr von Breusch.'

'And I prefer not to answer.'

By selecting this question and with the prisoner declining to reply, it verified von Breusch's guilt. 'Several days ago, Andros Stroud related in my presence what he knew of his father's death after being in your custody at the Chancellery. Stroud junior produced an order form written in a very neat hand. An educated man's hand, I would say. Verify if this scrawl resembles your signature please.' Reagent passed the page to a clerk court for the accused man's appraisal.

'Well, it does look a little like my signature. I would never have given a junior officer, other than my subordinate a form with a Reich letterhead. This must be a forgery. Probably forged by Stroud, I think you said.'

'Thank you, sir. You have satisfied my curiosity. In the recent conversation I can't recall mentioning a letterhead on a form,' the prosecutor decreed redeeming the page. 'I will now call my next witness to give his version of that event.'

Reagent then charged Andros Stroud to disclose what he'd earlier related regarding the topic under consideration. 'Herr Stroud, please disclose to the court what you earlier stated to my colleagues and myself in chambers.'

His testimony was consistent with his original statement. It was almost word-for-word and covered the full context of his father's life, until his death.

When he'd finished, Mr Reagent turned to address von Breusch. 'Now, after hearing Andros Stroud's testimony, do you still dispute the truth of his word?'

Rolf fumbled for an answer. How could he deny the man's testimony? Caught like a rat in a trap, he refused to commit himself.

'I declare this page was a written order you had given to Herman Stroud, this man's father, to follow. This writing and signature are identical with the data confiscated from your attaché case. Don't bother to deny it, Herr von Breusch.'

Stroud senior, like most of this Nazi's victims had been sentenced to death; never again to tread the passage of time. Unknowingly, his son's declaration had clinched the case and swayed the judicial panel's decision for the prosecution. It delivered another blow to defence counsel.

Next witness to the stand was Mr Kurt Baumer. Although his initial testimony was given "in camera" he, unlike Kendall, insisted on confronting the prisoner. A clerk ushered Kurt into court, after the room was cleared of all unnecessary patrons, including press members. Then proceedings got underway.

Even with his wonky heart trying to jump from his chest, Kurt triumphed by naming Rolf von Breusch as the Reich officer for whom his sister worked in Hamburg.

'May it please the Court; I now allow my witness to explain how his sister died and who ordered her demise.'

'Justice Reagent, time is edging towards recess. Direct your witness to make his statement brief and to the point. Is her death relevant to this case? I charge you.'

'It is vital, Your Honour. This relates to a question I withheld with judicial permission. My team and I consider this evidence relevant to set the record straight. It pertains to the accused man's mistreatment of his own family.'

'Very well, proceed.'

At this point, a caution was delivered for members of the legal and medical fraternities to refrain from speaking during this witness's testimony. Cameras were completely banned in the courtroom.

A clerk approached Mr Baumer with the directive to repeat the oath after him.

In a deliberate and sombre tone with right hand on the Bible, he did as instructed. The bloodcurdling look he received from von Breusch caused Kurt to falter and shudder, in fear of retribution if he recited his full name.

'Mr Reagent, your witness is exonerated from stating his name and the country where he now resides, as previously discussed in chambers. Please continue witness.'

Kurt shivered as frozen eyes pierced his own, which caused him to hesitate. The reason soon became obvious. 'I'm wary of disclosing what

I know of my sister's death. My adopted children were supposed to be with Heide. Instead, I'd driven that man's infants home. My deceased wife Anna, a qualified nurse … bathed and photographed their bruised legs and bottoms. Their father had beaten them with his leather belt or uniform cross-strap. If I accuse von Breusch my family and I may suffer the consequences …'

'Under extenuating circumstances due to your ill-heath, the court waives your right to disclose any further details. This signed affidavit is inscribed with all your particulars. I ask you sir, to verify if the signature on the document Mr Justice Reagent will place before you is your own?'

Kurt's bespectacled eyes accosted the document. 'Yes, that is my signature, Your Honour,' he truthfully validated its authenticity. His saltatory gaze tried to scan the words *Kurt Hewitt Baumer* twice. With blurred vision, a throbbing headache and shaky hands it wasn't easy. 'It certainly is mine.'

'State in your own words, your previous association with the accused prisoner. Speak clearly and be precise. Take your time to think, please witness.'

'Err … well sir.' Confused, Kurt paused. Then he recalled the order that his name would be withheld. 'My children and I are of German descent. I became aware of Herr von Breusch's existence one afternoon in May of 1942. My sister worked as a nurse in a Major's home, when I called to collect her. At first, she refused to tell me what had occurred earlier that day. I later learned that she'd been molested by Wehrmacht soldiers on her way to work. The major had also dispensed with her services. His two boys were in their early teens. The following day, she told me she might be employed by another Reich officer and named him as Kapitan von Breusch. I was curious how he came to know about her dismissal so soon. One morning week's later, I drove my sister to work and I caught a glimpse of von Breusch standing by his car, a new Daimler, I think it may have been.'

Members of the legal and medical fraternities buzzed with wild expectations. Their voices hushed in hope this witness, under duress, might reveal his name. Fever pitch hit the ceiling.

Senior Justice Strezlewski called the court to order. His "master's" voice had spoken. As spokesman he then instructed Kurt to continue.

Daunted by the hushed atmosphere, Kurt wrestled with his conscience. *How can I humiliate and dishonour my sister by telling the truth of Heide's attempted rapes by that Nazi criminal? They won't believe me. Her letter, I know Anna posted to Viktor Kassell, will prove von Breusch's guilt.*

Instructed to procced Kurt Baumer stood erect to speak. 'As I told you before, I am German. None of my family are Jewish, although I'm privileged to have known many people of the Jewish Faith. During the last war I worked with members of the Marquands and partisans in France and later in Berlin.'

'Sobeit! Keep your answers sharp and to the point, sir. Calmly continue please witness,' requested the prosecutor.

'The first time I met Frau von Breusch was in her home. I'd taken my sister her epilepsy medication, which she'd forgotten. My sister idolised both their children and they loved her. The morning she died we were taking the von Breusch children on a picnic. We found their mother very distressed. Erika had given the babes one of her prescribed tablets, and they were unconscious. In the car my sister said she'd seen the accused man push his wife down their stairs. She insisted that I take her home. She had forgotten her epileptic pills. My sister then insisted that I drive the babes to my home. She scribbled a note on my wife's letter. It stated that she'd witnessed von Breusch kill his wife. My wife was a qualified nurse who knew what tablets to give his children. She saved their lives. I left to collect their false documents and ours. An hour later I returned to my sister's home. It's all in her letter. You have it there, Mr Reagent. I need water, please.' Kurt Baumer clasped his aching throat and slipped a tablet under his tongue. Slowly the pain eased to a tolerable level. Unstable on his feet he collapsed, a clerk saving him from tottering backwards.

'You can't fall, sir. I'll hold you until this officer places a chair under you. There's a physician here, he will examine you.'

'My doctor is in the corridor. Help me to the door. He knows what tablets I'm taking. The pain down my arm has eased. I'll be okay, if I rest somewhere. Thank you, sir.'

Mr Reagent followed his witness into the corridor and spoke to Kendall. 'I saw Kurt place something in his mouth, it may've been a capsule. His breathing seems better than it did minutes ago. There's a private annex two doors down the hall. He can rest in there until court resumes. My secretary will be here with a tray of refreshments soon.

I asked Sam to select fresh fruit and sandwiches. Cheese and lettuce, or tomato on rye most of my clients prefer. Leave everything on the tray, Doctor Gwlynne. Samantha will collect it later.'

'Call me Kendall. I don't mind. It sounds less official than court jargon. I'll keep monitoring Kurt's vital signs until his palpitations settle. His pulse is fluctuating at a phenomenal rate. Unless he rests for at least an hour, I dread to think of the consequences.'

'Court isn't due to reconvene until two this afternoon. Let him rest here. It might be advisable for you to relax while you can. Kendall, do you still intend confronting von Breusch in court tomorrow. I know your sister does. Arneka is determined to take the stand. I agree with her wholeheartedly. She believes in people sticking to their principles.'

'A modern miss, she'll confront her nemesis with the dignity of a Queen in her court. My half-sister's a lovable rogue with a kind and generous heart. We don't always agree and our rows can be heated. In the witness stand she'll tackle von Breusch with decorum. She'll be direct and to the point in saying what she thinks of his brutality to her mother, Erika. It's all inscribed in the diaries Arneka kept from a child of seven until her teens. There's a postscript on the letter Heide scribbled in Kurt's car to give his wife. He told me Heide put the note in the children's nappy bag, after she'd seen him kill his wife, before they returned to get their coats. Kurt said he drove Heide home to fetch some pills.'

'Heide couldn't believe that von Breusch had killed his wife. She told you, Kurt that he released her braided hair and let Erika fall. The photos Heidi took verifies it.' He nodded. 'Erika's head collided with the marble floor below. The letter Anna posted to Major Kassell in Hamburg, and those two photos, will they all be introduced as evidence in court?'

'Yes Kendall, along with the two letters I received some time ago. One came from Düsseldorf, in Germany." Reagent sighed. 'The second was posted in Switzerland. I'm not at liberty to divulge the senders' names. I will at the appropriate time in court. Then you can browse through Heide's letters at your leisure.'

Unable to speak due to shortage of breath, Kurt Baumer pointed to his wallet on the table. 'I keep a copy of … her letters in there. You can read them later, if you like Kendall. Heide's signature has faded a bit, but it's legible.'

'Thanks Kurt, I might read it tonight in bed.'

'What caused my angina pains in court were the untrue comments von Breusch made about my sister. He must've known Heide resented him, after he'd tried to rape her in his office and again on the night he came home drunk. If Erika hadn't put the key in my sister's door, we all know what would've happened. God alone knows how that brute can sleep at night. I don't!'

'Men of his calibre lack a conscience. Their instinct to kill is too dominate and they don't feel remorse after committing an horrendous crime. It's a natural reaction to eliminate their enemies. Thank God, we don't live by that Nazi's code of ethics. Kurt, the shocked look on his face amused me, especially when you mentioned the letter Heide had written to Major Kassell. I naturally assumed Viktor was a close friend of your family. It'll be interesting tomorrow, to hear how Arneka challenges that Nazi dictator. She'll keep him guessing about how she and Chris avoided death in Hamburg. I'm anxious to hear how he'll worm his way out of the questions she's lined up to ask him. She hates him more than I do, and far more than my mother does. Gee, I must give mum a bell tonight. Remind me Kurt, to check the best time to ring home.'

'Daylight savings began back home early this morning, I think. The concierge at our hotel has a list of world times. That reminds me to phone Yeppoon to see if our house is still upright. The news bulletin this morning reported massive storms have annihilated cane fields and destroyed huge crops of bananas in North Queensland. I hope our town is still on the map. Last winter, hailstorms we copped, wiped out a great deal of our grain and vegetable crops. Graziers suffered the worst without food for their animals or families.'

Keenly interested to hear how this session was progressing, Kendall advised Kurt to rest and walked down to the courtroom. About to enter it, a firm hand gripped his shoulder. 'What the hell,' he began as a familiar face appeared in front of him. 'Arneka, you frightened the hell out of me. I've just left your father, he's okay. You didn't forget his warm overcoat and jumper …'

'No stupid, I didn't. Chris has them in his overnight bag. Where is Dad? I bought him fresh fruit and this hot coffee. Shit, I forgot the sugar and a spoon.'

Kendall directed her to the first aid annex. 'Both are on a table. I've left my locked medical kit and stethoscope in there for you to monitor his vitals. Here's the key, sis. He's resting at present. Keep me posted on his condition. I'll be in the first courtroom if you need me. Here's Chris, so I'll let you both care for your father. Don't overtire him, please. Kurt's next insulin injection is due in two hours. You'll find the list of his medications in my topcoat pocket.'

'Orders! Orders! Orders! Give me a hug and get in that courtroom, brother mine.' She laughed as Kendall tapped her bum. Their camaraderie and close relationship would carry them through the trials and tribulation they must all still face in coming days, of unimaginable hardships no person of a sound mind would wish to encounter in a lifetime.

A court usher directed Kendall to the only vacant seat. 'From here you can observe proceedings, Doctor. Neither Herr von Breusch nor the witnesses, or the legal panel can see you.'

'This rear seat is ideal. I may need to leave urgently. If a lady in a navy suit and lime blouse approaches you, come and tell me immediately, please sir.'

Justice Strezlewski ordered the court stenographer to strike off the Priest's last remarks. *Why did he do that? If I listen carefully the reason might be obvious. It's a bit hard to hear what they're saying from here.* Tempted to move closer, Kendall remained seated as the court spokesman addressed the man he detested most, his father.

'Herr von Breusch, your constant interjections and rudeness are disrupting court procedures. Remain silent, or you will be removed from the room. I warn you, do not force the judicial panel to hold you in contempt of court.'

Rolf glared at Strezlewski and disputed what the priest was saying. *I can't recall him being in or anywhere near my Chancellery office. If so, why wasn't he shot by one of my men. The priest's a liar and his testimony fictitious. That idiot must be one of the prosecutor's stooges. Is he falsifying evidence to make me confess to all the atrocities listed on their files? If so, the prosecutor's a bigger fool than this witness. Never would I be stupid enough to issue another priest's death warrant. There were hundreds of saboteurs who tested my patience at work. How this stooge escaped being shot, I can't imagine. And as for Kassell, I classed him a turncoat, an officer whom I neither liked nor trusted.*

Reagent nodded to the panel and quietly resumed speaking. 'In answer to your last question Father Riccardo yes, I have seen your deceased colleague's affidavit regarding a certain officer who worked at the Chancellery in 1944, and declared it as important evidence. It's still attached to the statement you posted to my secretary, just after the accused man absconded from Berlin. Before you continue sir, I would like you to clarify who these two photos resemble.' Reagent passed them to the priest.

The Italian looked at the snaps. 'They are both of your prisoner in the dock. I saw von Breusch standing outside his Chancellery office. I'd been to visit Father Ignatz in his cell, to give him Holy Communion and the last rites. We discussed the tragic death of our deceased friend, Father Kelly. Patrick worked with the partisans in Germany and helped a lot of our parishioners to escape the Nazis. We also helped British airmen, Gypsy and Jewish families, by supplying them with forged documents and passports. Two of our church deacons were plumbers. They transported all escapees to safe houses in Szczecin, north Germany.' Father Riccardo paused to review his notes. 'A local fisherman ferried them across the Baltic Sea to Falster Island. I believe they were taken to Odense where some were dispersed. Down the years, I have received letters from downed airmen and Jewish families from all points of the compass.'

Reagent then asked this witness to describe what had occurred after he'd spoken to his fellow priest in his cell.

'I left the Chancellery at two that day, a disillusioned man. I remember looking at my watch and thought, by this time tomorrow Father Ignatz will be dead. His execution in front of a firing squad was supervised by that man in the dock. This photograph of him giving the order to fire cannot be disputed. His driver had taken it. Sergeant Kohl disliked von Breusch's abusive attitude. His priest contacted me in Antwerp just before the war ended. I have great respect for that ex-German officer. Kohl wasn't afraid to disregard orders and he sent me this photo. I received it and his signed declaration through a diplomatic source. My only regret is not being able to visit both my friend's graves.

To this day I doubt if their bodies were buried. Most likely they were incinerated in a crematorium, to conceal the cruelty of a certain person in this room's guilt.'

Due to a prior church commitment, the panel excused Father Riccardo, who bowed to officialdom. He hurried from the courthouse to hail a taxi to reach the station on time. His scheduled train stopped directly in front of him. With his grey umbrella jauntily swinging he stepped aboard as its doors closed.

12 noon: In the courtroom Rolf von Breusch cursed his wartime driver. *If those miserable weasels in wigs exonerate me, I vow to hunt down and shoot Wilhelm Kohl. Not to kill. Just to maim. I'll make him suffer the pangs of Hell. Then I might put a bullet through his imbecilic brain. I never trusted him to carry out my orders, nor any other officer at work. The carpool drivers hated Kohl.*

Before the luncheon recess, Scharles noticed his client's red face had deepened and suspected his temper would soon reach a climax. Hesitant to query the reason, he chose to challenge Michael Reagent in chambers. With his entry barred by Reagent's clerk, his private secretary Samantha, he stormed back to his office cursing.

That young woman should be barred from all courtrooms in Austria. The length of her skirt is far too short and she's undignified for a legal secretary. Her legs are bowed at the knees and she acts like a street tart most days. Still, she may be a good bed-warmer on cold nights.

None of the defence counsel's idioms were correct. In fact, they were libellous. His jealousy derived from being criticised by the prosecutor and his team for misdemeanour in chambers on several previous occasions.

Scharles heard this witness repeating the oath gasp. His face had turned grey. Mumbling his words, he gained the strength to recollect his thoughts. Kurt Baumer began to relate how the situation had developed about the von Breusch children's coats, including the episode as he understood it, leading up to his sister's death. He also described how he and his wife had escaped with the accused man's children. Kurt concluded by disclosing what he and Heide's neighbour Karl, had seen from his car parked close to her house, as it burned fiercely on that fateful day.

During the course of Kurt's testimony, it was evident that his breathing had become laboured. Nonetheless, he described to the court his premonitions of his sister's death and how he feared for her life, especially if she'd gone with the von Breusch family to Berlin. Kurt related everything in detail even down to how he and Anna had reared both the von Breusch children as their own.

The look of disbelief on Rolf von Breusch's face showed his dismay. This knowledge confused his tormented brain. *If my kinder are still alive, this stupid old goat's confession will destroy my chance of being exonerated.* Rolf puzzled over why they and the son from his second marriage hadn't appeared in court. Known to that son only as Ian David Ross, he knew Kendall hated and mistrusted him. *I'm sure the lily-livered crud wouldn't have the guts to come here to expose, or accuse me of mistreating his mother on that winter's night in Yass, Australia.*

Reagent forthrightly praised Kurt Baumer for his diligent and honest testimony. Everything he'd said was consistent with his detailed declaration to ASIO, plus his written, signed affidavit was already in the panel's possession.

Excusing Kurt Baumer, the spokesman signalled him to remain seated in the stand. Strezlewski then calmly conferred with his learned colleagues. Notes exchanged hands between the panel members and a distinguished officer sitting adjacent to the dock. Nodding, the spokesman stood to address the court, but his eyes specifically pinpointed the ill man, Kurt Baumer.

'I will now personally address this witness. Sir, we the Judiciary has come to the conclusion that our court physician should attend to your health problems. The doctor in attendance has advised us that he prefers to keep you under close observation. Like us, he has observed that you sir, have on several occasions put something in your mouth. Possibly a tablet that dissolved under your tongue?'

Justice Strezlewski shuddered. *This man looks really ill. His features are pale and drawn.* Upon noticing the witness lunge forward, he pointed to the well-clerk. 'Quick assist that ailing man. You, you're the nearest,' he demanded as the court physician hastened to help Kurt Baumer.

Exercising his restraint, Strezlewski allowed time for him to be medically assessed, before instructing two court officials to escort him from the room.

'Just a few … chest pains, Your Lordships,' gasped Kurt, who managed to gulp several shallow breaths before his throat seized with pain. The dull throb travelled down his left arm and he supported its dead weight against his body. In a losing battle to survive, he tried to forge on. 'I've had the same pains before today. Yet … I've come through this … with pride.'

The doctor assigned to the court bent down beside him. 'Don't try to talk, sir. Let me listen to your heart.' The cold stethoscope disc pressed against his chest made Kurt shiver.

A physician's nod to the Judicial Panel indicated Kurt's breathing now seemed stable. He spoke in a convincing, gentle tone. 'Sir, please allow me help you down these steps. Once your feet are on the floor, these men will assist you to the first aid annex. Your own physician is here, I believe. He can take care of you in there.'

While being assisted from the courtroom, Kurt asked, 'I will be alright doctor … won't I?' Vaguely, he heard the physician's acknowledged reply.

In the fully equipped first aid room he gave Kurt a thorough examination. Finding his heartbeats were erratic, they confirmed his suspected prognosis. In his estimation Kurt was bordering on a coronary. Suffering from acute angina, usually a minor heart condition, these symptoms tied in with exhaustion and stress Kurt had suffered over recent weeks. His pulse and galloping heartbeats were developing into a deadly combination.

'I'm sure this attack you've suffered will be a huge warning. Take heed of that warning, sir,' the doctor advised. 'Nevertheless, it is imperative that you undergo a series of tests. We have fine modern hospitals here in Linz …'

'No, I prefer to have the tests done in Australia. Not here in Austria. I keep this bottle of Anginine tablets with me at all times. Well, they're old and perhaps they might need replacing. Oh, at long last the pains have eased and I'm starting to feel better.'

Scanning the label the physician nodded. 'Yes, these are well out of date. I'll prescribe you a fresh supply of tablets Herr Baumer, and have them delivered here while you rest. I shall be but a moment.'

'Wait a minute. I will be able to keep testifying? I can't die now.'

'In my professional opinion sir, I think so. If these angina attacks persist, you must seek medical attention immediately. You should take it easy and rest until you feel stable and your palpitations settle. With careful handling of your diet and with copious rest without stress, your doctor should be able to control your heart problems.'

'I appreciate all you've done for me, sir. Kendall, my doctor, warned me that if I testified it could affect my health.' Fighting his emotions, Kurt then stated, 'I can't die yet. My family need me and I've done my duty.'

'Is someone here with you today? If so, I'll call that person for you.'

'Yes, my daughter and son are here with Doctor Gwlynne. Kendall will be somewhere close. As I told you, he's my doctor while I'm here in Linz. You'll probably find them all outside. He won't be far away.'

'Ah yes, the lawyers and panel of this tribunal are acquainted with your physician. I believe he has given his testimony "in camera" to our judges.'

While Kurt's heart problems were being accessed by Kendall, Justice Strezlewski announced, 'Court is now in recess. There will be a stay of proceedings for two days. Court will resume on Monday next at ten.'

With the trial climbing towards a penultimate conclusion, this weekend's respite would be beneficial for all concerned. Everyone looked forward to a break after four weeks of relentless inquisitions and heartbreaking trauma.

A barrage of pressmen invaded the privacy of everyone who stepped from the courthouse building. Inside courtroom one, people tittered while collecting their raincoats, umbrellas and personal items. Within minutes the main corridor was cleared and empty.

The ghostly atmosphere resembled a tomb. Sombre reminiscences of the past that had cloaked the dock and witness stand in misery, now shrouded everything in a veil of mystery. Recollections of gut-wrenching memories were stilled, to be resurrected later, thankfully only for a short period.

Before leaving chambers, Michael suggested he and Sam should wait until the main thoroughfare cleared. In a quiet alcove they briefly conversed in soft monotones about the day's hearing.

'Well, that's a page turner for the ledger, Michael. My sorrow for Kurt Baumer is immeasurable. That innocent man certainly had a going over, in more ways than one.'

'Yeah, but not from me,' Reagent ratified solemnly. 'I sincerely hope this traumatic experience of delving into his sister's death won't forever haunt him. Kurt looked shocking. And to think that smug bastard in the dock with a satisfied smirk thrived on his misery. Oh, that Nazi bastard makes me furious.'

'At times his expression looked grave. I don't think von Breusch expected you to present those four photographs. Huh! That'll put a cap on his thinking for next session, especially if your previous witness takes the stand again. From what I know of the Italian's story Michael, it sure will capture the attention of those nosey show ponies of the press contingent.'

'No, it's not my aim to crucify the priest. If Scharles wants to play dirty, I don't have to, Sam. His idea is to denigrate my witnesses and crucify them by calling them liars. It's not my style. My plan is to let the truth be known to those media buffs, to get their bosses onside. It will help sway the "would be doubters" that those atrocities, gruesome in essence as they were, did in fact occur in wartime.'

Securing his briefcase, Michael added, 'Sadly Sam, there's still a hell of a lot of sceptics out there who for reasons of their own don't, or won't believe all those shocking events actually took place. Between you and me, I hope they hang this bastard. He won't cop a bullet. Whatever the outcome, I'll still fight my damnest to see von Breusch gets what he deserves.'

On leaving chambers, Michael and his secretary greeted some their fellow colleagues, one who remarked. 'I thought you gave a colossal shot in there, to bring the "B" down to size. Once we return to the hotel, we'll bring our notes up to date and run through them, prior to Monday's session.'

'Not me, Roberto. I intend to take my beautiful secretary to dinner this evening. And I've lined up a show for us to see. Don't ask …'

'About tomorrow?' queried his archive researcher, who'd butted in. Not getting a reply, he assumed it might have something to do with their hastily arranged trip. Reagent had booked their flight to Germany to interview a witness who worked with the Underground in the last war, and also a woman who knew von Breusch, before his fake suicide in Hamburg. Their flight to Berlin would be kept under wraps for the time being. Once his vital witness had consented to appear in court, all hell would break loose, especially in the opposing camp. Reagent smiled, contemplating the shock von Breusch would get after seeing his lady friend's appearance in court. This all depended on their replies and if they could arrange to fly to Linz, Austria, within the next few days.

35

10am: Monday morning court reconvened on time. In a bit of a hassle, after just driving in from the airport, Michael and his team of four reached the court house in time to run through their briefs. The trial was conducted in a similar vein as in the previous weeks. Two British airmen had given their testimonies regarding their misfortunes while in von Breusch's captivity.

Half an hour after proceedings began Reagent, handed a note, was forced to put his surprise witness on hold, much to his dismay.

'Oh, thank you.' Michael Regent accepted the folded clip from his deputy. Spreading it open he canvassed the text. 'May it please Your Honours; I have just been given an important note. I request time to investigate the contents.'

A noticeable disquiet channelled the room. Oohs and ahs, plus an episode of vibrant chatter moved the panel to declare "Silence", in stern tones.

With the racket abating, the spokesman civilly accorded, 'Request granted. In lieu of your attention being drawn momentarily from cross-examining the present witness, I now call a brief recess.' Glancing at his pocket watch, Justice Strezlewski declared, 'Court will reconvene in an hour.' The panel arose as people shuffled out to a forecourt, there to relax or stretch their legs.

In prosecutor's chambers a young UN officer in charge of the accused man's welfare informed Michael of a dire situation. It regarded his subordinate's find.

While sipping tea, Sam had made for those in attendance, Reagent accosted the junior officer about the declaration clutched in his hand.

'Corporal Bedler, are you aware what this note implies? Could you possibly be mistaken about Herr von Breusch's intention?' Reagent's offsider observed the officer's reactions, as his secretary took shorthand notes of their discussion. 'Are you certain his intention was to bribe you with this gold fob watch? You must be positive. These accusations are serious in context and imply deadly overtones,' decreed the deputy prosecutor.

'I am positive of both, sir. The documents you have just scanned were given to me this morning by the accused man in his cell.'

'What was his motive, do you think? And, did he make his reason clear?'

'His reason was exceptionally clear. I promised to give that memorandum he's been working on to the press. I lied. It knew it would be wrong to allow him to blab his war history to their bosses. I let von Breusch think I was going to keep the notes until he advised me otherwise. "When the time is right, I'll let you know", were his actual words. He also gave me this, Sir.' A brown suede pouch lay in Bedler's palm. He passed it to Mr Reagent. 'Mm, a gold butane lighter.'

'A nice bit of work.'

'Yes Michael, it looks rather expensive. There are intertwined initials on the front. The inscription on the back reads "M.S love R.von B." Sam flicked the flame. The glare she got from Reagent caused her to replace this priceless bit of evidence in his hand.

'Sam, I wasn't referring to the lighter, although, it doesn't look like one a man would carry from choice. I meant the way his deviate mind works. Are you willing to testify, Corporal Bedler?'

'Certainly, sir. I feel dirty, having touched those things.' Wiped hands gave the impression that he was trying to be rid of filth from his fingers.

'There's a basin in our toilet. You're free to use it,' Reagent nodded. 'On your return, I'll ask you to sign this affidavit my secretary has drawn up. Now I need a moment alone with your superior officer. Thank you for your diligence and your honesty, Corporal.'

Now it all makes sense to me. Bedler smiled while Reagent was conferring with me, his superior. General Passfield appraised the prosecutor's opinion of Bedler's verbal declaration. Sam then spoke to the deputy prosecutor. Together they quietly discussed the latest data, received from another junior officer.

Justice Beck, an advocate in his own right, agreed with her notes. 'This news is a positive bonus for your team and us. Wait till our yokefellows hear this. Oh boy! Will it set the hounds among the turkeys.'

'Yeah,' noted Reagent re-joining them. 'I reckon it will. Finished Sam?' he asked looking over her shoulder.

'Michael, this copy of the corporal's declaration is done.' She reefed the pages out of her portable typewriter, and separated the duplicate and triplicate sheets then handed them to her boss. 'All they need is Bedler's signature.'

'I'll co-sign that document with the same pen,' the general stated. Until this issue surfaced, he hadn't realised what he'd heard of their conversation was important. *Ah, now it all makes sense. I suspected von Breusch of being an arrogant and treacherous individual, a fiend with less scruples than a gutter snake. I disliked his arrogance in court.*

The general collected his cap, baton and brown leather gloves off the table and hurried to his vehicle. The icy glare he gave von Breusch and his counsel would freeze the balls off a brass monkey. Michael an observer, smiled.

2 pm: In the courtroom, proceedings were quite heated. Reagent addressed Mr Justice Strezlewski, who had asked the reason for such a long delay. He clarified it by saying, 'I have in here a signed affidavit from another prison guard who is willing to testify. May I present him now, Your Honour?'

'Your request is a little unorthodox at this time. I need to confer with my colleagues and sight his affidavit before we grant your request.' The panel momentarily discussed it. 'Mr Reagent, call your witness to the stand,' Strezlewski nodded, handing the folder back to floor clerk. 'We are ready to hear his testimony.'

Bellemy was duly sworn in. He explained how he'd seen the initialled lighter in Bedler's hand. 'I know it was given to him as a bribe to secure von Breusch's unofficial release by leaving his cell door unlocked. Tonight, after light's out he wanted Bedler to commit that felony. I heard Lewis say he wouldn't comply with the prisoner's demand. Definitely not before his replacement came on duty. He seemed convinced my colleague had lied. Then von Breusch offered to give him an unframed valuable oil painting to convince him. I heard him say to Bedler, "It's in lieu of payment. And the title is, *A Nude in Love.*"'

The chief prosecutor read out what both Bedler and Peter Bellemy had stated to his legal team in chambers. Passing their affidavits over, Reagent then asked Bellemy to verify if the signatures were theirs. Peter confirmed, 'Yes that's mine there, sir.' He pointed. 'Bedler's looks a bit like scribble.'

'Now please describe in your own words, what *you* actually witnessed last night outside the accused man's cell, also what you told us in chambers.'

Peter Bellemy attested that he had assumed Bedler never intended to unlock the cell door to release their prisoner. 'After signing the nightshift book, I reported the bribe and what I'd heard to our superior officer.' On counsel's advice he verbally enunciated what he'd said in chambers. 'I told you, at four I finished my rounds on that floor, signed the time sheet and then went home.'

'You sneaking rat, *die schweinehunde*,' shrieked Rolf at his accuser. 'You and Bedler promised not to let my memoirs out of your sight. Now is *not* the right time. Months of my hard work wasted, down the drain and for what? I warned you pair of traitors twice. You're both as thick as two planks,' Rolf von Breusch bellowed even louder. His eyes followed a hand as it rotated towards the two junior officers. 'You sneaky bastards. You're like all the legal misfits in this room.'

Reporters scampered to get their scoops first online and on the streets in record time. The proceedings thus far proved worthy of this terrific mainliner.

A constant hum of twittering voices echoed throughout the unsettled courtroom. Sensibly, Strezlewski ignored them, until their owners quietened.

'Pass those pages to the well-clerk. We the panel, need to review that entire document Mr Reagent. We should have accessed that data this morning.' This done, the folders were retained for their detailed perusal at their leisure.

'I request Your Honours for the prisoner's previous declaration to be struck off the record,' defence persisted. 'My client is so distressed he doesn't know what he's saying. In his confused state, he is likely to admit to any accusation the prosecutor puts forward.'

Directing his gaze and voice at the defence counsel, Strezlewski retorted, 'I'm afraid it is a little late for that, Mr Scharles. Your client has committed himself. In his own words he has disgracefully defamed this court and all within these walls. By his rudeness, the panel is denying your objection.' A regal, non-verbal gesture passed for the prosecutor to proceed.

Reagent responded curtly. 'Now, what say you prisoner? You have betrayed your own counsel by allowing your private diary to fall into one of the guard's hands. Written in your own script, there's no way you can deny the document's authenticity. My,' Reagent issued an atoned sigh, 'what more conclusive evidence can the legal representatives of this tribunal expect?'

With discrete smiles, the judiciary panel made their individual notations. They were unanimous in their brief assessment of the unlogged document. The personal dossier written by von Breusch would take time to assess in detail.

To be impartial in judgement would now be impossible. The panel, as sworn members of the judicial system, were obligated to see this trial was fair and equitable, be it now ever so difficult. The criminal in the dock had all but secured the final knot in a noose. All questions he'd responded to were of his own volition.

With this afternoon session drawing to a close, the panel's decision to call more witnesses was deferred until tomorrow's session. By then they would've perused the accused's dossier and responded accordingly.

The morning session would bring an avalanche of memories, born of tragedy, yet heralding the truth. The most poignant stories tendered would bridge eons of history with current woes. The ebbtide had turned in favour of the righteous, leaving liars to grovel in the dirt beneath society's feet, where they belonged.

Taking the prosecution stand and after acknowledging the panel, Reagent declared, 'My next witness will not be the one listed. Before I call that witness to the stand, I request the court's indulgence to read two signed depositions.'

Permission granted, Michael Reagent began to quote from the letter in his hand. "Wounded in the neck by a stray bullet, I had to bail out of my aircraft, doomed to crash. After an hour of trying to keep from drowning in the North Sea, I was rescued by the captain and crew of a fishing trawler. Landing in Cuxhaven two of the crew took me to their local hospital. Several months later in 1944 the Underground arranged to take me to a safehouse on the outskirts of Potsdam in Germany. From there I was hidden under bales of hay until the wagon driver delivered me to St Xavier's Church. One of their parishioners carried me down into the catacombs beneath a religious grotto. I stayed there until I was well enough to leave, just before Christmas that year. An elderly priest, Father Ignatz and his colleague Father Riccardo, with the help of two plumbers and some local people assisted my escape through to Sweden in January of 1945.'

A short break followed then Michael resumed speaking. 'With the Panel's indulgence, I shall now read an affidavit. It is similar in context to the previous letter, a signed and dated deposition from another British flyer, brought down over Hamburg in Germany late in 1943. They have both been verified as evidence. Both the pilots were destined for "special treatment" in Mauthausen Prison Camp. Because of ill-health from war-related injuries, neither retired officer is capable of attending this tribunal in person.'

'The letters on file we have duly noted,' declared the spokesman. 'Call your next witness, Mr Reagent.'

Summoned to the stand, Mr Quinton Marciano, guided by his walking cane shuffled from the small annex set aside for selected witnesses to courtroom one. In desperation he tried to knuckle his horrific thoughts in silence. His left eye focused on rows of people until aided, he reached the witness stand.

A hostile witness waiting in the corridor to be called, looked at the crippled man who'd hobbled past him wearing a flesh-coloured eye-patch, bellowed at the usher, 'Why the delay. I should be next, not him.'

A thin scrawny man in his mid-fifties, who once possessed the physique of an athlete, now slightly bent in stature, repeated the oath. His almost sightless eye, strained and with a glint of wearied sadness, analysed the expressionless mass of faces.

Quinton flinched as his gaze settled on von Breusch's unshaven features. Memories of this Nazi's harsh abusive voice, who while signing his death had ordered him to be shot in the Chancellery quadrangle, ran through his mind. Quint shuddered on recalling a split-thronged whip that had laid bare the flesh on his back and shoulders. Harassment of being threatened by von Breusch, his abuser, had left a lasting effect on his nerves. Now his stable eye reassessed Reagent, as a hint of a smile appeared on both corners of his lips.

'Mr Marciano, within a day of my team arriving in Berlin, I phoned you and visited you in a small Henning Dorf apartment. Then with your permission, my clerk and I escorted you here to Linz in Austria. Is that not correct, Sir?'

'Yes, it is Mr Reagent. I sat between you on the plane until it landed.'

'Please tell the court what you were doing, when we first spoke to you?'

'Not much, pottering around the flowerbeds with a garden fork in my hand. Sadly, I can't see to write or do close work now. It annoys me that I can't read. I lost my sight, or most of it, in a cell towards war's end in Berlin.'

'Take your time and in your own words describe how the loss of your eyesight occurred. If you become stressed, stop. No one in this courtroom minds a brief respite for you to collect your thoughts, or to catch a breath.'

'It's like this. With my German priests, I 'elped people to reach safe 'ouses. Now our beloved Father Ignatz and Father Kelly have both gone, Your Worship.'

'By gone, do you mean the two priests are deceased.' Michael smiled at the terminology of his official status.

'No. They were murdered, along with many of our friends. Murdered by that man there.' Quinton pointed to the dock. 'Oh, von Breusch didn't do them killings. The guards followed 'is orders. I know, 'cos I was next in line to be shot dead in the Chancellery quadrangle below 'is office.'

'Think back and try to remember clearly, if you can.'

'I don't need to think. It'd be impossible to forget that man's face.' Quint surveyed the accused prisoner through his magnifying glass. 'We, all except Father Ignatz, were captured by the Nazis, not long after our friends had left with George in their truck to Sassnitz, in north Germany. A dedicated man, George Brady wanted to return to Ireland, there to study in the priesthood.'

'Why did Nazis arrest you? What happened to Father Ignatz? Define how he died and why. Were there consequences of you all helping pilots and other refuges to escape Nazi tyranny in Berlin, Germany?'

'Father Ignatz blessed a wooden cross in our church and gave it to George just before 'e boarded the plumbers' truck. A couple of empty cement bags covered George who huddled in their toolbox. The two airmen they collected on their way up to Sassnitz. At that Baltic seaport they all boarded a lugger going to Denmark. From there George went back to Ireland, so the letter I got six months later said. Sir, it's in your hand.' Quint shaded his weak eye, by covering it with one hand. 'My mind's not the best since I was tortured by that fiend's men.' A nod again indicated Rolf von Breusch.

'Now witness, describe what you know of the priest's death. This is important to establish what actually occurred on that day. And for the legal teams in this courtroom to access if it does concern the accused man.'

'After I escaped from Berlin, a partisan told me Father Ignatz was shot by a SD officer. I suspected von Breusch had signed his death warrant, 'cos I'd seen SD on the cuff of 'is sleeve. And I saw 'im sign mine. I heard about this trial in a bus. I knew then it'd be about von Breusch. A friend showed me a photo, a big one. With my magnifying glass, I recognised whose face I'd seen earlier that day in a newspaper, Sir.'

Quint paused to unravel a conglomeration of muddled ideas running through his mind. 'In them days von Breusch wore a grey uniform. A Captain's, I think from memory.'

Rolf gasped on being shown the enlarged photo in Reagent's hand.

'Confirm for the records, is this the same officer you recall seeing in that paper and in his Chancellery office. These two original miniatures were in your jerkin pocket when you were arrested by the Gestapo in Xavier's church grounds.' Reagent then asked the witness to verify them. 'My records show that you, Mr Marciano gave a miniature to George Brady before their truck left the church.'

'George put the stamp-sized snap under the lining of one shoe, just before the plumber's truck left for Sassnitz. The stoker who worked on their engine gave it to me, to pass on,' Quint replied, ignorant of the fact that this elderly priest would be called next to testify.

'May I, sir …'

'Yes, by all means use your magnifying glass. Don't rush. You'll need this to see the miniatures clearly,' Reagent passed him a pocket torch. 'Then one of these guards will assist you over to the dock. There you will be able to observe the accused man at close range.' The judicial panel nodded their approval.

'Yes, it was Kapitan von Breusch who questioned me in the Chancellery. You can see SD on that uniform cuff. A guard clouted me with the butt of a rifle, but I never said how we got George away. For 'elping 'im to escape, we were sentenced to death without a trial. Just with von Breusch's signature on a scrap of paper. It wasn't only George the plumber's truck drove to safety. They loaded it with fresh produce from the preparatory garden, blankets, freshly churned pats of butter, eggs packed in sawdust and tinned supplies before they left St Xavier's Church. Our priests, nuns and some parishioners 'elped dozens of families to escape from Nazi tyranny. It wasn't wise to be on Berlin streets in them days.'

'We know that. How you were blinded is what we need to clarify.'

'Gestapo thugs stopped and searched my car. Then they searched me. I got angry with the Black Shirt bullies. It weren't the Brown Shirt Brigade. They were disbanded not long after Hitler came to power in 1933.'

'Yes, I understand that. You were saying about the Gestapo arresting you?'

'Yes, they did. One of their officers gave an order to drag me out of my car. Then they threw me in a truck crammed full of dishevelled men. Mine was a decoy vehicle. Another car had taken three English airmen by a different route to a safehouse. Because I refused to tell them where they'd gone, a Gestapo thug bashed me from behind with something hard. Probably the butt of 'is rifle. I awoke next morning in a prison cell.'

Quint wiped the sightless pit where his left eye had been, with a cloth. He then put the eye-patch down to cover his sightless socket. 'That day I was taken to the Chancellery and dumped in a tiny cell. In an upstairs office von Breusch questioned me for hours. I still refused to say where George was, because I didn't know the route our truck 'ad taken. Two guards dragged me back down to that filthy cell. One guard used a hot poker to burn my naked buttocks. Sir, I gave you the photo of my scars.' Reagent nodded and held the snap aloft. 'A young guard who'd taken me for a shower, had taken it. I think 'e felt sorry for me.'

'Objection! Scars do not mean maltreatment. I have a few myself and they were not inflicted …'

'Enough, Mr Scharles. I will not warn you again. Another outburst and your deputy will take over. Have I made myself clear?' demanded

Strezlewski. 'You have been constantly heckling this witness while he's trying to speak.'

'Your Honour, I understand you perfectly.'

'Instruct your witness to proceed, Mr Reagent.'

'Mr Marciano please say what you intended when you were rudely interrupted.'

'I wouldn't tell them roughnecks how George escaped then a guard bashed me senseless. One of von Breusch's men threw acid in my eye and laughed. I screamed and passed out in pain. I didn't remember anything for days. I awoke almost a week later in a different Chancellery cell with my face swathed in bandages. Father Ignatz was dead. And Father Ricardo was wounded, so the guard who took my photo said.'

Busier than bumblebees in flight, the court hummed to life. The wooden bench paid homage to Strezlewski's abandoned gavel, which he seldom used.

'The accused man never again threatened you with death?'

'No, sir. One night I bribed the guard who'd taken my photo, to smuggle me out in a laundry truck, full of dirty linen. Because a stooge pimped about our church being involved with the Underground, two of that man's stooges,' Quint pointed to von Breusch, 'had arrested our priests. Someone ratted, but I didn't sir. There were lots of our parishioners who hated the Nazis. It may've been one of their family members who contacted the Gestapo.'

'I have no more questions at present, for this witness, Your Honours.' Regent bowed to the judicial panel, then returned to be seated alongside his secretary and their research team.

'Your witness, Counsel for the Defence,' ordered the spokesman. 'Take note Mr Scharles, the panel has been extremely lenient with you this day. Be warned. Keep the questions short, and your voice moderate.' He acknowledged the spokesman's ruling.

Scharles began cross-questioning Marciano, whom he'd ridiculed and chided, while Quint tried to think of the correct answers to an array of important questions. The defence counsel's meagre effort proved futile. With affidavits and depositions already noted and read, Terri Scharles found little comfort in quizzing the Italian. He did, however, find him a hostile witness who objected to every proposal he put forward. The Italian priest seemed relieved when a halt was called to proceedings.

Scharles was inclined to believe Quinton. *This prosecution witness has more or less proved that von Breusch has been feeding me lies. Now the truth must be revealed. No lawyer, brilliant or otherwise, appreciates being made to look a fool in this or any other courtroom.*

Furious over being deceived by his client, Scharles spoke to von Breusch in a firm, dismal tone of voice. 'If you continue lying to me, Rolf, I'll refuse to be your lawyer. Your erasable and uncontrollable temper has been unbearable. If you don't curb your tongue and speak civilly, you'll cause my disbarment. I'm damned how your mind functions at times. I require truthful answers from you before the next session. Now I must re-examine every scrap of evidence and ask endless questions of you, if I'm to succeed in winning this case. No more indecisions and lies if you expect me to win this case, Rolf.'

'You can resign. I *am* quite capable of taking on my own defence, Terri. Leave now, or continue as I instruct you do in this damn courtroom.'

Going by previous testimonies, there was little or no likelihood of Scharles either salvaging his reputation or keeping the hangman's noose from tightening around this client's neck.

The following morning unlisted and important issues were raised; this delay detained proceedings for three hours. With the time approaching two, Reagent began to question his next witness, an Irish ordained priest.

On entering the witness stand Father Brady was defiant to confront his former accuser of sabotage. This moment he had sweated on for decades. Now George's anger had subsided to a degree, he felt gratified in righting the wrongs of his protagonist's criminal deeds.

'Before I name this witness, I will endeavour to establish one thing. I challenge the accused man to verify if he's ever seen my witness before today.' Reagent smiled. 'Herr von Breusch, tell the court if you recognise the priest who is about repeat the oath?'

'Not really, his face does look a little familiar. What's one face among the crowd of nosey onlookers packed in this room.'

'By giving me a sarcastic answer, you've justified my question.' Satisfied Reagent walked to review his briefs and bending low he said to his secretary, 'That silenced the quilled tongues. They're stiller than a grandmother's crutch.'

'Shush Michael,' Sam whispered in ventriloquist style, 'you'll earn a rebuff from his nibs. Don't forget the spokesman is a wizard at reading lips. I overheard him mention to his clerk of having a deaf sister-in-law.'

Reagent stood upright and smirked on readdressing this witness. 'Now Father Brady, please state the reason why you spent more than twenty hours incarcerated in a freezing cell, below the Chancellery building in 1944.'

George briefly narrated the reason for his capture, the misery he suffered in that cell for longer than a week as a prisoner, why he had exchanged clothes with his friend Father Patrick, and how he escaped from the Chancellery.

'Kapitan von Breusch ordered me ta be shot and Patrick took me place before his firing squad. Outside the Chancellery a car full of partisans collected me. They were all dressed as priests.'

With his opposition ready to embrace the floor-well, Reagent looked across at von Breusch in the dock. The smirk on his face no longer existed. The look on his scrawny features was one of disaster. And of total disbelief.

Rolf von Breusch looked through Reagent as though he never existed. *Will that arrogant prig in the dock answer my previous question? Evidently, he has recognised the Irishman, who he blatantly denied knowing. I'll ask him again. He needs to unequivocally state why he gave orders to have this man beaten and then shot by his junior officers. I am not aware what he'd done to deserve the beating. These records were verified.* Reagent seethed with anger. *I WILL hear von Breusch admit the truth of what actually occurred in his Chancellery office. And why he used a riding crop to strike my client, George Brady across his left cheek. The scar is still obvious, even after twenty or so years.*

Puzzled over where he'd seen the witness in the stand before, von Breusch frowned, *His face looks familiar. Not unlike the priest whom I watched being shot by our firing squad below my second floor window around forty-two. How did this man escape? Did those idiots shoot the wrong prisoner?* He pondered long and hard to think where he'd seen this priest before and under what circumstances. *No, I'm never mistaken, so it can't be him.*

In retribution, the devout Catholic at last hoped he could make him feel some remorse, before paying for his crimes. *For as long as God allows me to breathe this fresh country air, I pray this evil Nazi will never be able to harm or murder another human being.*

'Father Brady, I asked you a question and you apparently didn't hear me. In your own words, define what occurred on the day that you were taken into custody by Gestapo officers in Berlin,' challenged the defence counsel. 'Also explain your first encounter with my client, Herr von Breusch.'

'As I said, in mid ta late 1944, their officers arrested me by the railway marshalling yards. After extensive questioning they took me to the Chancellery where their guards threw me inta a tiny cell and brutally beat me. Later that day they dragged me up to be interrogated by von Breusch. I refused ta tell him why I'd been with them gangers. He accused me of thievery, crimes I hadn't done and called me a liar. All because I wouldn't dob in me friends fa sabotaging the tracks, he struck me across the face with his riding crop. Angry, he signed me death warrant. Then his thugs threw me back in that filthy cell.'

'What happened then, Father? Speak clearly. Your answer is important.'

His finger angled at the dock. 'While I clung ta life, ready ta face his firing squad, a local priest, Father Patrick Kelly came ta give me the last rites. He and I worked as undercover agents with the Urban Underground in Berlin. Father Patrick, me dear friend from Ireland, undressed and forced me ta change inta his clothes. I felt ill and wanted ta vomit, 'cos I didn't know what would happen.'

'Clarify one more thing please, Father. Why did Gestapo arrest you in the first place? When you told them neither you nor your stoker had sabotaged railway tracks in or near Berlin, you lied, didn't you? My client knows you did, as do I. Written evidence I have proves you were both the saboteurs of those railway tracks and viaducts, as my client adamantly declared earlier.'

Bewildered, he hesitated and couldn't respond. Confused by all this intense questioning, Father Brady seemed overwhelmed and quite distressed.

Defence Counsel Scharles fumed, his flushed face deepened over what he assumed was a purposely orchestrated delay. *I'm sure the witness is wasting time just to fabricate more lies. What other excuses is he concocting now? He seems so bamboozled that he can't remember how to tell the truth.*

Reagent interjected. 'Mr Scharles, I object. Are you confusing my client by repeating the questions you asked earlier? You have distressed him to such an extent that he can't think logically. I suggest you desist and let him rest.'

'During his previous testimony, when briefly questioned by you, the prosecutor, I found his answers vague on all points I asked,' declared Scharles, who had goaded the witness, while checking the paraphernalia in his hands.

'Might a man … be allowed ta finish? Then if ya let me, I'll be tellin ya the truth,' pleaded the anguished witness. Father George clutched his chest as a severe pain travelled down his left arm. Feeling dizzy, he soldiered on regardless without disclosing his physical discomfort. Stressed to the limit due to mental torment, his mind fluctuated on the brink of despair and he felt as though he would physically collapse.

Aware of the priest's anxiety, Justice Strezlewski intervened. 'Defence Counsel, desist stressing this witness in order to extract what you think may be the truth. I and my panel colleagues suggest that you cease persecuting him. Rephrase your question, and moderate your anger.' The spokesman's patience had reached saturation point. He disapproved of Scharles' arrogant manner which he considered abhorrent, given the priest's exhausted condition.

'Your Honours, I apologise if I have distressed the prosecutor's witness,' Scharles smugly concurred. 'It wasn't my intent.' He then addressed Father Brady again. 'I want you to clarify the truth. I think you have deliberately lied and refused to answer my requests to clarify whether you and your accomplices were arrested for sabotage. You incited a riot by obstructing justice because you did commit sabotage. This statement wasn't pulled from thin air. My client, the accused man, has signed an affidavit verifying that fact. Remember witness, you are under oath.'

'I don't care what von Breusch signed. And no, we didn't incite a riot, as ya crudely put it. We were taken by force. Me part was ta pass on messages that's all. My stoker and I did not sabotage any railway viaducts or the tracks.'

'So you never, at any time, used explosives to blow up train lines, after your engine had crossed the points.' The defence's increasingly loud demand echoed around the courtroom, now hushed. The atmosphere was electrifying.

Kendall walked into the room just in time to hear Scharles abusing Father Brady, who was being subjected to ridicule by this arrogant defence counsel. 'Excuse me Sir,' he challenged the nearest court officer, 'what has he been saying to upset my patient? I have Father's injection ready to administer. He's due for this drug. Please ask Justice Strezlewski or his judicial colleagues, if I can speak to them in private.'

'Doctor. I shan't be a minute. Please be seated until I return. I cannot interrupt the panel. They're in the middle of deliberating at present.'

Furiously, Kendall scrunched one fist over the delay. *This is urgent and that fool ought to use his eyes. Even from here I can see Father is gasping. The pains in his chest must be excruciating. If I don't, or can't administer this drug in his arm now he'll collapse. Bugger standing here. I'll interrupt the panel and insist on giving him this heparin injection.*

'When arrested by the Gestapo thugs … I wasn't anywhere near our engine. We only stopped in between stations to save the lives of people in cattle trucks on their way to concentration camps. We only helped to sabotage railway viaducts and tracks, whenever possible after we'd travelled over them.'

George Brady found it difficult to breathe with the stabbing pain in his chest. It penetrated his back and then dissipated within seconds. The intense tightness in his throat made it hard to swallow. Probably indigestion, he thought. Nevertheless, he ignored the pain to clarify what had occurred.

'You admit there was just cause for your detainment?' Scharles queried.

'No, not then there wasn't. T'was me job ta drive our engine and pass on information to our contacts in Berlin.' Feeling dizzy, Father grabbed the witness stand for support. Talking slowly, he plodded on. 'Things were pretty crook in the towns we went through. Poor and underfed folk had no hope of surviving cold winters, without our or the Underground's help. We made food and coal drops while guards were on their meal breaks. I did what I could ta block the enemy tanks and trucks from reaching their front lines.'

'It sounds as if you and the railway gangers had deliberately provoked the guards. They were following a lead, so I'm informed. A positive lead, that led the Gestapo directly to your engineer on that day. More than a coincidence wouldn't you say? That's the real reason why they arrested and transported you all to their Berlin Headquarters. There *you* refused to cooperate and wouldn't tell their officers what you had done. Not satisfied with your lies, they handed you over to Captain von Breusch to be interrogated.'

Gasps from several sources were heard across the courtroom. It appeared as though this witness was on trial and not the accused, by this acrimonious questioning by defence counsel. To lay observers it sounded as though Scharles was endeavouring to break the priest's spirit.

'I object, Your Honours. Defence Council is deliberately baiting my client. Can't you, Mr Scharles, see the man in that witness stand is ill and cannot cope with your demands? If not, it is plainly obvious me and to everyone in this courtroom.'

'Sustained,' Justice Strezlewski decreed fervently. 'I have warned you more than once today, not to badger this or any witness. You have totally overstepped the bounds of protocol. It is you, Mr Scharles who should keep your questions brief. The witness is not on trial. Your client is. Remember it.'

Scharles acknowledged Strezlewski's response. *His reprimand of me was unwarranted. How can I get the truth without cross-questioning this priest? If I can rip his testimony to shreds, Strezlewski and the panel won't get another chance to ridicule me again. I hope they overrule Reagent's next question. I'm determined he won't win this case. I will.*

Rolf von Breusch looked through squinting eyes at the witness. 'I remember you now, priest. You threatened me and refused to answer when I accused you of being the instigator of those gangers who sabotaged our bridges and railway viaducts. You deliberately lied to me, in my office,' Rolf bellowed. 'The guards who dragged you back down to your cell, were not mine. They told me they had whipped you until you collapsed. I never gave that order. Their superior officer did on that day.'

'Herr von Breusch, refrain from bellowing and badgering the witness. If you continue with your unwarranted and outrageous accusations, I will order your withdrawal from this courtroom. Do you understand my demand?'

Rolf nodded and desisted. Aggravated, his eyes blazed with fury as they glared at the distressed witness.

Instructed to proceed, Father Brady paused before continuing. He sighed with relief as this bout of excruciating pain had eased until it was tolerable. Taking a shallow breath, he spoke in barely an audible tone, 'They threw me inta a tiny cell at the Chancellery. Two guards flogged me until I collapsed. I awoke later that day soaked in a pool of me own blood ...' George's shaky hand embraced the witness stand.

'Your witness looks ill. He is exonerated from further questioning by you and the defence counsel, Mr Reagent.' The chief clerk rushed to catch Father Brady who collapsed. Kendall also rushed to his aid and together they laid him flat on the floor. Kendall tried to resuscitate his patient. With tears in his eyes, he felt the priest's carotid artery and shook his head. Judging by his actions, the judicial panel knew this witness couldn't be revived.

Immediately courtroom one emptied, an officer covered the deceased priest with a sheet. Kendall followed the stretcher bearers through to the first aid annex. With the quell of endless chatter still ringing in their ears, Scharles and his team collected their briefs. Reagent's team had left the courtroom.

He and Samantha stood a-gasp outside the door. 'Go and buy me a brandy in the member's bar. I'll join you there, once I've spoken to Strezlewski in chambers, Sam. I won't intrude on Kendall's privacy. He needs time alone with his friend. I can't believe how Scharles pushed him to the brink of despair. That foolish quibbler will never learn the rules of decency, not while his arse is pointing to Hell.'

'Arrogant men like him don't know the meaning of compassion. I think *he* thrives on greed and I dislike his uncouth manner. He doesn't know the meaning of dignity. The snide look on his face equalled von Breusch's smug scowl. Their aloofness floored me. Scharles could see the priest was in agony, yet he still persisted in persecuting him. I felt like jabbing Scharles in the eye with the heel of my stilettoes.

'It wouldn't have shut the pompous prig's big gob. Sam, you hit the correct button by calling him uncouth. My terminology for him is an uncouth bastard. Here's Miss Baumer. She's probably looking for Kendall. Excuse me, Sammy, I'll be tactful telling her of the priest's death. Then I need to speak with her half-brother. I won't be long. See you in chambers. It'll be difficult for anyone to comprehend how his death occurred so suddenly.'

Bordering on tears, his secretary nodded. A cotton handkerchief worked overtime. Sam wondered how Miss Baumer's father would accept the sad news. *I know he suffers from congenital heart disease. Kendall seems a dedicated physician. Oh well, I can't stand in here miserably thinking. Michael will think I'm lost in this massive building, if he doesn't find me in chambers.*

Samantha balanced her briefcase, files and coat down the corridor until she met Miss Baumer, who was in tears. 'Come with me Arneka, and I'll make a cup of coffee. We can drink it in the recreation room, it's quiet in there and no one will bother us. Where's your father? He and Christian might like a cup. I can ring down and order lunch or mixed sandwiches.'

'I'd love a cup of black tea, nothing to eat, thanks Miss … I've forgotten your name. I feel so distressed. My father's taken the news really hard. He's lying down in the first aid bay. I won't disturb him. I think Kendall and Chris are with him now.'

'That's understandable. I honestly don't know why the defence lawyer accused your friend Father Brady of vandalism during the last war. He did what he thought was right to save all those unfortunate people from starving. Michael was furious with Scharles. Most of our team detested his arrogance especially when von Breusch gloated … I keep forgetting who he is …'

Arneka interjected. 'That's okay. None of my family call that beastly Nazi a relation. I remember the savage beatings he gave us kids in our home. He murdered our mother there. It's all noted in your files. He placed innocent people's lives in jeopardy by signing their death warrants in Berlin. Our nursemaid, whom we called Button, told Dad that. Heide was also murdered. Not by von Breusch. He ordered her death by sending his goons to firebomb her home and car in Hamburg. The smug bastard thought he'd killed us kids. I can't wait to face von Breusch in court. I'll stick to the oath. I will accuse him of murder and attempted murder of both our mother and Heide. I hope he cringes as I confront him. Your boss has our written depositions, including Kurt's and several letters. One was written by Heide, my father's sister to his wife, Anna. Heide also wrote a letter to Major Kassell. He was our closest friend in Hamburg.'

'Arneka, call me Sam. Mike and I've seen those letters. I gather by you getting all that guff off your chest it's made you feel better. Please try to eat something, or you might faint. I need a drink and I'll order one for my boss. It's been an extra-long session in court today.' Samantha used the phone and placed an order for refreshments to be sent up, including the brandy for Michael. 'The canteen's sending us a plate of mixed sandwiches, a jug of coffee and a pot of black tea, Arneka.'

'While we're on the subject, Kendall and his mother dislike von Breusch immensely because of his cruelty. We all knew him as Ian Ross back home in Aussieland. Kurt, my adoptive father and the Marquands in France had a code of silence and loyalty. Dad says it was the same in the underground. I typed his deposition. The French partisans helped us to escape Nazi tyranny in Hamburg. Otherwise we would've all been murdered. Nobody could walk the streets without fear of being shot, or shoved in cattle trucks then sent to concentration camps. Kurt feared we kids would end up being skewered on some Nazi's bayonet. The partisans signed our documents Kurt had forged. It was his greatest feat in wartime, although he won't admit it. I hope he doesn't have another angina attack or heart seizure. None of us expected the priest to die. It occurred so suddenly. Kendall told me about his death a moment before you did.'

Samantha referred to her notes while sipping tea. *Everything Arneka has just said tallies with her adoptive father's deposition and my records. She looks so pathetic sitting there pensively thinking of the past. Her honest appraisal of their wartime traumas and brutality by that Nazi astounds me. Michael and I admire her tenacity and strength of character. Her testimony on the witness stand will be interesting. I'll leave this page of court procedures on the table for her to read. It'll make things a lot clearer and easier for her to understand our legal jargon.*

Samantha stopped on seeing Justice Strezlewski approaching her boss in the corridor. Michael nodded for her to continue onto chambers.

'I'm here on behalf of our panel. Michael, please pass our condolences to Kendall. I can't find him or Mr Baumer and his family. They seem to have disappeared. Have they all left the building?'

'No. Miss Baumer's in the recreation room. Her brother Chris could still be in the men's room. I'm going to see Kendall in the first aid bay. Edward, I was extremely annoyed by von Breusch and Scharles' attitude after my client died. They both sniggered. Scharles made offensive remarks about him, in my hearing. I had to control my temper and I felt like kicking his arse.'

'I know what he called the priest. Have you forgotten my lip-reading talent, Mike? I repeated it word-for-word to the panel. Leonid Pyotr almost took a fit. Bob Spiller feels the same as we do. Ingivar Brage's face turned a deep shade of an overripe plum.'

Scharles called the deceased priest an overstuffed, drunken bum and a fucking old liar, which disgusted me. Michael's frown deepened.

'None of us condoned his despicable behaviour. I confronted and warned him to respect the judicial panel and all future witnesses. Court is adjourned for twenty-four hours, in respect for the priest's passing.'

'I'm glad you ticked Scharles off. He's an arrogant pompous prick. The local morgue rang me. There won't be an autopsy. My client and your physician have both signed the death certificate. One of their vehicles will be here soon to collect the Irishman's body. Sad business. Now if you'll excuse me, Robert. I'll see if Kendall or the Baumer's would like a hot drink or refreshments.'

'Doctor Gwlynne looked absolutely grief-stricken. I spoke to him in court. How's his sister taken the news of their friend's demise? She and her brother Christian went through unadulterated hell in Germany. I re-read their depositions this morning in chambers. They all suffered the indignity of being manhandled and threatened by Nazi hooligans in Hamburg. Herr von Breusch has a lot to answer for his war crimes. The official files we received from Berlin of his service in the Wehrmacht reads like a monstrous tale of horror. None of my colleagues can fathom how his brain functions. Off the record, Mike. At times he acts like an egotistical megalomaniac. Yet he's as sane as you and me. It's a ruse to incriminate a witness who challenges his authority. Being a strategist, his knowledge of controlling rioters and saboteurs is used as a ploy to confuse everyone. Have you noticed how he uses his eyes as a tool to intimidate his opponents in court?'

'Yes, and I agree with you Bob. If I don't get moving, the hearse-driver will be knocking on our doors. That'd be a turn-page for the snoopy paparazzi. They're lined in dozens outside this building with their cameramen perched on cars, ready for a mainliner for tonight's rabble rags.'

'Don't panic, Michael. I'm pleased we arranged for the disabled man to see Father Brady before they took his body downstairs. A curtesy to say his farewell to an old friend. A hearse will be here soon and it'll pull in the rear entrance.' Strezlewski tapped Reagent's shoulder and returned to chambers.

Instead of lingering in the corridor, Michael hurried to see how Kurt Baumer was feeling. He moved one foot and heard a familiar voice behind him. Turning, he faced the French advocate. Before he could speak, the Russian spoke in perfect English. 'We want you to pass on our condolences to the Australian doctor and his friends. It came as a shock to witness the death of that Irish priest today. Tell them we feel for their loss over his sudden demise.'

'I won't forget, Leonid. Are you ready to leave now? If so, I can give you both a lift in my car to our hotel. My clerk and I will meet you in the basement carpark in approximately ten minutes. I tell my clients to use that entrance.'

The judges declined his offer; they'd booked a taxi to the Metro. Shaking hands with Reagent, they left the building by the same entrance.

36

10.30 am Tuesday: Court convened with its members and witnesses in solemn moods. Memories of Father Brady's sudden demise shocked everyone. Kendall spent an hour rewriting the illegible scribble of his deceased patient's medical history. It concurred with the court physician's preliminary version. This morning they compared notes, before he delivered them to the local coroner.

11.15 am: As court resumed Reagent began, in a more sombre tone than his arrogant counterpart. 'I apologise for the unnecessary delay, your Honours. It was warranted, due to the distressed state of my next witness. I shall now call her to the stand. The lady has requested that her surname be withheld, for health reasons. I re-read the judicial panel's sanctioned reply, and have duly noted it.'

The defence counsel smiled at his client in the dock, asserting his right to reconfirm some answers later, if necessary. He seemed elated with his efforts to extract the truth from von Breusch. Scharles had deliberately tried to derail the priest's testimony in the previous session.

He suspected Reagent had filed a request to have him disbarred and ignored him completely while signing the daily register. Scharles detested Michael and was tempted to block his entry into court. This defence mechanism failed dismally. Samantha dropped her wet umbrella and Scharles tripped over its handle. He landed arse-up on the courtroom floor. His obnoxious attitude accompanied by indelicate swear words didn't enhance him to the spokesman nor his colleagues, who were talking not a metre away in the hall.

Struggling to his feet, he gathered his pink-tied folders, glared at Sam and angrily pushed her aside. 'You should be more careful, young lady.

The next time you cause an accident someone may knock you flying with a clewed fist.'

She ignored his rudeness and proceeded behind her boss and their team to their seats. 'Oh, that prig infuriates me Michael. It was an accident. I tried to juggle these memos, files, my brolly and attaché case. Something slipped and you all saw the results. He's a supercilious old grump. I detest Scharles. He's a pompous man. He thinks he's God's gift to all secretaries in this quadrant of rabbit-warrens and passageways. I hope he falls down the stairs and breaks his wrinkled neck before he chastises me again.'

'My, my, displaying temper? We all agree he's a pompous ass. Now, let's settle down to business. I need to skim through my notes before this session gets underway in earnest. Although I must say, our learned wigs enjoyed seeing him arse-up in front of strangers and witnesses signing-in for today's hearing.'

How could Father Brady not misconstrue all questions put to him by a legally qualified ignoramus. Scharles egged him on until he collapsed and died. Michael will rip all unproven evidence he puts forward to sheds. My boss and our research team have studied his client's last statement in the dock and they're ready to tackle him head-on. I doubt if von Breusch will readily accept the panel's deliberation. He'll most probably deny them the pleasure of seeing him hang. Before Scharles fell, I heard Strezlewski say they will be fair, equable and unbiased in their judgement. His marital crimes will also be taken into account. On Arneka confronting him in court, he'll get the biggest shock of his miserable life. He deserves to be punished for the despicable way he treated them as kids.

Michael nudged her hand with his pen. He suspected she was daydreaming. 'Don't go to sleep on me. Our learned panel are seated. They nodded. We all responded. You didn't. You stood up and looked as if they didn't exist.'

'I couldn't sleep last night. Pains in the tummy. Just now my mind drifted back to yesterday's tragic occurrence. I'll listen, with my eyes closed. Nudge my arm again, the moment von Breusch enters the stand.'

Reagent tapped her hand within an instant. Sam's eyes flickered then closed. Michael surveyed the listed witnesses on her clipboard. *I hope my next witness has arrived. If not, I'll call my surprise witness to the stand. I know she and her son are here. An exceptionally busy orthodontal specialist her husband may not come. I've spoken to him on the phone a few times in their Prague home.*

Michael prodded Sam. 'Defence counsel's client is in the dock. Keep my seat warm. Once von Breusch is settled, I'll throw a few doozies his way. I will, however, keep my questions polite and short. He looks a bit dishevelled this morning, a rough night sleeping on a hard mattress with threadbare blankets, I suppose. Here I go Sammy. Look lively, or you may cop a gruff rebuff or a frown from our black robed, red-collared colleagues.'

'Herr von Breusch, before I ask you an important question, I would like you to scrutinise this miniature photograph. Have you seen it before and where? Also do you recognise that insignia on the officer's cuff?'

'Of course, I recognise the SD insignia and myself. What are you trying to imply? I'd like to strangle the idiot who took that photo without my permission. Huh, he's dead probably. All Reich officers looked alike in uniform. It could've been taken in Potsdam. I attended an international conference there with my superior officer. What's it to do with you and this tribunal? I do not and have not denied that. Define what you meant, Reagent. If not, I refuse to answer any more of your imbecilic questions. Use your brain, or is it the size of a dried pea burnt to a cinder?'

Rolf von Breusch received a harsh scowl and caution from the entire panel. Leonid Pyotr, the Russian also scowled. Brage Ingivar, the Swedish judge glowered. Robert Spiller, the English advocate clenched both his fists as did Hérbert Pautenel. Mr Justice Strezlewski, a tall man, stood upright as the gavel thundered down onto its wooden block.

'Herr von Breusch, I countermand that caution. You will be re-shackled then handcuffed to that dock rail. One more outburst of temper and you shall be removed from this room. I am *very* tempted to charge you with contempt of court. Moderate your language and your tone of voice.'

Rolf drew in a breath to expand his lungs. Fuming, he cleared his throat and spat on the well-floor. 'Let one of your pet donkeys clean that up. I won't. Am I to stand here listening to the prosecutor's drivel? I will respond to my own lawyer. At least he knows how to respect and not threaten his clients.'

'Proceed, Mr Reagent.' Exasperated, Strezlewski wiped his hands on a clean towel his clerk had passed to him.

'I will read aloud this letter which I received from a woman in Düsseldorf. She describes her brother, a clone of yourself, whom you claim you never murdered in your home in Hamburg. She has enclosed in the letter a small sepia photograph of him, taken around 1944 in his factory uniform. I think they will both astound and interest you, Herr von Breusch.'

As Reagent read, the furrows on Rolf's forehead pleated and his ruddy cheeks deepened to a rich puce. 'Lies from a liars pen. Hans never wrote, or posted that letter to his sister. He was to drive his car to my home. He phoned me, because my directions weren't clear. I told him the correct route to follow. He then left in his car outside my gates. I honestly don't know why he returned. My lady had gone home, long before Hans arrived that night. I'd given her gifts for her birthday. She promised to ring me the next morning.'

'I believe this letter spells the truth. It contradicts what you just said about Hans Selig leaving your home. He didn't return, or leave in his car at all. You murdered him to confuse Reich investigators. This letter from your children's deceased nurse proves that you ordered and instigated her murder. You bribed Gestapo thugs to carry it out. They firebombed her home and car with live grenades. All because she had twice rejected your sexual advances.'

The French advocate whispered to Mr Justice Strezlewski. 'Bob, will you object if Spiller or I direct both counsels to approach the bench?'

'Not in the least. I'm almost hoarse from chastising von Breusch. Do so now while Scharles is collecting his briefs. Call them both to the Bar, Spiller. Hérbert, you can be his spokesman, if an argument ensues. I need a leak. A few minutes break won't disrupt the legal fraternity. See you all in five.'

Reagent gathered the three letters and photos then walked over to the bench. Terri Scharles took his time approaching it. His eyes focused on von Breusch, who shrugged both shoulders.

'As you are aware Michael, your client is ill. His son and Doctor are with him at present. His daughter is also in the first aid annex. We have exonerated him from attending court again. His son is undecided about testifying. His daughter still insists on taking the oath. The letters in your hand we have read. Our initials are on the reverse side. Remember, mention no names in court. Refer to the younger Baumer's as witness X and Z. They are listed under those …'

Scharles interrupted Spiller. 'What's this about witnesses and digits? A harsh look from the spokesman, who'd just returned, warned him to listen, not speak. 'You can now both step back from the bench. Reagent, have you finished cross-questioning this witness?'

Michael confirmed he would after speaking to Kendall about his ill patient. *Snoopy old Scharles will be all ears if he could hear what I need to ask them.*

'May I intrude on your privacy, Kendall? How's Mr Baumer feeling now? His face looked gaunt and turned grey with pain in court.'

'He's asleep. The drug I gave him took effect instantly. Arneka's fetching another light blanket from the warming closet in your bathroom. Kurt's relieved he won't have to face von Breusch again in court. Once was enough.' Kendall looked at the trinket clutched in Kurt's limp fingers. 'That gold locket belonged to von Breusch's first wife. Under Erika's miniature, there's one of him in uniform taken in their Hamburg garden by Heide, Kurt's deceased sister. Anna, his wife kept the locket until she died in England. He gave it to Arneka some time ago. She's never worn it. The one she's wearing today belongs to my mother. As you know, Mum was his second wife. She gave her the locket as a good luck charm before we left home. I know Kurt wanted to show von Breusch the one of *him* in *his* Reich uniform taken in their garden. Well, that's gone by the board now.'

'I'm sure Kurt Baumer won't mind if his daughter, Arneka produces it in court. The miniature could, as evidence, clench our case against von Breusch. Ask her if I can borrow the locket please, Kendall? One of my research team will run off a copy. I promise to return the original photo and her locket to you. Now I must hurry, court reconvenes in twenty minutes.'

Reagent spoke to his next witness outside the courtroom door. 'I knew you and your son were here and ready to face the bar. A court usher will call you to take the stand.' Michael addressed his witness by her married name, still undisclosed to the defence counsel's client. 'Your testimony will undoubtedly surprise von Breusch. He seems to think you died towards the latter part of World War Two.'

'Sir, that's why I accepted your proposal to testify. My husband's flight is scheduled to land in twenty minutes. It depends on how dense the traffic is. Our taxicab stopped at every traffic light, that's why we arrived here later than planned. You met my son earlier in chambers.'

Reagent grasped Denby's hand. 'It'll be surprises all round. I might have a special one for you and your family later. Now, if you'll both excuse me, I'm overdue in court. There's a cafeteria downstairs, if you use this card your meals will be half-price. Let an usher in a blue uniform know where you'll be, in case you are called, please Madame.'

She nodded, collected her handbag, gloves, grosgrain hat and navy topcoat off the seat, then followed her son to the downstairs lift.

'Lost your umbrella again, Mother?'

'Oh dash. I left it on a seat while speaking to that usher. Fetch it please son, while I order our lunch. Don't be long Denby, or your soup will get cold.'

About to retrieve her brolly, he noticed a woman in distress. 'Have you lost something, my dear. I'm Denby Kuper. My friends call me Den. I've come to collect my mother's umbrella. She always forgets things.'

'I'm Arneka Baumer.' She handed him the brolly. 'My father's ill. He's lying down and I've come to get another warm blanket. This usher fetched the last one. Denby, are you here as a witness for the prosecution, or the defence counsel?'

'My mother is a prosecutor's witness, Arneka. I'm just here to boost her morale. Dad is due to arrive within minutes. Mum knew Herr von Breusch in Hamburg. He gave her a gold lighter for her twenty-second birthday, I think. She dropped it in the snow outside his front door on her way to the car. She'll be delighted to meet you and your family. Can I see you tonight? Where are you staying?'

'At the Metropole Hotel. We've rooms on the second floor. Den, I'll meet you in the lobby at eight. That's if my father's okay. My brother, Kendall gave him a new Nitro-lingual spray this morning. His heart reading, an ECG was okay, thank goodness. Kendall's his physician while we're here in Linz.'

'Terrific news, Arneka. Here's my father's professional card. You can buzz me on this number.' Denby jotted it down. 'I must run. Mum will think I'm lost. We're having lunch in the café downstairs.' Den squeezed Arneka's hand and after kissing her fingertips he hurried to the lift.

She stood in awe of this man with courteous manners. *Kurt will be pleased to meet his parents. I wonder what nationality Den's father is. His mother will probably be of German extraction, if she lived in Berlin. A blanket can't warm itself.* Arneka ceased fantasying and accepted one from the court usher.

7 pm: Dressed ready to catch the lift, she popped in to see how Kurt was feeling tonight. 'Kendall, how's my father? He looked really ill on his return from the court. I'm going down to meet a gentleman in the lobby. His parents have an apartment upstairs. And he's offered to take me out to dinner.'

Kendall frowned. 'What gentleman? You have no friends here. Well not to my knowledge. Does your father know him? Arneka you should be cautious and pick your friends carefully. I'll come down to meet him. If I don't, Chris will.'

'You're my junior and he's my elder brother. So don't dictate to me Kendall. I'll do what I please. Take care of my father. I'm going now, *without* you. Why must you two always be spoilsports where my friends are concerned. I don't need a chaperone, not at forty. I should be back here around ten. Leave the door on its latch. Better still, loan me your keys. I won't lose them.'

Kendall threw his set to Arneka. Her red eyes were because all this rumpus had unleashed horrible wartime memories that they, the Baumer's needed to forget. Her adoptive father's signed affidavit and their family's details were recorded. She couldn't understand why Kendall was so pedantic about her casual friends.

Wouldn't he love to know who Den is? Kendall's worse than Chris when he snoops into my private life. He knows damn well to who this locket I'm wearing belonged. One snap hidden under its heart-shaped band I treasure. The other person I detest. The Nazi dictator wielded his cross-strap with fury across our small bottoms for breaking his new glasses back home in Hamburg. I often rub the scars on my bum and feel like killing that beastly creep.

Now, I'm going to enjoy tonight, and forget all those horrible memories. They won't spoil my dinner with Denby. I loved how he kissed my hand and respected my privacy. I deserve a pleasant evening without those two whingers ear-bashing me. Every time they do, I end up with a whopping headache.'

37

10am Thursday: It teemed rain as Arneka and her brother Christian entered the courtroom, now well underway with the accused Nazi refusing to cooperate or answer questions lobbed at him in quick secession.

'Herr von Breusch, why the hesitation? You know why I'm holding these miniatures in my hand. As I've stated before, they were taken of you on your front lawn, by a young woman who cared for your children in Hamburg. That same woman you ordered murdered by thugs under your control. They shot her, tossed grenades at her home and then firebombed her vehicle. All because she refused to submit to your unwarranted seductive charms.'

'Lies, pure supposition. Neither you, nor anyone in this room can prove who took those photos. Heide wouldn't have dared to take one of me, in or out of uniform without my consent.'

The prosecutor looked over his glasses at von Breusch. 'Without a doubt I can prove they were taken by that nurse with you leaning over your Daimler. You shaded your eyes to prevent the strong morning sun blinding them. Here you are shown dressed in a Dress Grey uniform while talking to a mechanic. Your vehicle wouldn't start and that mechanic was your allotted chauffeur, Sergeant Kohl.'

Michael smiled as his next witness approached the witness stand to repeat the oath. He queried if a particular snap was a photo of him talking to von Breusch in front of an ungaraged vehicle.

'That is me, sir. The engine had a blocked valve. I cleared it with a good hard blow and it started. I drove Kapitan von Breusch to work in

Hamburg most days. His wife was standing on the top stair when I left their home. I never saw Frau von Breusch again.'

Defence counsel glared at Reagent. *What stunt is he going to pull now?*

'Is this sepia photograph of you, taken in a Wehrmacht uniform?' Reagent withheld the other two snaps.

Rolf scowled. 'All officers looked alike in uniform. Why target me?'

'I asked, is it you? Not another officer. Remember, you are still under oath von Breusch. Verify if these two other photographs are of you bending over the body of a woman. It says on the back, she was your wife Erika. If you refuse to reply, I have another witness who will verify whether they are of you or not.' Reagent held back, not wishing to show his hand at that moment.

'You don't know if they're doctored. How can you prove they weren't me?'

'It's proven the three snaps are not fakes,' interjected the spokesman. 'They have been verified by two independent sources. Proceed Mr Reagent with questioning this man.'

'I cannot, Your Honours. I have Father Brady's signed affidavit here. I asked what physical violence was perpetrated against him in the Chancellery cells. It says that he 'was struck across his face by a riding crop held in Kapitan von Breusch's hand. The incident occurred because I, George Brady denied being the instigator of causing German viaducts and marshalling yards sabotaged by his stoker and himself after their troop transporter had crossed them. Every incident is written in detail on his affidavit you have all seen and initialled.' Reagent then said, 'That gallant Irish priest speaks from the grave to condemn the accused man of those atrocities. Please try to understand, he wanted his statement to verify the truth of what had happened to him in that Chancellery cell. Persecution is never pleasant to hear or witness. Nor is being condemned for something you were not guilty of perpetrating. Father Brady served his partisan friends well and suffered immensely for doing what he could to save the lives of innocent refugees, Jewish people and families ready to be taken to concentration camps in Germany, Austria and Poland and every country the Nazi dictators occupied.'

Feeling calmer, Michael quoted in a congenial voice, 'I was subjected to a great deal of cruelty by officers under your control. All because I refused to betray my friends, some of the finest men I'd worked with in the Urban

Underground. There were no trials. Yet we were all sentenced to death by shooting, our death warrants were signed by you, Herr von Breusch.' After a brief respite, Reagent afforded in a lower more composed tone. 'Might I add, the deceased priest owed you sir, a huge debt of gratitude.'

The court again buzzed with profound disbelief after hearing the priest's affidavit being read from the grave by Michael Reagent. *If this bastard disregards the truth that I've just quoted, he will be embarrassed by seeing my next client. It will place him in an awkward position and he'll cringe. I know he intends to disclose something vital about his Germanic past. How will he dispute the woman's testimony? It's described in her signed deposition. Boy oh boy, are we in for an interesting time, listening to her denigrate him from the stand. I can't imagine her being dogmatic or rude.*

'Michael, were you thinking about your next client to take the stand? I asked because she seems such a passive and kind person.' He nodded.

In a sharp contrast and in an impressive tone Strezlewski stated, 'The panel will have this courtroom cleared, if more noise echoes from the press gallery. I hope our point is noted.' He then affirmed in a severe voice. 'I charge the prosecutor to proceed. The entire world should hear the deceased man's report on that brutal Nazi regime. It will be horrific listening of a courageous man who can no longer speak. His epitaph challenges us all to be upstanding. Every word he's written will be publicised and filed to prevent the rejuvenation of the Nazi Party, which will forbid the freedom of innocent peoples this world over. To prevent one crime being committed in their name, will be the Irishman's legacy.'

All heads were bowed in memory of a priestly man who'd sacrificed his life to save others. To incriminate this Nazi was only a secondary factor. In a revered tone Reagent continued, 'George Brady found his faith in God because of your cruelty.' A nod indicated the accused prisoner. 'After he walked from the Chancellery, a carload of priests took him to St Xavier's Church. In the catacombs down in its bowels he met an elderly priest.' A sigh expelled from Reagent's dry throat allowed him to finish quoting. "Father Ignatz was a dedicated man who raised my spirits. I, George Brady swear if I am allowed to live in my beloved Ireland, I promise to take my vows." Reagent then added. 'I greatly admire the deceased man's ability to succeed with his vow of chastity. He proved to us all that life cannot be snuffed by wilful, savage individuals who thrive on hate. I shall now retire to reconsider what I have read and quoted to this court and my esteemed colleagues.'

Excessive strain had denied this middle-aged lawyer the majestic triumph he'd hoped. Head bowed, Michael returned to his seat to deliberate on a lot of unanswered questions. A splitting headache caused him to withdraw from the courtroom. He nodded to his peers. Tomorrow could wait. He needed time to get some rest. Sleep had eluded him its pleasure in the dawn hours.

With fixed gazes the panel stood to appraise their fellow and his team leave. No one envied this man who had extracted the truth from a criminal's life history, a story so horrific it consolidated every word the priest had written.

Rolf von Breusch looked dumbfounded and astounded when he recalled having seen the photo of himself bending over his deceased wife's body. The focal point of the snap was the red fire-truck that had belonged to his infant son Kristian.

Who gave Heide a camera and who developed the photos? I don't recall her being anywhere near our house before I released Erika's braided hair. She couldn't have been. She'd taken our kinder to her home. That's one reason why I ordered her death. That bitch of a nurse only lived ten kilometres from our Hamburg property. There's also the letter and photo from Erika's oma, it went missing on that day. I may've burned them. I can't remember now.

This problem magnified as the day wore on. Every word of his memoirs lacked substance. The truth he distorted to read as factional events. Rolf von Breusch's overtaxed mind tried to construct a picture of the day's advent in court. His remembrance of accessing the circumstances surrounding all three photos taken of himself in their Hamburg garden were haunting him.

Free of those heavy handcuffs and calamitous ankle-chains I can relax to read the pages of my completed memoirs in peace. His peace was short-lived as a guard announced his lawyer. *What's he want? Come to gloat? He's the last person I need now. I explained the negatives and photos were reproductions of another officer. Heide couldn't have taken them. I'd like to know what else that blasted wig for the prosecution has in mind. I mistrust him more than I do Scharles. I paid him well to smuggle in the oil painting, my father's gold fob watch and the lighter Maddie gave me. He's a fucking numskull if he thinks I'll knuckle down to his damn will. He'll earn a bullet if I'm exonerated and his empty brain will my first target. Likewise, so will that faggot Kohl, my treacherous ex-army driver. All the opposing team's evidence is concocted rot. Nothing is factual.*

A guard with his rifle ready to use stood beside the cell door. 'Must you stand there listening to our conversation? Get lost for five minutes. I need peace and privacy. Stand well away from that blasted door. Or are you too dense to understand an officer's command.'

'It's alright Rolf. I can't stay. I'm due in courtroom four now. I'll see you first thing tomorrow, with his signed affidavit. He's promised to be here by ten.'

'He better bring the letter I asked you to procure a week ago. You can't mistake his limp. It's quite pronounced. Or it was in army boots. Keep those nosey press bastards well away from here. They're bloodhounds ready for a kill. They gathered in the corridor as if they were instructed to lynch me. What time do you expect the court to reconvene tomorrow? I'll need time to dress in this navy suit, blue shirt and tie you brought in. This rough prison garb is revolting. It's baggy and most uncomfortable.'

Scharles slammed a note on the cell table. 'Read that. It has the time and all the data you'll need to finish your dossier. Do you want me to bring in a hot breakfast? I know you dislike the unpalatable gunk they serve here, Rolf.' He nodded. Scharles also nodded for the guard to unlock the cell door.

38

Reminiscing, the court physician sighed while signing a copy of his report on Father Brady's sudden death. *What a tragic price to pay for truth and his freedom. That saintly man fought the demon of Hell and won by leaving his signed deposition, which the prosecutor read in court this morning. Unfortunately, Herr Brady died before either Kendall or I could reach him.*

Shortly afterwards, as everyone returned and had settled, the Senior Justice sadly enunciated. 'It is this tribunal's solemn duty to inform our learned colleagues, and all dignitaries attending this court that Father Brady's body will be flown to the Holy City later today. His Holiness is sending a delegation to escort his casket to Rome. From there he will find his final resting place in Innes, Ireland.'

Strezlewski paused until the twittering abated. 'Now you are all informed of those proceedings, duty calls this tribunal to order. Call your next witness to the stand, Mr Reagent. Before you do, we members of the panel request everyone to be upstanding for a minute's silence, in respect for a courageous and devout man, all free peoples of this world should be proud of. He forfeited his life for the price of justice.'

Hushed, every breath taken was expelled silently. If a mouse could whisper it would sound similar to a clamorous shout in the stillness of this silent courtroom.

Nobody observed von Breusch, with elbows balanced on unshaken knees, clutched his head. He imaged a man defeated. Not by earthly intervention. Rather by a Heavenly hand. Truth had conquered defeat and truth had won.

With revered silence permeating this icy room, the presiding judge waited until all present had paid his or her respects. To attract the attention of all assembled, Strezlewski announced in a pious voice: 'I am indeed humbled by Father Brady's declaration of faith. And in the brief time I have known him, I've come to admire all he stood for. By denouncing all evil, he has set a precedence to all unbelievers of his principles in times of adversity. I think I can speak for everyone present on this account. That priestly man proudly declared his faith; denounced the wrongs and the perpetrators of those wrongs, stood alone, in pain. He has, through you quoting his last written words, repeated his agonies Mr Reagent, at the cost of his life.'

Kendall and Quinton, who'd been talking to Mieze, his countrywoman of whom he'd grown so fond, were called in from the witness annex for this brief period of silence. She had just commented on the hubbub stirring from within this courtroom. Kendall knew the reason.

Peacefulness is borne out of necessity and simplicity. One should never say or think never. It doesn't exist. An unexpressed thought means nothing to a weak-minded individual. The strong shall inherit the peace. Thank God I can face my Maker with an innocent conscience. Kendall checked the time of death he'd written on the coronial report. It matched the court's physician. *We both were right in recording the time of his death. My wife's lifelong administrator of their faith.*

He closed the priest's medical file and left the room. In quiet repose he stood to contemplate on the cherished life of his deceased patient. Quinton Marciano stood beside his friend, and pensively waited for him to speak. 'Kendall this lady is my dearest friend. Mieze, this is Doctor Gwlynne. You asked me earlier who the man wearing a red-lined grey suitcoat was. Now you know.'

'Hello sir. Quinton met me at the local hospital. We've known each other for over a decade now. I've come to Linz for two reasons. After giving my testimony, I'll be admitted to the same hospital. My feet were brutalised by that officer's men in the Chancellery late in forty-three. You know who I mean. He smiled every time the red-hot poker burned my toes. In his quiet domain he abused me and called me a liar, because I refused to divulge how I had escaped from the machine gunner who fired at random. Wounded, I fell back on the bodies of women with babes in their arms and young children in that forest. I saved myself by climbing over wriggling people who were dying. The faces of dead people were twisted in agony. Their lifeless eyes scared me. Each person looked different to their neighbour. Men in army uniforms dragged some of the bodies away and threw them in blazing fires. Their clothing was burned to ashes, so the owners couldn't be recognised. Two young boys found me lying pooled in my own blood the following morning. Now you know why my feet are deformed and I can't hobble, nor walk more than a few paces without pain.'

With lowered head, Kendall held her hand. He couldn't believe this woman in her mid-thirties had suffered so much in her short lifetime. He shuddered on realising who had caused her endless distress.

'Once the tribunal is over, I'll be flying home to Australia with my friends. Mieze, I have asked your friend to come with me. I purchased a property on the Gold Coast last week, with two cottages and a watermill. It also has a fully equipped stable. Quint, you could groom our two horses. If you choose to come Mieze, you may prefer to do some of the cooking. I'll pay you both well. Then you won't have to worry about medical bills. You have a month to think over my proposal, Quint. Please read the brochures and you'll see what a viable proposition it will be for you both.'

'Sounds exciting, Kendall. I'll let you into our little secret. We're going to be married in two days here in Linz.'

'Congratulations Quint. You and Mieze must be thrilled.' She smiled at Kendall, who then said, 'I'm flying home to Australia with the Baumer family soon. I'll be calling in here on my way back to Ireland. You can give me your answer then, Quint. You love horses, I believe. No mucking of stables or hard work for either of you. I'll employ the people I need to do the hard yakka and difficult jobs.'

'Yes, we both do. I worked as a groomer in my youth. Mieze and I might have some good news to tell you, on meeting your wife when we see you in a month's time, Kendall.'

Kendall's prime concern was for the Baumer family. Kurt, his elderly patient blatantly refused to see a cardiac specialist in Linz, so his options were limited. 'He may, in time come around to my way of thinking. Arneka's worried and so am I about his health problems. I'll inform Kurt's physician in Queensland and he'll probably meet us at Brisbane airport. Now I better phone Brianna in Innes. My wife will be anxious to hear what arrangements are in the pipeline for Father Brady's funeral in Ireland.'

Arneka had just drawn up a syringe to give her father his heparin injection when Kendall stuck his nose around her doorway. 'Why the huge smile? Your gob will stay that way, if you don't stop grinning sport.'

'Don't comment until you've read this cablegram. I'm going to be your boss. Brianna and I have purchased a rural property in Mount Tamborine with a terrific view of the entire Gold Coast.'

'You sneaky twit. How will you be my boss? What surprises are you conjuring now? It says here there are three houses, a set of stables and a surgery on this property. Who's going to occupy them all, I want to know, Kendall.'

His gleaming smile broadened. 'Dorian and I will be partners in our new practice. He's arranged all the paperwork and his father's backing us. He's given us 500 grand to stock the surgery and it has an annex where Dorian will perform surgery on our patients, if necessary. And he's a qualified surgeon now. His main field is obstetrics. How's that for a surprise, sis?'

Absolutely gobsmacked, Arneka couldn't find the words to express her delight. 'Doran's a fully qualified quack now? Wow! You've stunned me. I can't wait to meet him again. I'm over my childish crush, but I still care deeply for him. He may've changed his mind towards little ole me. Do you want me to be the head nursing sister in your Tamborine practice? You certainly know how to bamboozle a disbeliever. I'll jump with joy, if it pans out and we're all living on your property. What's Bridy think of your proposal? Bet she's jumping hoops in Ireland. By the way, when are you bringing her home to Aussieland?'

'Once you and your father have settled into your new home on our property, then I'll book my flight to Ireland. There's still Father Brady's funeral hanging in the wings of time. It could take quite a while to arrange in Innes. He's to be a martyr in Rome. I should know something to confirm that, either today or tomorrow morning. These things take time to arrange, Arneka.'

Mouth agape, she couldn't think of something to say. 'It's all too much to collate in one sentence. Gosh, a martyr? Well he deserves it. I admire the way he handled being browbeaten by that bastard's spiv. He sure gave them a mouthful, in his polite manner. Just you wait to hear what I confront him with in today's hearing. His smug churlish ego will suffer and so will his arrogance suffer the embarrassment of trying to figure how Chris and I are still alive. I'll be straight as a dye when I needle him to tell the truth of our mother, Erika's death. He won't be able to wheedle his way of answering that. The truth will come out, as they used to say in the olden days. I won't turn the screws to make him confess everything he perpetrated against us as kids. Heide and Erika will be looking down from Heaven to boost my strength while I'm testifying.'

Arneka and I both know her father may not be well enough to attend today's session. We best leave it up to Kurt, he'll decide that.

'You haven't fooled me, Kendall. I have a fair idea of what you were thinking. Of course he will. Chris will take good care of our special father. He's even bought him a fleecy-lined coat to wear in that bleak courtroom today. I've sorted his warm socks and thermal underwear. Kurt's feet and body won't freeze in the taxi. Let's tell him your bonza news. I'll give his something different to mull over in court.'

A puzzled look on his unshaven features said more than he could say in a lifetime. 'Do you want me to sell our home in Yeppoon, Kendall? Dad intended to sell long before we left Australia. Your new property sounds a profitable venture for both our family and yours. What will your mother and Pierre think about your proposal? Pierre might offer to be your financial backer in the project.'

'No, Kurt. Dorian's father's a millionaire. He's promised to be our backer and has proved it. There are sizeable funds in our joint bank accounts. The main clause on our contract is for both to sign all cheques. It will eliminate confusion and superstition. We both get on fine. I trust Doctor Payne and he does me. A mutual trust is best in every business venture. Do you agree, Kurt?'

'Yes. I'll telephone my estate agent in Yeppoon later today. He's already approved the sale of our old weatherboard. Kendall, it'll be a huge relief off my mind to sell. We should make a good profit on the sale, our home being only a short walk from the beach and shopping centre. Thanks for the offer to rehouse my family on your property. It'll save us all excessive costs and payment to those greedy house-hunters and sellers.'

'With our housing problems sorted, Kurt it's time to get ready for court. Now I've given you this heparin jab, Chris can assist you to shower. Arneka's put warm clothes on your bed. I'm off to shave and don my glad-rags. Can I borrow one of your ties, Kurt? Mine are packed. Your navy one tones in with my grey suit and this blue shirt.'

Dressed ready to leave their apartment, Kendall glowered on hearing a knock on the door. 'Shit, who's that? If we don't hurry the taxi bloke will be tooting its horn. I'll answer the door, Arneka. You and Chris can walk your father out to here. He doesn't need your help. He's quite stable on his feet this morning.'

Kendall was astounded to see a strange man with a bunch of roses in his grey-gloved hand confronting him in the hall. 'How may I help you sir?' He twigged. 'Oh, you must be Denby Kuiper. Please come in. I'm Kendall Gwlynne. My sister, Arneka will be here in a moment with her father.'

'My parents are down in the car. My father offered to drive you all to the courthouse this morning. We have umbrellas, so you won't need yours. The weather here in Linz is so changeable, one never knows what to wear. My mother's looking forward to meeting Arneka's father and Christian. Let me carry your attaché case, while you fetch a raincoat, Doctor.'

'The beautiful rosebuds smell divine. They have a delicate perfume. I'll put them in this vase. Arneka can rearrange them on our return from court.' Kendall tossed the dead sweet peas in a bin. 'She's just changed the water, so they won't wilt. Here they are now. What should we call you, Denby or Den?'

'Den's okay. I prefer it to the poshness of Denby. My mother, Madeleine must've been drunk while naming me. I'm joking. She's a tee-totaller. Coffee drinker on freezing nights. Dad puts a tipple of rum in his hot chocolate.'

Kendall smiled. *Well, that's one thing I won't need to ask. So much for a lecture on the family history. Kurt's walking cane, where did I put it? Ah, it's here in the umbrella stand.* 'Hurry on Arneka, you're holding up the works. Your friend's family can't wait in a freezing car while you dawdle.'

'Get lost, Kendall. We're ready and you aren't. Going out in those comfy slippers, are you? Chris, this is my friend, Den.' The men shook hands and she looked at the item in her hands. Her brother's grey brogues fell from her fingers.

Comfortable in their rented convertible, she dug him in the ribs. 'Den, your mum's breathtakingly beautiful. She's a classy dresser. I bet her clothes are from some courtier's chic salon in Prague. And I love how your father's honey-tinted moustache twitches if he disapproves of some misdemeanour you do. His hair has a golden glow, slicked back with Brylcreem. It not unlike yours Den, parted on the side. In Aussieland we call red hair a carrot top.'

Kendall tapped her knee. 'Shush Arneka. They'll hear you. Keep those rude comments to yourself. Or you'll embarrass his parents. Denby, I can't apologise enough for my sister's rude remarks. Please accept *our* apology?'

'Don't worry,' he whispered back. 'They didn't hear her. I've forgotten what she said. Neither of you know how we're related. I can't tell you now. I will before we enter courtroom one. It'll be the greatest surprise of your lives.'

Den hasn't fooled me. Kendall and I both know who he meant. So does Kurt. He's free of pain. The heparin worked well to ease it. His breathing's normal and there's a hint of pink in those gaunt cheeks. Dad still refuses to see a cardiac bloke here. I've talked myself hoarse trying to persuade him. Oh well, after today's session, we'll be free to book our flights home, thank goodness.

Quietness ruled as a smartly dressed, veiled dark-haired woman in black entered the witness stand. 'Madame, will you please remove your veil, or tuck it under your cloche. Either will suffice.' She raised the veil high enough to disclose her vivid blue eyes. They peered at her son's father, standing erect in the dock.

'Madame, do you recognise the accused man?'

'Yes. Mr Reagent. The last time I dined with Herr von Breusch was in his Berlin home. I'd received a gold cigarette lighter with our initials intertwined on it from Rolf for my birthday. Unfortunately, I dropped the lighter in fresh snow near my car. With my arms laden with gifts, I failed to notice it wasn't still wrapped in paper. At his dining table, I commented on the fine china.' She again looked at von Breusch. 'You replied that you disliked violets and told me to keep some items, which I did. I still have the one piece setting. There are clusters of forget-me-nots on their decals. Twice I tried to tell you that I was pregnant with your child. You fobbed me off by lying. Now you know. I will not disclose the child's name or its gender. Nor must you, sir.' She nodded to Reagent. 'Our child is now in its mid-forties.'

'All the rubbish you've stated are lies. I never lied to you.' Cautioned not to disclose this witness' name, Rolf glowered at his counsel.

'You never expected to see me again, did you Rolf? I believe you assumed that I had died in the last war. Well, you were wrong. I took a photo of you in your lounge room with one hand on your beautiful Steinway. I played the piano that last night we dined together. You were beastly to me on a couple of occasions. I have, however, had the misfortune of meeting people, listened to their tales of woe and heard them describing under oath how your thugs had maltreated and threatened their family members. I doubt if there's many people still alive who you sent to the Gestapo. Photographs cannot lie, unlike some people whom I dislike in the extreme.' Witness Q nodded to the panel who let her finish speaking.

However, Reagent intervened. 'Witness, on your arrival at the von Breusch home that night, what did you see or hear? It is important for you to be precise and to answer truthfully.'

'Kapitan von Breusch took my coat and gloves then kissed my cheek. We dined with soft music playing on his turntable. I commented on his china, even though I knew he detested the aroma of violets. That stems from his mother wearing the perfume, I think. Rolf told me to keep a setting for one, which I did. Just before I left his home the telephone rang in his hall.'

'Did you overhear the conversation,' Reagent smiled. 'If so, tell the court what you heard.'

'I think it was around ten o'clock. Rolf called the person he spoke to Hans. And he abused him for not knowing which shoulder the cross-strap was worn on. I assumed they were talking about a Reich uniform. I later checked his wardrobe and found his old blue-grey uniform was missing. Then I left. On the way to my car I dropped one of my gifts. I didn't hear the lighter fall, because it must've landed on soft, fresh snow. I then drove home.' She hesitated. 'On recall, I now remember the tie of Rolf's robe caught on my wristwatch. He wore nothing under the gown and I saw his full-frontal body. I think I commented on him being naked.'

'I have no more questions of this witness. Your Honour. Yes, I have. How can you prove Herr von Breusch fathered your child, who is now around forty, I believe you said.'

'Last summer, my husband ordered a DNA test. The results verified Rolf was and is the father of my child. The laboratory compared it with a lock of his hair, which I'd removed from his comb and put in my locket. Those few strands proved an identical match with our child's blond curls who, as I said is now a mature adult.'

Letting her veil fall to hide weary eyes, Madame Q stepped down from the stand, nodded to the panel, collected a few things off a nearby seat and with her head high left the courtroom. A slight hum could be heard in the wake of her silent footfalls.

'Well mother, you look extremely pleased. I commend you for tenacity to confront the dragon. He looked gobsmacked while you were speaking. I smiled at you calling me "it". I don't think he has a clue to my real identity. It'll keep him wondering whether I am male or female. I thank God that he didn't try to denigrate or humiliate you. I quite expected him to belittle you on that stand. I'm sure you kept the rotter intrigued and under that magic spell you create to hoodwink most lawyers. It worked. Come on, Dad's waiting in the car and we're all starving.'

'I need to see and speak to Arneka. You haven't told her about our little surprise, son? Where did you put the parcel?'

'Don't panic, Mother. It's here in my satchel. I've just spoken to Arneka and she's accepted our offer. Her entire family will be our guests tonight. I'm going back to that courtroom to see her confront the dragon face-to-face and head-on, were the words she used. Arneka has a peculiar way of saying things if she's annoyed. Will you tell her tonight that we're related? I thought it best if you told her, rather than me. We can't ruin a terrific friendship before it develops into something magical. I really treasure Arneka's company. She has a great sense of humour. I like both her brothers. Christian's only fault is he seems to derive pleasure in tormenting her.'

'I can't imagine Kendall doing that. He's a gentle-natured man. I think he would protect her against all troublemakers. I'm pleased you genuinely care for her. They're returning to Australia soon, so her father said this morning.'

In their hire car on the way back to their hotel, nagging thoughts of his wife's beloved priest's death were persistently reoccurring in Kendall's unsettled mind and had triggered a headache which wouldn't cease. 'If this raging pain doesn't soon ease, I better find a local bloke to give me an injection.' Kendall refused to self-administer any drug or illegal substance into his arm or bum, which would be impossible. Nor would he let Arneka do it. Only once on the continent had this occurred. He now recalled the female physician's name. Stepping into their apartment he dropped the key. 'Damn it. I better give her a tinkle. Where did I put the capsules she prescribed? They worked wonders. I'll look up the drug's name in my *Merck's Manual.* No, a good night's undisturbed sleep should remedy this problem. Cold compresses on my temporal lobes and top vertebrae haven't eased this wretched pain. I feel as if my head is ready to explode.'

'I noticed a glimmer of light under your door and heard you muttering. What's wrong, Kendall? You look ghastly. There are dark circles under your puffed eyes and your cheeks are scarlet. Running a temp, I bet. Gosh your forehead's burning. Into bed. Don't worry about stripping off. Your clothes will survive, but you won't if you don't get a decent amount of sleep.'

'What are you doing, Arneka? Put that receiver down now.'

'Ringing the hotel quack. I don't want to wake in the morning to find you dead, stupid. Kendall, if I don't phone him, you'll be in hospital within an hour, or in the morgue. I'm going to fetch Kurt and Chris. You might listen to them. Because you won't listen to me, or accept my advice. You're a fool.'

'Don't disturb your father. Chris has just come in. He'll be exhausted, after searching this town's boroughs for your father's lost briefcase. Taxi drivers don't give a damn, if lost property isn't returned. Most of them prefer to make a good profit for a full night's work. I'll be fine, if I can get twelve hours of undisturbed sleep.'

'If I make you a hot mug of milk and honey, will you drink it? Don't up and chuck it. And it will be my pleasure.'

'No. If I drink anything except a sip of iced water, I *will* vomit. Will you fetch me another cold compress from the small fridge? Wrap the crushed ice in a hand towel. There's a clean one on the bathroom basin. Then please check on your father. He'd due for another heparin jab soon.

His temperature was normal an hour ago. I've probably picked up a flu bug in court this morning. Every second person kept on coughing. Who knows? It's not impossible.'

'Don't lie to me. This ain't no damn flu bug. Your temp's off the score board. Shove this damn thermometer back under your arm. If it reads high again, you're going to hospital. No arguments. I refuse to listen and now I'll pack you a small bag.'

After checking on her father and administering his required drugs, Arneka returned and sat by Kendall's bed all night. In the morning his temperature was below normal, he had no headache and seemed anxious to face a full day in court.

This was her day and Arneka sweated on challenging the man she detested. She dressed in her favourite suit. Donning a salmon, self-striped silk blouse and grey skirt with a fishtail-fluted rear pleat and low heeled, deeper grey shoes, she then assisted her father to dress. Kurt's ablutions were done with Christian's aid. A thick navy jumper, overlaid his sombre grey shirt. His tie was pale blue. Black shoes over woollen socks made him feel like dancing. A soft grey Fedora kept his bald head warm. A multicoloured alpaca scarf wrapped around his neck kept the icy winds at bay. A black leather-covered wooden walking stick completed the outfit.

'Get a move on lazybones. Taxicabs don't wait for stragglers. Nor do buses. What's holding you up? And I don't mean a pair of skinny legs. Kendall, how's your nogging feeling this fine, bleak morn? You snored the pigs to market all night. I'm glad you took my advice and let the quack give you an injection. I was sitting beside you most of the night. Except when I poked my nose in Kurt's room. He was sleeping like a drunken bum after a night imbibing on whisky. I know he doesn't like drinking grog.'

39

10am: On approaching the courthouse by its rear entrance Kendall whispered to his sister, 'Arneka, in the witness stand you must not look at his eyes.'

She frowned. 'Why? I think they're evil. They spew fire more than any dragon could. He will *not* browbeat me. I'll face him with the strength of an ox. I'm ready to go in there now. I've just been called to give my testimony. Wish me anything, but not good luck. I don't believe in luck. It's superstition. Good fortune is a far better word to use, I think.'

'He uses his eyes as a tool to intimidate opposing witnesses. Look at his bald noggin. Your slang, not mine. Then it'll have the opposite effect. He'll try to antagonise you, by throwing a furphy in every word you say. He is evil. Downright evil. So remember my caution. Otherwise he will make you suffer some indignity and you'll be the loser. And I don't have to spell out the results of that to you. He'll try every lurk in the rulebook to twist your words solely to bamboozle you.'

'Got it all logged in my mental computer. The cogs are turning and they'll keep on churning. He has a ghost of a chance of outwitting me. I'll make you and Kurt proud of me on the stand. I hate courtrooms. Most of the time they're bleaker than a cupful of crushed ice. Chin up, mate. Because mine's higher than this hall's ceiling.'

As they all filed into the lift Kendall winked at his sister. 'No gags. No frivolity. Be serious. It'll throw him off guard. Look at Kurt, Chris or me, if your courage wanes. We'll be sitting in our usual seats towards the back of courtroom four, not one.'

'Why four, this morning Kendall?'

'Security reasons. It's a smaller room. Besides this whole building will be in lock-down within twenty minutes. They're taking no risks and there have been two bomb scares already this morning. We'll all have to walk through a scanner, so Reagent just told his research bloke.' Her name came over the intercom. 'That means you, Arneka.'

'Miss Zed? Have they changed my name?'

'Yes, you're his final witness. Look lively and be ready to face the dragon. Have you got it in your jacket pocket?'

'No Kendall. I gave the folder to our spiv's boss outside the lift on our arrival. The bigwigs will probably have scanned its contents by now. Kurt looked absolutely miffed on finding the folder tucked under his suitcase. I read the contents, before I gave those pages to you.'

The electronic scanner buzzed. 'Oh shit. My bracelet. I better show it to this bloke. Otherwise they might strip search me. Gosh, how embarrassing.' Arneka fished it out of the inner pocket of her overcoat. 'That's all I have on me. Unless you want to examine my earrings. They're silver kookaburras.'

The official laughed. 'No Miss. You're free to pass though into this courtroom. Leave your handbag in this tray. It'll pass through our scanner and you can collect it in there.'

Kendall knew the ropes and emptied his pockets. 'Sis, they're taking no chances. This lot are very thorough. Remember my advice. Don't look at the dragon's eyes,' he whispered in her ear. Feeling a little hesitant, he watched Arneka approaching the witness stand.

In a clear voice she repeated the oath. Her eyes focused on von Breusch's eyes. Her impenetrable stare made him look away. His gaze focused on everything, but her glum features. *What is this bitch trying to do? She won't browbeat me, with her glares. Who is she? I can't recall seeing her before. Probably one of the prosecutor's stooges ready to lie, or concoct some cock and bull tale to accuse me of. I'll soon bring her down and make her cringe.*

Arneka deflected her gaze on hearing Reagent speak. 'Miss, do you recognise the accused man? State how, why and where you have seen Herr von Breusch. Speak clearly and do not be intimidated. The panel will then address you.'

'My first recollection of seeing him was as an infant of three in our Hamburg home. Although I have disowned that man, I am his daughter. He treated our mother abominably and beat my brother and I unmercifully. Our nursemaid, Heide tried to protect us on many occasions. She also tried to keep us out of his way, by taking us for walks in the forest, or around our garden ...'

'She's lying. My kinder died with their nursemaid. I know, because I drove past her home. I needed to give Heide some more documents plus a book of fuel coupons for their journey to Berlin later that day."

'May I, Mr Reagent?' Arneka pointed to her left shoulder. 'I won't undress. I ask you Herr von Breusch, did your daughter have a teardrop-shaped birthmark on her neck?'

Rolf stiffened and stood erect. The look on his face was indescribable. Teeth dug deep into his bottom lip. He was ropeable. 'How do you know she had a birthmark on her shoulder, just below her left ear? Tell the court what colour it was? No, you cannot. Because you are a liar.'

Arneka eased her blouse down a fraction. 'What colour is that. Unless you're blind, you and everyone in this room can see my birthmark is ruby-red. Now deny I am not your daughter. There were a lot of compassionate and honourable Nazi officers. Major Kassell, the father I adore, knew Viktor well. The paper's Heide photographed in your office prove your intent to kill innocent Jews you persecuted and sent to their deaths. Everything you wrote on those Reich letterheaded pages will confirm what I have just stated.'

Justice Strezlewski showed the pages of open documents to his legal colleagues in the room. 'Reagent will quote what you, Captain von Breusch wrote. Be warned, most of the text in these copies are extremely gruesome. Horrific in context. Michael quoted, "To save our precious resources, fuel, manpower and perishable foods, I suggest all our rail transporters full of Jews, gypsies and undesirable dissents be taken to sidelines. There the cattle trucks can be opened and sprayed with Zyklon B, or some other deadly gas and each car then sealed. The bodies can be easily disposed of in lime-pits, then sprayed with cement. This method will eliminate the need to dispose of them in numerous concentration camps littered throughout our occupied countries. Think over my proposal? Let me know your decision at the earliest convenience, I remain your most loyal colleague, a dedicated officer of our beloved Führer".' Reagent then added 'This letter is addressed to W Freck. It's text also refers to the German Propaganda Minister, Herr Goebbels.'

'Tell me one thing?' Reagent asked von Breusch. 'Whose signature is written on this letter?'

'It's not mine. I never wrote to Wilhelm Freck. I did know him, casually.'

'The writing and signature on this letter are identical to all documents we received from an archival officer in Berlin. They are yours Herr von Breusch. They compare favourably with documents confiscated from your tan satchel.' Strezlewski looked at the court clock. 'I and the judicial members of this tribunal will now retire to deliberate. Court is adjourned for twenty-four hours.' He nodded to the UN armed guards. They followed the usual procedure, only this time they escorted von Breusch to a security van, not back down to his cell.

Arneka looked at Kendall. 'What happens now?' She raised one hand then looked at their counsel. 'Where are they taking von Breusch?'

'How would I know? I must speak to Reagent. You can listen. No, walk with Chris and your father to the lifts. Wait for me there, I shouldn't be long.'

In a huff she pouted. Holding her father's arm, she guided him along the corridor. 'Darling, is Kendall coming? I need my injection soon. He looks tired and drawn. I hope he's not feeling unwell again. The last bout of headaches really caused him so much pain. None of us want to fall ill now. I'm looking forward to flying home. God alone knows when that'll be, Arneka.'

Kendall caught them just as four UN guards entered the lift. 'Did you think I wasn't coming? Everything's arranged for our flights home. Reagent assured me that we'll be surprised on arriving back at our hotel. No more rotten news. Good for a change. I'm at a loss to know what he meant, Arneka. I forgot to sign-out my medical kit. Oh, he or the court doctor can initial for it.'

'Kurt's really exhausted. Kendall, he needs his usual heart jab. Let's get out of this freezing lift. We all could do with a solid hot meal and a hot coffee.'

'Excuse me, Miss Baumer. We're to escort you all to our vehicle. It has dark windows and air-conditioning. I parked it alongside this rear door.'

Arneka gasped on seeing the black limousine with a grey leather interior. 'Kendall, look at the silver emblem on its bonnet. Gee, those seats sure look comfortable. We'll be travelling in style back to the Metro Hotel.'

'Miss Baumer, we're not going to your hotel. You will all be staying at the airport hotel. All your luggage will be in the executive suite on its seventh floor. The bedrooms have spectacular views of Linz central. That magnificent panorama sweeps right down to our coastal beaches.'

'Sir, I insist on you driving us to the Metro. I don't want strangers mauling my delicates. Besides, we need to say farewell to our friends. They're leaving at noon today.' Arneka glared up into the rear-vision mirror.

She copped a dig in the ribs from Christian and shut up. Their father had fallen asleep. Kendall disapproved of her demands and rude comments. He suspected the real reason behind their sudden change of hotels.

Feeling remorseful over ignoring his wife's plea to fly back to Ireland after three months separation, he sighed. He could feel a strong pulse in Kurt's wrist. 'It has improved, Arneka. I feared his pulse may be threaded like it was this morning. Another heparin jab may not be needed, if his heart's rhythmical beats keep this regular.'

Entering their new abode Arneka rushed to the toilet. 'I'm busting and I've cut it fine.' She scowled as feminine fingers dropped a packet in her carry-all. 'Who's there and what did you just put in my bag?' No one answered. The main door shut quietly. She closed the toilet door and washed her hands. Intrigued, she unwrapped layers of tissue paper. Arneka gasped in disbelief. 'This fine porcelain I saw on a cabinet in the Kuper's apartment last week. Den told me his parents were flying to Prague this morning. I must've heard wrong. These violets and forget-me-nots are beautiful. I think they're hand-painted.' She ran a finger over the silver-rim of the fluted-edged saucer. Holding up the cup she could see a glimmer of light through its opaque porcelain.

Outside she saw both her brothers talking to a man near the ticket booth. Ready to intrude on their privacy, she heard a familiar voice behind her.

'What are you doing here, Den. Miss your flight, or has it been delayed?'

'Neither Arneka. I switched flights. I promised Dad I'd be his stand-in at a dental seminar in London. Not to speak, just take notes. I'll use my pocket dictaphone to record the lectures on modernised medicine and instruments. Mum left you a small breakable parcel. A friend of yours promised to give it to you. The lady talking to Kendall and your brother, Chris. Or she was a minute ago.'

Mystified, Arneka tried to think who Den meant. The penny dropped. 'I can't see her face. I wonder how she is.' She half-turned and Arneka stood there mystified. 'How come you're here? You're the last person I expected to meet at this airport. Where's your special man?'

'Don't I rate a kiss, or at least a cuddle, Arneka?'

'Yes, you do Gigi. I'm sorry, seeing you I almost dropped my carry-all. It has breakable porcelain inside. The china is delicate and translucent. If you hold a cup up to the light you can see your fingers through it. I bet Kendall was pleased to see you here. Where's Pierre?'

'Don't step back, or you'll tread on my husband's shoes.' Gigi's face beamed as her son approached them. 'I followed you into the powder room Arneka. A well-dressed woman asked me to give you the parcel and I carefully placed it in your totebag. Kendall wanted me to be a surprise, that's why I didn't speak. Just left. Can I see the gift, please? The lady was in an awful hurry. Her flight to Prague had been called. She also handed me this letter to pass on.'

Arms full, Arneka pointed to the seat. Relaxed and excited she proudly showed her *special* mother the three-piece gift set. 'Gigi, this dinner set is what our mother Erika used to serve all her delicious pastries and cakes to visitors. I can still remember the first time Heide came to our front door in Hamburg. She looked intimidating in her nurse's uniform with a red-velour cape swishing in the wind on a summer's day. I could only reach her fingertips, because I was so small. She hugged us, twinkled our noses and said they looked like tiny puffy buttons. From then on we called her Button.'

Tears welling in her eyes, Gigi couldn't express her love for this charming young woman who'd stolen a little corner of her heart. 'Darling, you must've loved your mother dearly. This tribunal has been a great strain on you and your family. I know Kendall senses your pain. He admires your tenacity and the strength you displayed while testifying. So do Pierre and I. It can't have been easy confronting the dragon. I often think how you used to say Heide had named that evil man von Dragon. It suits him to a tee. Let's not drift on to such a maudlin topic. It's a magical day and we shouldn't spoil it by thinking of him or the past. The past no longer exists. It's tomorrow we should be looking forward to. Oh, I have some terrific news to tell you. Kendall, through Peter Bucknell, has prearranged all our flights home to Aussieland, your words.' They laughed. 'I think he might have some special news to tell us very soon. That's all I can say, at present.'

'What about Pierre? Is he coming with us? I dearly love his family. Well, in a way, we're all one big contented family now. Mercedes is a darling. She's caring and so kind to everyone she meets in this whopping big, cruel world. Her beautiful smile and gentle nature reach beyond the Moon and back. If she can't help someone in trouble, she wouldn't harm or hurt that person.'

Arneka began to shake violently. Her carry-all dropped between her knees and then toppled onto the floor.

'Kendall, come quickly. I can't hold her much longer. Her eyes look glassy and she can't stop shivering.'

Christian reached his sister first. Pierre followed with Kendall close on their heels. 'Give me a hand, Chris and we'll lay her on this seat.' He felt for a pulse. 'Fetch my medical kit please Pierre. It's on the floor beside your cases. I quite expected her collapse. Arneka's been overdoing it and the stress of standing for hours have taken a toll on her nerves. Buy a bottle of water from that kiosk, please. Her pulse is shallow and fluttering. A warm rug …'

Pierre threw their travel rug over Arneka's inert and shivering body. 'Will your sister be all right, Kendall? She looks so pale lying there. Gigi, pass me your alpaca coat. It will help to keep this little lady warm.' He tucked his woollen scarf over her shivering shoulders. 'There ma petite, that is yours to keep.' Pierre bent and kissed her limp fingertips. 'Arneka, we all need you to keep us free of trouble. Your papa is here. Open those pretty blue eyes and look at him.'

Kendall supported her neck. Accepting a wet hanky from Kurt, he pressed the damp cloth against her hot lips. Her eyelids fluttered and opened. 'Don't try to talk, sis. Just lie still and sip this crushed ice. It'll cool your throat. Take it slowly. Don't gulp it, or you could choke. That's right, easy does it. You'll feel better in a minute.'

A security guard approached Christian. 'How may I assist you, sir?'

'My sister collapsed. We need a wheelchair. She can't walk out to the main carpark. The gentleman bending over her is our doctor.'

'No trouble sir. Will this one do? Someone left it here and it's blocking the lift corridor. Would you like a pushchair for the elderly gentleman talking to her, sir?'

Gigi intervened. 'Excuse me Christian, my husband's Linz chauffeur is ready to drive us to your hotel. Pierre and I still have to book somewhere. And your father is tired.'

'Thanks Gigi. One of those chairs is for Kurt, the other is Arneka's. She still looks a bit groggy. We better get moving before she collapses again.'

'Kendall thinks she's battled this illness for some time. I admire her strong willpower. She'll rally once we get her to bed. He and I both feared this might happen. She's usually the strongest one in our family.'

Noon: Settled in their hotel, Gigi paid the Baumers a visit two floors below, to see how Arneka and Kurt were feeling. 'Pierre and I are quite perturbed over Arneka's collapse, son. How's her father now.'

'They're both fine. Arneka's asleep. Come on in Mum. Where's Pierre?'

'He's lying down. Pierre has a full day tomorrow. Off to London again. Something about importing more Mercedes or other types of cars. It means I'll be shopping to fill in time here in Linz. No, I might come back to Australia with you. I'm anxious to see this huge property you've bought on the Gold Coast. If you need a financial backer, Pierre's offered to go guarantor for you and Brianna. How is she Kendall? I bet she's missing you. Your absence has been just over three months now, son.'

'Bridy's fine. I rang her last night and spoke to her parents. Mum, I miss her terribly. They've made arrangements for Father George to be buried in the church crypt. He's being honoured in the Vatican first. His body should arrive in Rome today. A huge honour for an innocent man who braved evil and won. We might stop off there on our way home from Ireland.'

Gigi had a spur of the moment idea. 'I want to visit the Svenssons in Yass. Do some sightseeing with them on the Gold Coast and tour though the Whitsundays, before returning to the continent. Pierre will be away from Paris for a month.' She knew neither her son nor the Baumers could travel anywhere until the tribunal was over. They all knew this might occur within days. By then all their flights would be booked.

9 am: By resumption of court the following day, Michael Reagent accompanied by his secretary and research lawyer, deserted camp so to speak. Vital news had come in from a reliable source in Zurich. Fresh data the Mossad's representative was holding related to a bank in Winterthur some twenty kilometres northeast of the city.

Leaving his deputy prosecutor Justice Beck to take over his stint in court for this next session, Reagent, accompanied by a UN observer met his contact at Linz Airport. Together they boarded a private aircraft bound for Zurich. On touchdown their Israeli contact joined their group. After a private discussion over lunch, the four men drove across town to their designated hotel.

After a brief respite, their first port of call was to the local politizei, where with little trouble, the Mossad's agent was supplied with a legal warrant.

Michael had previously listed Doctor Gwynne's affidavit for Monday's proceedings. This allowed everyone concerned to relax over the forthcoming weekend.

Kendall, having already disclosed all knowledge regarding the case "in camera", considered his duties done by declaring the truth. His signed and witnessed declaration describing his life in its entirety was tendered as evidence, in his stead.

Verbally, he'd described his early childhood years while living under the prisoner's roof. Also disclosed was his memorandum, which related to maltreatment of his mother and himself at the hands of Ian David Ross, who he had only known by that name. In truth, he'd declared nothing more than ASIO and the AFP documents already tendered in court showed.

The judiciary in their wisdom had graciously vowed neither to divulge either Kendall's family name, nor his professional standing in the medical field. Nor was their country of origin disclosed. This was one story the paparazzi had missed out on.

With completion of all signed documents, plus verbal information Kendall Gwlynne was discharged of any further responsibility, his sojourn in the city of Linz, Austria had officially drawn to a close.

The tribunal and court adjourned for a two-day recess. Provision was in progress for the prisoner to be transferred to an unknown location after sentencing. Only the legal appointed panel knew which city this would occur in Austria. A security move had prevented roughnecks and Nazi sympathisers from rioting, or undesirables causing untold trouble. Nazi splinter groups were keen to know where von Breusch's final port of call would be in Austria.

40

Settled in their Winterthur Hotel, the prosecutor's legal entourage were on time for an appointment at the Verkehrs-Kredit Bank. The manager, Herr Sigvard Nandor reluctantly confronted with the awkward dilemma of producing his client's confidential safe-deposit box, relented on sighting a warrant for its release.

In a room adjacent to the vault, Nandor, in the presence of Reagent and his team opened a box with his client's key. With gloved hands Michael carefully examined the brochures, manuscript and documented letters enclosed and discovered they were all written by Rolf Max von Breusch dated August 15 1944. 'Well now, what we have here? More undisclosed evidence to prove that bounder's guilt. These documents show how he whittled away his first wife's inheritance. Not that they will be included in the pending verdict. Still this untouched fortune should go to his dual families. No, I don't think so, Robert. I think Kendall and his half-brother might class it as being tainted money from ill-gotten gains. Such a huge amount could be a ruminal for aged victims of the Reich. A fund to rehabilitate some of the Jewish organisations that have been struggling to help thousands of people who'd suffered the indignity of being punished for crimes they hadn't committed in wartime.'

The manager pinpointed the owner as a disgruntled man who seemed rather irate over something he couldn't quite fathom. 'Perhaps someone may have stolen something our client valued. All the officer said was that all his documents would be collected later. Because he, the box-holder had paid its fees well in advance, I never disputed the reason.'

Recognising the officer's surname, Reagent knew it was his opposing team's wealthy client. The notoriety this indicted German officer had received caused worldwide press reviews, which induced Nandor to seek his own notoriety by producing new evidence. He'd realised the dishonest felon, who now owed his bank substantial funds, was in fact Herr von Breusch. To clarify his reason for contacting the Austrian authorities he relayed his initial dealings with him.

'The judicial panel in Linz will indeed be interested in these documents, even though they have already deliberated. Herr Nandor, where do I sign for everything we've confiscated from this box, owned by Herr von Breusch. You understand sir, all we have accomplished at this bank is confidential. Your bank will be reimbursed for funds owed. I guarantee that. This letter signed by you, requests the judicial panel of this tribunal to release these accumulated funds to the Jewish Mossad.'

'Well yes. Their organisation always needs funds to combat a mass of problems that treacherous Nazi regime left after the last war. Please sign both the sheets there.' Nandor pointed to a cross. 'Sir, this duplicate is yours to keep. We will retain the original. Now, if you and your colleagues will excuse me, I need to attend to my other customers.' He returned the safety deposit box to its correct slot, closed and locked it, then the vault. He followed his four foreign guests to the bank entrance.

Reagent sighed on re-entering the car. 'You all realise that Herr Nandor's signed letter resolves us from being charged with contempt. The judiciary guys will have to make the correct decision, regarding this whopping amount of confiscated dosh. I can imagine the look von Breusch gives his counsel on hearing this news. Not that it'll do him much good, dangling from the end of a rope.'

'My sentiments exactly, Michael,' responded his legal researcher.

Released of any further responsibility, Kendall Gwlynne turned his attention to procuring seats on a commercial flight out of Linz to Heathrow. Their luggage was packed ready to leave their hotel. A commercial aircraft with Pierre and his manager had left for Oslo an hour previously. Their journey two days later would continue onto London and then to Paris.

Kendall, his mother Gigi and the Baumer family were anxious to hear the Judicial Panel's decision, after deliberating for over forty-eight hours. Kendall attended to Kurt's medication well before leaving to hear the verdict. Meanwhile his mother dropped in to see how Arneka was feeling this morning. 'It's such a divine day I thought you and I could stroll down by the river. I made sure your beautiful china wasn't damaged yesterday. I wrapped it in my alpaca coat Pierre had placed over you, Arneka. His topcoat would've been too heavy and we didn't want you to freeze laying on that bleak seat.'

'Thanks. I've rewrapped the delicate china in my undies. Yes, I'd love to stroll with you in this warm sunshine, Gigi. We can't be away long. Kendall's ordered a taxi for ten this morning. I'm eager to hear the verdict. The panel of judges were all genuinely sorry about Father Brady's death. Their spokesman and two other bigwigs spoke to Kendall. I think they all felt a bit guilty that it had occurred in their courtroom. Accidents do happen. It was tragic. Your son and I always made sure he got his medications on time. Let's get out of this bitter-cold corridor. That lazy-ole sun won't keep us warm, if the weather turns wintery.'

Anxious to hear the judicial verdict after their long hours of deliberations, Quint and his wife Mieze were seated in the courtroom long before Kendall and his half-sister's family arrived. They had all entered the locked-down building by its rear entrance.

Silence reigned as the spokesman read the verdict. Arneka looked at Kendall who held her shaking hand. The stern look on Strezlewski's face gave them a slight clue to what he would soon announce.

'Keep looking at me, sis. Don't let that rotter's eyes penetrate yours. We all know what he deserves and may God grant us our most ardent desire. It will only be minutes now and all your traumas will be over. Look how passive your father seems. Kurt is stoic and he'll willingly accept their decision.'

Ordered to be up standing, Herr von Breusch smiled and nodded to his counsel. Herr Scharles shuddered, he feared hearing the verdict.

'This day we have taken into consideration the shocking treatment of your first wife's premeditated death. Reams of proven evidence cites you treated her abominably. And you also tried to murder your children. That failed. However, you then successfully instigated their nursemaid's murder. These are minor incidences compared with the hundreds of innocent people you sent to the Gestapo, after signing their death warrants.' Edward Strezlewski donned a black-fluted cap. 'This panel now sentences you to death by hanging for your war crimes. From this building Herr von Breusch, you will be taken to a prison somewhere in Austria. At the Governor's pleasure of that prison, you will be hanged by the neck until a medical officer declares you dead.' The spokesman removed his black cap, then declared, 'As of this minute this tribunal ceases to exist. We have read and judged every bit of proven evidence and our final decision was fair and just. May God have mercy on your soul.' He addressed von Breusch and nodded for his guards to remove the prisoner. Bowing to his legal colleagues, Strezlewski collected his files and left the courtroom.

'I sincerely thank God that their decision was fair. It's all over for us and we can move on with our lives, now Arneka.'

'Yeah. I hope a guard in that prison has the guts to give that Nazi bastard a thorough flogging, before they hang him. I'm glad none of us will have to witness it. I'm anxious to catch our plane home to Australia. What time does our flight leave tomorrow morning, Kendall?'

'Eight-thirty. I'll book our taxi tonight. That'll make sure we arrive at the airport well before six. Mother's likely to dawdle. Pierre's warned her to be prompt and ready by four. He misses Gigi terribly. Still, he knows she'll be safe until she reaches the Gold Coast. She'll phone the Svenssons tonight. She's hoping they can come to Queensland soon.'

'What day do you expect to fly up to Ireland? I know Brianna must be missing you, Kendall. Her priest's funeral in Innes is scheduled when? And don't forget to give our love to Brianna.'

Christian joined them as they were ready to leave for their hotel. 'I've decided to fly home to Yeppoon as soon as we land in Brisbane. I should get a transfer to a hinterland school. That'll alleviate some of the burden from you, Kendall. Dad is my responsibility and I can take him for trips along the Gold Coast beaches. He loves dining in good restaurants and picnicking on riverbanks, or by the sea. It'll also put some colour back in his peaky cheeks. Then he won't be a burden on you, your wife or Arneka.'

She frowned at him. 'Christian, what an awful thing to say. Dad's never been a burden to you, me or Kendall. It'll be terrific if you're living close, then you can take him on short fun-loving drives. This brochure says there's lots of places to enjoy a day by the beach, or walking through forests above our new home. I won't mind driving you both on my days off in your practice. I'm looking forward to working with Dorian.'

Kendall looked at his sister. 'Have you sent your resignation to the matron of Yeppoon Hospital, Arneka?'

She laughed and held the page up for him to read. 'I'll post this letter at our hotel, before we leave first thing tomorrow.'

After proceedings concluded in the Linz Courthouse, Mr Reagent spoke to his legal colleague, Justice Strezlewski in their hotel foyer. 'Last weekend I had the good fortune of attending a small town of Winterthur in Switzerland. At a Reich bank my legal team and I discovered, in front of a Jewish Mossad agent, an independent observer, this extremely interesting dossier. One I'm aware you will be mindful of, Edward.'

Strezlewski read the text and smiled at Reagent. 'How interesting! I gather you mean von Breusch, the criminal we just sentenced to death. The specific contents of this document are fascinating. I knew some documents in that city were missing.'

'If you will Bob, examine the writing on this page. Please read each line carefully. It relates to his signature on a document that he signed there over five years ago.' Reagent passed it to Strezlewski.

'Wow! This should have been declared as evidence during the tribunal. It's unbelievable to think a man of his intelligence could write such drivel. It does implicate him in a gigantic string of murders that he committed in war-torn Germany. My God, this reads like a thriller written by Conan Doyle in the last century. How could a civilised man commit such heinous crimes, while of a sound mind?'

'I don't know the answer to that question. It looks as though von Breusch planned and executed the murders without flinching an eye. In this light you should have no difficulty reading the entire script. It seems he deceived us. He led everyone to believe he was bankrupt. As you can see, he lied. There are thousands of pounds in Germanic currency in that bank account. Money stolen from his victims over long periods. The eight-digit rows of figures listed there once belonged to his first wife, Erika. What do you suggest the bank manager should do with those funds, Edward? None of this money will benefit von Breusch. Not where he's going.'

Strezlewski thought for a moment before responding. 'Michael, I can name over a hundred organisations that are struggling to retain enough funds to keep them viable. The Jewish Mossad would be the best ones to ask. This staggering amount of money could feed and rehouse a thousand orphans. Kids whose parents died in concentration camps. It wouldn't be blood money, if used wisely. I'll inquire and let you know. It could fund fifty fully equipped hospitals in and around Linz, even further afield. Aged hospices for those unfortunates who suffered the indignity of being imprisoned in those wretched camps in Romania and Poland.'

'The ink looks so faint and old. Who can tell whose signature this is?'

'I suggest you read the last line, where I've placed a marker. Then answer. If not, I'll read it aloud.'

'Yes, that's a sensible move. Definitely that's his signature. I have seen von Breusch sign lots of declarations during these past three months. I am, however, pleased the tribunal has drawn to a close. I thought you and your companion judges should see this initial bank draft.'

Michael and Strezlewski both knew the text of this dossier was of an extremely disturbing and urgent nature. Bank managers the world over all knew the importance of reporting huge undeclared funds. They were also cognizant of not letting such large amounts be wastefully spirited away without putting it to some positive use.

Gathering his briefs, Michael and the deputy prosecutor headed for their chambers, followed close at heel by their secretaries and legal teams.

The judiciaries' eager eyes focused on this dossier, each page in turn was examined then passed onto them entering their warm chambers. 'We find all documented facts on the sheets of a fanatical nature. Knowledge the words of which are capable of inciting unrest, enough to regenerate a Nazi-type regime, or induce idiots to try and seize power over our youth. After duly noting all this data it will be passed to the UN's Security Council, to be classified top secret. Therefore, you are all sworn to secrecy. I'm afraid this information is incredibly serious for us to handle.'

'May I come in, gentlemen?' requested a senior UN delegate standing in the doorway.

'Please do, Frances. I believe you're well acquainted with my colleagues,' the spokesman said and caught the observer's nod. 'I was just saying this data should be in your capable hands. Not ours. You'll find this dossier extremely disturbing. The contents are deadly.'

Accepting the dossier from Strezlewski he fanned its pages.

'Getting back to our original problem. Might I suggest, I throw in a ruse to fool those media barons. They'd have a field day with this tale of destruction and bigotry.'

'What do you propose, Justice Pyotr?' inquired Strezlewski who observed his companion flipping through this important document.

'With your permission, I could state all these drafts of von Breusch's must be scrutinised in-depth before their contents are revealed. I jumped the gun,' Leonard Pyotr smiled. 'And I won't object to being politely reprimanded for my misjudgement.' The group of five sniggered. 'Seriously gentlemen, I'll state in my report that perhaps this data requires lengthy inspection, before we disclose it publicly. How's that for rejection of my goof?'

'Might work! You could try it. Be interesting to hear what our superiors in head office think. What say you my esteemed colleagues?'

All five agreed. Neither they nor their teams wanted or needed adverse publicity hounding them or criticising their ideas. It could easily produce repercussions not yet imagined. Even the defence counsel gave merit to their idea, thus he accepted their decision.

12 noon: Reagent left the Linz Courthouse by its rear entrance, to face his own demon. A woman whom he once loved and respected. *I'm determined to win this case. My son deserves someone who loves him dearly. Not an old crow who will destroy his future. I intend taking him on a wonderful holiday touring through the Swiss Alps. He can catch up with his schooling later. I may have been a little hasty in my judgement of his mother. Still, I can't let her ruin his life. Jacob needs me.'*

In truth, he won the case and was awarded sole custody of his son. There was no stay of proceedings. This was a cut-and-dried case. He had, however, considered the risk of von Breusch's financial documents being publicised would be a catastrophe for everyone concerned with that case. This was unimaginable to Reagent.

Having analysed the folder's contents a UN delegate placed a temporary veto on disclosure of its details. 'This entire dossier must be presented to our Security Council for a final analysis before any sanction can be given. The investigative teams will have to ascertain whether von Breusch was working alone. This is our main concern. Perhaps there may be other factions implicated in his treasonable plot.'

With completion of their tête-à-tête, the judicial panel disbanded and went their own ways. Leonard Pyotr and Brage Ingivar headed to the local train station. Robert Spiller, the British QC, hailed a taxicab and offered Frenchman Hérbert Pautenel a lift to the airport. Edward Strezlewski chose to walk back to his hotel to collect his packed cases before heading to a bus station to take him to Vienna, where he intended to meet his American wife. The couple planned to travel all through Europe, before flying home to Philadelphia. This allowed them time to enjoy a brief holiday.

Preparing for tomorrow in courtroom six Mr Justice Beck, a new wig in Linz, intended to send the affidavits he just received from the Australian diplomat, Mr Peter Bucknell and the British charge' d affair, Henry Powell, now delegated to the British High Commissioner. Their affidavits substantiated all data relating to Quinton Marciano's request for visas for his wife, Mieze and himself to migrate to Australia. Apart from those, signed depositions came from Powell's personal secretary, Mr Homer Ellis and Doctor Gwlynne, both of whom had nominated the newly married couple to receive their limited visas to reside in the country. Homer Ellis had signed a form stating that his previous dealings with the Italian and heard about his wife's war injuries from several sources. The initial one came from Monsieur Bouvier and his Welsh wife, both of whom now realised the couple's importance in her son's life. Gigi encouraged Kendall to put in an application for this handicapped couple to be accepted to reside in Queensland. The second source had originated from a Prague dental specialist who praised their strength to battle adversity in Germany during the last war. Arneka and Gigi also approved of her son's decision.

With tears mounting in her eyes, Arneka bade farewell to the couple. 'I know you will love staying on my brother's property he bought in the hinterland, which overlooks the Gold Coast. I've seen a photo that a friend of his took last week. It's beautiful. Until we meet again, I wish you both God's speed and good fortune, Mieze. Your husband seems contented with Kendall's offer. Quint might lose those dark shadows beneath his eyes once he relaxes.'

'May I call you Arneka? It's such a pretty name. It reminds me of a summer's day that I spent in Cologne with my mother, before war broke out in Germany. Bombs fell and demolished the chalet, so we left there. We travelled by night in hay or fruit wagons bound for markets until we reached a small town of Bergamo in northern Italy. That's where I first met my husband. A year later, I saw Quint buying fruit at market there. Even though I was crippled, he took me to his mother's home. Both our parents are dead now, so I know how your brother must feel over the loss of his wife's priest. Father Ignatz saved Quint's life and lost his own after the Gestapo raided St Xavier's Church. My husband says he misses Father Kelly who exchanged clothes with a friend in prison.'

Arneka confirmed that she had known Father Brady and how he had died a week earlier in the courtroom. 'I remember von Breusch saying that his first wife's death was an accident. He declared, under oath, that he'd had no hand in her death, as his defence counsel had incorrectly assumed. That Nazi rogue led everyone to believe that she'd tripped over Christian's toy fire-engine, and fallen down the stairs in our Hamburg home. I knew he was lying. I've read Heide, our Hamburg nanny's letter. She always told us the truth, Mieze.'

Mrs Marciano couldn't find the words to convey her sorrow.

A Nazi salute and clicked heels were ever-present in Arneka's mind as she recalled von Breusch stating in court: "Mark my words, the Hitler Youth will be rejuvenated again by today's youngsters. Believe me, from Germany it will spread all throughout the world. I reiterate again, that I am NOT GUILTY of anything. Only of performing my duty as an officer in the time of war".

Arneka gathered her hand luggage and coat off the seat and watched this woman struggling to walk ten paces to meet her husband. 'Aren't we fortunate Kendall, we can walk without dragging our feet. My heart goes out to Mieze. She and Quinton have suffered, battled and braved that Nazi octopus and won.'

'Octopus? You usually call von Breusch the dragon from Hell. It sounded weird when you called him an octopus.'

'Waving his arms around like he did in that freezing courtroom, he reminded me of something floundering in a sea, or caught in a fisherman's net. What does the criminal's name matter, when you know what he was capable of doing in Germany? He thrived on his victim's misery and murdering people.'

'Nothing. If we don't hurry, we'll miss our flight. Mum's already gone aboard with her hand luggage. Let's help Chris get your father moving, or we'll all be left standing here, twiddling our thumbs.'

Safely aboard and seated Arneka looked across the aisle at his mother who looked fragile and tired. She yawned then nodded off to sleep. This woman who cared for her son and his wife Brianna, also took a great interest in her and her family. Finally, as their aircraft cruised high above the clouds Arneka drifted into a gentle slumber. Nobody noticed, least of all her father and brother who were talking to Kendall. He did however, notice Kurt's head drooping and smiled. 'Your father will fall asleep soon, Chris. It must be the hum of this plane's engines. I hope I don't fall asleep, because I'm interested in seeing some of the scenery far below us. That's unless we strike some form of turbulence. There's a storm brewing in the clouds above us. Gee I hope we don't fly in or through it. Growling thunder will awaken everyone in this smoky cabin. I hate people smoking and spitting their germs all over their neighbours.' Kendall wet his handkerchief in a cup of cold water and held onto his nose.

Twenty-four hours later their aircraft touched down on a wet tarmac in Brisbane. The sun broke through a dense, clouded ceiling as they passed through customs. With nothing to declare this group of five were waved on by a stern-faced officer. 'It's terrific to feel solid earth under our feet again, Gigi. Here, let me carry your cases. Kendall can't, he's helping my father to keep both feet on this slippery walkway. And Chris's arms are laden with our bags. I never thought we'd reach here alive, after striking that whopping hail storm over the Timor Sea. Someone high above the clouds must've been taking care of us. Thankfully, our plane didn't crash in those monstrous white-capped waves far below.'

A barrage of welcomers rushed to greet them on leaving the airport confines. 'Well, fancy meeting you here,' stated Peter Bucknell. Dee, his wife nodded in agreement. 'There's another couple who arrived on the plane before yours. They've also come to the Gold Coast to see your new home, Kendall. If I tell you who, it will spoil their surprise.'

'Don't be a spoilsport, Peter. Who were you just referring to?'

'Us. Bjorn and I still have to collect our luggage from the turntable. Don't I get a hug, Kendall? Where's Brianna? We thought your wife would be here with you. She's not?'

'No Kirsten, my son will be flying to Ireland within a few days to bring her home. I never expected you and Bjorn to come up here. I must've mistaken what you told me before we left Austria to fly home. Let's get moving or our limo driver might get impatient. Pierre paid and ordered the limousine and its driver for us, before he left Linz for Oslo. Oh, it's wonderful to see you both again. Our next trip will be up to Mount Tamborine where we'll see my son's new home and his and Dorian Payne's surgery. Are these your cases coming though now, Kirsty? I recognised their spotted straps.'

'Yes kiddo. Ooh, we're so excited. It's seldom we have a holiday free of worry. Sam and his crew from Abergeldie were crutching our mob of sheep. And Edna his wife has promised to care for my spring garden. I've sown asters, dahlias and we should have a good crop on all the fruit trees. The pears are finished now. I am sorry that we'll miss seeing Brianna again. Never mind, there's always next year, if we are fortunate enough to get another break from Yass and the drudgery of raising two massive mobs of sheep.'

Gigi sighed with relief once their car left the main road and turned in through her son's farmyard gate, painted grey. His roadside drum, a letter box looked sparkling in sunlight. She loved the summer months and looked forward to going for a swim in his dam. It looked clean with sun reflecting the trees in its water. Pumps kept filtering the water to keep it clean.

'At least we won't have to go far for a swim, Kirsty. I'm a bit hot, but not unpleasantly so. I'm sorry you'll miss seeing Brianna. Still, she'll be here soon. Kendall's booked his flight to Ireland and their return trip home. Oh, for a good soak in a cool of scented water. I'll use some of your rose perfumed crystals in my bath tonight. Lunch first. I'm interested to see what Harriet, our elderly cook has prepared for all of us. Probably cold beef, a salad with fresh fruit on the side and greens from her garden.'

'We miss her cooking with Helen on Jalna. Her sister's a boon. She never minds doing any jobs in moderation. I won't let her do heavy work I wouldn't do myself. She keeps the house spotless. Her daughter Jasmine makes delicious cakes and puddings for our meals. The shearers still get on well with them both, as you know.'

Kendall sighed. 'Come on you lot of dawdlers, or you'll miss seeing the pink and grey galahs being fed. The black red-crested cockatoos are already tucking into their grain. Gee, I sounded like you, Bjorn.' They both laughed.

'I'm ready for a cold beer and a few nibbles to munch on while we're resting on your front verandah swing, Kendall. Good cool spot on a hot spring day.'

'Yeah. First I better say hi to my colleague. Doctor Payne probably thinks I forgot to buy him a gross of cottonwool and swabs in town. That's why I asked our driver to stop at the chemist on our way home. Robert will be your chauffeur while you and Kirsten are staying here. I hope your small flat was clean BJ. I can't imagine they wouldn't have done a good job. Harriet and her daughter are both particular in their work. After a good rest why don't you, Kirsty and Mum, walk down to the village. It usually takes Dorian ten minutes. If you dawdle, it might take you all a bit longer.'

'Yeah. That sounds a good idea mate. I need a cool shower first. Phew, I didn't think it'd get so hot at this time of year. Summer's around the corner, ready to make your fruit trees bloom.'

'See you in ten, BJ. I'm going to have a shower, before I rest.' Kendall looked at the empty beer bottle, saluted Bjorn and hurried indoors. Feeling refreshed and drying his hair, he walked to the kitchen where his mother and Kirsten were enjoying iced orange juice.

About to speak, he jumped when a hand gripped his shoulder. Turning he got a shock to see Bucknell. 'I thought you and Dee were lying down in your flatette. Sure is warm. We must expect it to grow hotter as the days lengthen. Still, there's a cool breeze wafting through the gums outside those louvre windows.'

After their noon siesta, the Baumers and Kendall joined their guests strolling through the antiquated, quiet village. Tagging behind them came the Bucknell couple while Arneka held her father's arm over rough patches of unsealed roadway.

She then marched ahead of Kendall, nose held high like she'd done three months earlier on their arrival in Canberra. 'Your mother will never forgive us, if we forget to buy the right talcum powder at this chemist. She'll scowl if we get the wrong brand, other than her usual one.'

On their return Arneka undid the chemist's bag and discovered the girl had given her rose-scented talcum. 'Good,' she sighed. 'I'll leave it on Gigi's dressing table. I don't want her to pay me for it, Kendall.'

Busy packing his case and suit-folder ready for an early departure in the morning, he didn't bother to respond. 'Arneka, can I borrow your small travelling clock. Mine won't be suitable on a plane or in hotels.'

She tossed it to him, plus its box. He knew how to set the time and its alarm. It went under his six freshly-ironed shirts atop four pairs of trousers, all different colours and ties to match. Lastly, he folded his warm gown and two thick jumpers and laid them over the lot. He closed and locked his brown leather case. His shaving kit would fit in alongside the spectrophotometer in his medical bag. He looked at Kurt standing in the doorway. 'The idea of you having colour blindness is nonsense. Your vision and eyes are perfect, Kurt. What makes you think they weren't okay?'

He thought for a second and then responded, 'My headaches seem to be getting more frequent, Kendall. You've eased my mind greatly, thanks. I'm really glad the tribunal's over. Now, if you'll excuse me, I'll finish dressing.'

Perturbed, Kurt refused to believe his headaches were a fly-by-night episode. 'I suppose being a doctor he knows best. Now, where did I put my navy-blue tie? I know, it's still on the rack in my wardrobe.'

'Hurry up, Arneka. Your father's ready to leave and so am I. Planes don't sit on a tarmac waiting for late passengers, you know.' Kendall collected his luggage, ready to put in the taxi. His overcoat and hat he placed on the tan-leather suit-folder atop his gloves. An uncoiled umbrella rested against the lot. Tapping his brown brogues on the hall floor, he looked at his wrist watch. 'If she and Mum keep on nattering, I'll miss my flight from Brisbane to Ireland. 'It's getting late and the taxi has arrived, Arneka. Are you two coming to the airport, or not?'

'Yes we are, Kendall. I needed to get a light jacket. Now we're ready to leave, son. We'll follow you and Kurt to the taxi, crotchety bum.'

His plane left the international terminal on time with a full complement of passengers aboard. Within two hours it flew over Darwin and headed to Singapore, where it set down and picked up passengers bound for London. Air turbulence over Jakarta caused some to feel nauseous. Not Kendall though. He's taken the precaution of downing two Quell capsules, which eliminated that trouble. His judgement was spot on, due to an earlier flight. All precautions taken and no gastric upsets were his idea for a comfortable flight.

He dozed until the Captain's voice came over the intercom. 'Within ten minutes we'll be landing at Heathrow. I hope you have enjoyed your flight with British Airways. We wish you all the best in London. Those passengers wishing to commute interstate or to Ireland, your flights leave from runway four in one hour this morning.'

Kendall retrieved his hand luggage from the locker above his head. Then he assisted a young Burmese lady to collect her bag. Sitting down he relocked his seat belt, closed the magazine and put it back in its rack, behind the person in front of him. His hat and all-weather coat were ready to slip on as soon as they touched down on runway two at Heathrow. A quick walk saw him retrieve his luggage from the turntable. It took him fifteen minutes to arrive at terminal four. Showing his gatepass to a hostess, Kendall boarded the plane bound for Rineanna Airport in County Clare, Ireland.

'What a relief everything went smoothly. Now I can relax and read a magazine. The meal smells good. Boy oh boy am I hungry. I found the food on the previous flight dry and overcooked. A hot coffee will go down well. Thank goodness, there's no smoking on this flight. At least I won't choke from cigarette smoke.' He ceased mumbling as the drink-hostess approached his seat. 'I'll have a black coffee, please. And an apple pie with a scoop of vanilla ice-cream. My lunch order will be …' he scanned the menu, 'Roast beef rare and vegetables with gravy, thank you.'

'Will you desire a dessert, Doctor?'

'Yes thanks. I might have the fruit salad with a small dob of cream. Oh, and black coffee. No, please make that a mug of hot chocolate. That's all thank you, Madame. What time will it be served?'

'Luncheon is at noon. Dinner is served at six. Morning tea's around ten, depending on the turbulence we encounter. Afternoon tea will be served at four, sir. The menu may alter, if we hit stormy weather over the Irish Sea. Enjoy your flight. There's magazines in those other racks, if you prefer something medical or educational to read, Doctor Gwlynne.'

Kendall decided to doze rather than read. His medical magazine, *The Lancet* would suffice should he become bored. Falling asleep was a pastime he enjoyed now he was free to dream. No worries of the past could rob him of this luxury. To forget the trauma over recent weeks, stretching back three months could now remain in the past. Lots of more pleasant dreams awaited him with Brianna in her home town of Innes in County Cork, Ireland. The only minor disruption of their happiness would be attending Father Brady's funeral.

Seven agonising hours dragged by before Kendall saw Brianna standing alongside the custom's bay. He nodded to her while an anxious officer checked his documents. Free to go, Kendall raised his arms in relief.

'Hello darling, I thought that officer was going to search your cases.'

'No Bridy. He only looked at my entry visa and passed it back. My signed declaration he barely glanced at.' Kendall gave a hearty sigh. 'It's terrific to see you again, Brianna. At one stage on the plane it hit turbulence and I thought my number had come up. Thank goodness, it diverted to a different flight path without jostling. It scared me, only for a moment. Let me hold you and don't talk. There'll be plenty of time for chatter later. Did you receive my last letter about Father's proposed interment in Rome?'

'Yes I did, yesterday. Our church here has arranged his burial tomorrow. Thank God the weather forecast's fine. No drizzle or rain to spoil his service. Our new priest is going to say mass at nine. That means we'll all have to be up and dressed by seven. My mother sends you her love. Dad's been busy polishing the car. Oh, I've missed you, Kendall. Letters aren't the same as speaking to the one you adore in person. Come on, we'll collect all your luggage and then go to the car. Dad parked it close to the first exit door. It's chilly in this terminal. Let's hurry.'

They nattered all the way to O'Braily Street, their family's new abode. 'How was your flight, Kendall? I dislike flying in small aircrafts.' Her father looked in the rear-vision mirror. 'I dread it when hailstorms strike our locality in steamy or inclement weather.'

'We struck quite a bit of turbulence, followed by rough weather over the Irish Sea. Otherwise the flight was okay. I slept most of the way until we touched down at Rineanna Airport. Thanks, Mr O'Shea.'

'Not Mr please, Kendall. Call me Bob, or Robert. I detest formalities. My wife prefers Marg. Not Marjorie. She reckons it's an old-fashioned name. It'll be good having another man in our home. I'm tired of being surrounded by women who keep nattering all day.'

Brianna frowned at her father. 'We don't chatter all day, Dad. If we have something to say to you, we say it to your face. Forget this bickering, or you'll spoil our day. We'll all need peace to cope with the funeral.' Determined to stick by Father George's famous quotes, she spoke one from memory. "I will fight until my last breath and make you welcome in our Lord's house. The knowledge He bestowed on me I will now impart on some disbelievers. Take each day as it comes, with the grace and chastity of those who fear evil." How did that sound, Kendall? You probably recall him saying those words. Reach to the sky and pray, Father often sprouted if in a good mood. History belongs to the believer, not to the unknown rogue who doesn't care about his neighbour. There, I've quoted most of his compassionate sayings.'

'Thanks for the historian's history lesson, Brianna. Kendall knows what Father's favourite sayings were. Please help me in with his ports. Your husband looks as if he needs peace. Come my girl, let's not dawdle.'

Kendall's reflection in the bathroom mirror looked ghastly. 'Gosh, I need a shave and a hot shower. Or her mother will have a fit to see a raggle-taggle visitor appearing at their table this evening.'

Brianna, who walked in the bathroom, heard him speaking and laughed. 'Mother says she's unpacked your heavy suit-folder. Now, she wants to know if you'll object to her unpacking your two small cases, as well?'

'You can unpack them. There's personal items in the smaller one. Things she might object to Bridy. In the bigger one there are presents for both your parents. I won't be long dressing in these fresh warm clothes. Then I'll come and survey what you remove or unwrap from both my cases. Early to bed tonight.'

He just looked at her and they both burst into gales of laughter.

'Take my watch and put it on our bed, please darls. I can't afford a new one having spent a lot of dosh on our new homes and the surgery. You'll love living in the hinterland above the Gold Coast. I shan't let you into the secrets Mum and I have planned for you, Bridy. Is this towel mine?'

She threw him a clean one. 'Use that one. The others I used to dry my muddy feet on. There are more clean towels in the warming cupboard. They should be ready by dinner. Hurry darls, I'm anxious to give Mum her present. You can give Dad his later. I put your good suit on our bed. Dress is semi-formal tonight. They have church friends coming to dinner. See you in a tic. Dry your hair in our bedroom, Kendall.'

Startled by his nakedness, she closed her eyes. Turning she almost tripped over the hall mat. He smiled, but didn't comment, just looked vacantly at the ceiling painted a delicate shade of aqua. Grabbing her used towel to cover his naked pelvic region, Kendall gave her a sheepish smile.

Brianna sniggered on leaving their bedroom. 'That'll teach him not to parade around in the nude. Mum would've taken a fit, if she'd have walked in our bathroom and found him starkers. Oh well. So much for a bit of fun.'

A familiar voice echoed from the kitchen. 'Brianna, I'm serving dinner. Please ask your husband to hurry. Hot meals dry out, if you both aren't ready to sit down at the dinner table. Your father is already seated and slicing the beef-roll.'

By the time everyone had finished eating, a sudden gale turned a pleasant night into a freezing zone. Brianna shivered and cuddling her husband's back, she stroked his shoulders. 'I'm frozen. There's a doona in the wardrobe drawer. Will you get it please, darling?' He wrapped it around her and she responded by kissing his bare back. 'I have something exciting to tell Kendall. My doctor here in Innes sent me for an ultrasound last Friday. You couldn't guess what he said. I'm not three months ...'

'It's questionable.' Kendall felt her swollen stomach. 'I think you're closer to five months pregnant. How far did he say you were, Brianna?'

She gave a long unimpressionable sigh. 'I'm glad you're not standing Kendall. Cos you'll get as shock. In four months you'll be the father of twins. A pigeon pair. We're having twins. A darling boy and a sweet little girl.'

'That sounds incredible. No wonder you're excited. I still find it hard to believe. Twins? You certainly know how to astound me. A father of two babes. I can't wait to phone Gigi. You won't have to wait I'm phoning tonight. Did you bring the ultrasound films home, Bridy?'

Brianna laughed. 'Yes, they're in a plastic sleeve Doctor Marle gave me. You can look at them and my x-rays tomorrow. They don't have legs and can't run away, darling.' She began undressing in their room. Clothes were folded and put away.

His mind lingered on the growing township of Mount Tamborine with its older-styled houses, milk bars and a grocery store, set among rows of trees in the main street. The barbershop with its red and white striped pole and pottery moustache-bowls, leather razor strops advertised for sale in its front window. Avenues of conifers, private homes being converted into cafés, restaurants or business houses. Lawyers and solicitors attending to their client's problems in small professional rooms, their shingles swinging wildly in the breeze. In the distance waves roared as their white-tipped crests crashed on the Gold Coast seashore. There maybe a few horse-drawn drays delivering produce to local markets as shown in Dorian's last brochure which he recollected reading as well as buses taking people on tours.

Kendall thumped the bedhead. 'Bridy, I forgot your birthday last November. My mind must've been distracted by that damn tribunal. I'll buy you a gift at Rianna airport, before we leave.'

'Don't bother. I haven't forgotten yours. I have a belated gift for you, one you can't see. My secret will in time be revealed.' Brianna laughed and pretended to frown at her husband, who thought her antics hilarious. 'What mischievous scheme is that devious mind of yours conjuring now. Let's go to bed, I'm exhausted, it's been a long bleak day with rain weeping down off the escapement. Innes is almost submerged in snow tonight, the tops of vehicles are covered by impacted snow. This room is freezing.'

'Only because our bunker is out of wood. Dad'll order more tomorrow. You must expect it to be cold. This is 1975. My parents will probably be well-funded in later years. My mother's cancer drug cost a lot here in Innes. They have no medical insurance; you know that Kendall.'

He longed to be home again. He climbed into bed, kissed his wife's cheek and turned over. He needed a good night's sleep to face the ordeal of Father Brady's funeral. Their plane tickets with their reserved seats sat on the dressing table. Tomorrow they would be stressed after attending her aged priest's funeral.

Unguarded and unprepared for love, his temper rose. 'Not tonight darling, I need some sleep. Jetlag leaves one frazzled. You know tomorrow will difficult for us all. We need a good night's rest, before the funeral.'

'You promised to love me tonight. Why the change of heart, Kendall? I wasn't going to tell you my secret yet. I want some private time alone tomorrow with Father George to tell him of my special secret.'

'Lower your voice, before you awaken everyone in the house. Brianna, I'm overtired. If I can't sleep, then I refuse to attend the funeral tomorrow morning.' Kendall knew this little fib would make her see reason.

In a huff, she turned over and cried. Exercising her right to be silent, she thought, tomorrow's another day, he may feel differently then.

9 am: Slowly the hearse cruised to stop outside their church alcove. Bunches of flowers glistened in every available space on the pathway. An array of posies containing roses, buddleias, spring orchids, blue and purple delphiniums and multicoloured carnations lay close to the chapel steps. Their sensual aromas welcomed each guest entering the nave.

A quadrant of men in black supported the coffin on their shoulders until they reached the low altar. Kendall, with Brianna's father, the chief pallbearers, lowered Father Brady's casket at an identical time as their two companions. They then returned to their original seats.

Brianna sat in the back row, in case she needed to make a quick exit. Her mother sat alongside friends in the first row of seats. The service proceeded with a large echelon of Catholic dignitaries seated in the next and following two rows. A choir of youthful voices sang *The Lord's Prayer* followed by *Ava Maria*. Then came the eulogies, written and spoken by three priests from the Vatican in Rome. Latin in context, some of their words sounded unprofessional for learned men.

Doctor Gwlynne spoke of Father George Brady as a kind and beloved friend. His wife's father, Bob talked of their childhood days, riding together over hill and dale in County Clare; how George loved saddling his mare Dolly while his mother, Molly feared her young son might fall and break every bone in his fourteen-year-old body.

12 noon: After the final eulogy, the four men regrouped to carry his coffin down a short flight of steps. Father Brady had left instructions for his body to be buried, not interred in the church crypt.

Brianna was standing under an arbour of full-blown and budding roses as Kendall sauntered over to her. 'Why did you leave the church before Father's ceremony was over? I was frantic when I couldn't find you anywhere, Brianna.'

'It's ironic that a person has to die, before another two are born in this huge, wide wonderful world.'

'What do you mean? That sounded odd coming from you, Bridy.'

'It means Kendall, as you know in four months you will become the father of twins. I hope and pray our boy will have dark hair, like yours. I shall name him George Ken after you and Father George. Our girl's name is Adelle. Not Idelle after your mother. And not Gigi. It sounds as if you're calling a horse to come and eat or be groomed.'

He laughed. 'The correct pronunciation in French for Gigi is Chee-chee. I'm sure my mother will be wrapped if we name our baby daughter Adelle. Some people in Paris misspell her middle name. It should be Adelle, not Isabell or Idelle. The priest at their wedding made that mistake.'

'Can I have that minute alongside Father George's grave tomorrow, darling?'

'Yes. We can say the prayer he taught you Brianna, at school here. I know you're tired and need to rest. Let's go home. Here's your father's car. Be careful entering it. Don't trip on its running board. I'll follow with your coat, bag hat and gloves. I thought what you said enlightening. We should write that poem in our babes' birth diaries. I bought two in London at a nursery clinic, before I caught my plane to you. Pretty clever for a quack, as Arneka keeps on calling me.'

They recited the prayer silently in their car. Brianna removed a handkerchief from her pocket. The hearse was leaving, she and her cousin were stressed. The media in attendance rushed to wire and phone their news barons, in order to be the first story to hit world newsstands. Bold-lettered headlines plastered on bulletin boards read, "Brave elderly Irish warrior buried with the regal honours of his church". They were similar to bulletins that were flaunted on newsstands and banners waving in the breeze on a frosty August morning, after the death of von Breusch in Linz, "ONE MORE MURDERING WAR CRIMINAL BITES THE DUST. Ex-Nazi officer hanged in front of officialdom in our city's prison, the capital of Austria".

EPILOGUE

At Rineanna, County Clare's airport, business had taken a busy turn. Aircraft were leaving at a phenomenal rate. Hovering above the airfield were six to eight-seater Piper Aztecs and other small aircraft stacked in turbulent conditions, behind much larger commercial airliners waiting to land.

Wind sweeping across the river's isthmus, deflected its fury to the lower eastern shoreline. Shannon's estuary fortunately had escaped the full blast. The distant hills were inundated by this unpredicted stormy weather.

From the airport it was a twelve-mile journey by road to Innes. Travellers from all nations had descended on this small town's privacy. This funeral had outdone every other event since the last war. A couple of Orangemen waving banners were determined to disturb the peaceful hamlet, this being a non-Catholic area. However, the Church was inundated with townspeople who had prepared for an influx of strangers invading their territory. There wasn't one growl from these gentle country folks, after setting up large marques on their church lawn to accommodate an influx of righteous visitors on this sad occasion. Twelve noon was the time designated for the solemn event to begin. Their hometown priests were listed to conduct Father George Augustus Brady's funeral, according to their Irish tradition.

Brianna, at her wits end, was trying to cope with her husband's irritability, due to trauma and stress after battling through the recent tribunal. Death's sting had left its mark on him as well as her. Its repercussions bounced off her immediate family and devastated everyone in their town, now full to overflowing with strangers of all nationalities.

Kendall surmised his wife would crumble over the tragic death of their dearest friend. Since his arrival, he could only imagine the sorrow of everyone in their household. At a loss to explain his own sorrow, which Kendall found hard to express, his solemnness increased tenfold, yet he tried to remain calm for his wife's sake.

It seems so unfair that this caring and devout man, having gone through such terrible times, survived, only to have his life ripped away at a trial, which he should never have attended in the first place. The cruel cloak of injustice has denied him the privilege to be here with us today. Kendall floundered in misery while awaiting their hire car to arrive. Brianna's parents were on their way to the church in a friend's car. Her father had misplaced the keys of his aged vehicle.

Gigi and Pierre unfortunately couldn't attend the funeral. They were in the process of unpacking in Paris. Pierre had just returned from London, Gigi from Australia and they both were anxious to hear how Father Brady's funeral had progressed. Gigi, tried to settle Amber, ready to whelp her first litter of puppies. Mercedes had cared for the pup in their absence. Raoul up until today, had taken Amber for her daily constitutional in a local park, before leaving for his tutorial at the conservatorium. Their entire household was preparing for this lovable, yet roguish bitch to whelp.

'Darling, may we have good news by morning to tell my son. Pierre it's a wonder Kendall hasn't phoned to let us how the funeral went this morning. Oh, I keep forgetting they're on a different time-zone in Ireland.'

'Ma chérie, your son will telephone you before they leave Innes. Probably tomorrow? Do not be upset. I have a feeling that they may have something to tell us. Still, we will know something soon. Believe me, Gigi dearest.'

'How's my son bearing up?' she asked Pierre, filing her nails.

'I am worried,' he confessed. 'Kendall seemed quite distant when I spoke to him last evening. He sounded listless, tired and overstressed, Gigi.'

'He's hardly spoken to us since the tribunal ended. Brianna says he's sleeping in their second bedroom at Innes. I'm afraid he's retreated right back into his shell. Kendall seemed extraordinarily fond of her priest. I don't mind admitting Pierre, I have some idea how he'll be feeling, after this funeral.'

'Gigi, do not forget he and his friend Kurt have suffered greatly in recent months.'

'Oh, I know. Haven't we all. I do know Kendall's anxious about me. The difficulty is I've never been able to cope with a friend's death before. Don't get me wrong. He respected Father Brady and he must be missing him dreadfully. Kendall, by buying into the practice with Dorien, has secured a wonderful future. They both have. I can't go on much longer watching him being morbid, like he is at present.'

'Once today is over, I'm sure Kendall will settle. It must have been difficult for him being there when Father Brady died.' A gentle squeeze of his wife's hand boosted her courage. 'Thank you, Pierre. Because my son's a doctor, it makes no difference. People have feelings and must learn to cope with their own misfortune. I'm positive Kendall will survive. I'll chat to him on the phone tomorrow, or tonight in Ireland.'

Gigi felt like crying, yet the tears wouldn't flow. *It rips my heart apart to hear of him suffering in distress. It's seems as though he can't bear Brianna to be near him. Something must be wrong.* Her frown deepened. *If it's good news we shouldn't worry. Otherwise, I don't want to hear more sad news.*

Sniffing, she tried to smile until the phone rang. 'We'll be stopping off at Orly tomorrow. Can Pierre, or Raoul collect us, Mum? Our plane's scheduled to land at nine, depending on the weather. Brianna and I have some news to tell you. Good news. No, wonderful news. We can only stay for two days. Mother dear, no fuss and no party. Tell you more then. Bye darling. Love to Pierre and Amber, your faithful watchdog.'

Feeling composed, Gigi stood flabbergasted. She dropped the receiver and hurried to bathe before their guests arrived for dinner. 'Pierre will be too busy to collect Kendall and Brianna from Orly. Perhaps you Mercedes, or Raoul might collect them.'

Pierre intruded upon their friendly tête-à-tête. 'Excuse me ladies, I need to speak with my wife for a moment. Where is my black bow tie, chérie? I cannot find it anywhere and it does match …'

'What would you do without me, my fastidious man? It's under your suitcoat on our bed.' Gigi calmly noted and bustled off to see how her stepson might be faring. Earlier Raoul couldn't think or concentrate on where to find the bow of the new violin. He looked in the piano stool. 'It's usually in my violin case. Somehow it must have fallen, or

I've lost it at the conservatorium.' Another hunt and he found it lying under the music stool.

Hanging up the phone, Kendall recalled having spoken to Quint and his wife outside Innes Catholic Church. The couple had intended to stay with Brianna's family, but changed their minds at the last minute. They preferred to bed down at the local pub, it being a short walk to the church. With little persuasion, he decided to accept Kendall's offer for them both to fly to Australia after the wake.

Joined by her mother, Brianna chose a seat beside a young lady, whose husband was one of the pallbearers. Feeling a little queasy, Brianna moved back closer to the church doors to where throngs were gathering.

'Would you believe I was christened in this church,' she'd whispered, touching the arm of Mieze. 'We probably would have married here, if Kendall hadn't insisted on me staying with his mother at Yass, in Australia.'

Mieze nodded, distracted by the magnificent stained-glass windows above their heads. They mesmerised her. 'How is your husband feeling now? I caught a glimpse of him on our arrival. Quint seems distraught over the loss of his and my beloved priest.'

'He'll be fine, thank you Mrs Gwlynne. Your husband, Quint and I met at that beastly man's tribunal in Austria. Sorry, today is not a day for ugly reminiscing. Today, I must remember the kindness he showed to our ageing priest, who suffered immense pain, after being tortured in the Chancellery. Quint's shown me photos of his grotty cell. Mine wasn't much better. Huge rats and other vermin ran up its filthy walls. My bed consisted of two threadbare blankets, no sheets or pillow. My torn coat became a pillow. I had no shoes. They were discarded by the Gestapo thug who ordered me at gun-point, to climb up into their truck.'

'Kendall told me about the horrific times you and Quinton both suffered there. Now, if you'll excuse me, I'm going outside for a while. I can't breathe in this stuffy air. The deacons have closed all these windows. Why is beyond me to query.'

Mieze opened her prayer book at the chosen hymn and nodded to Brianna.

Huddled in the brick alcove, she uttered, 'I want to ask Kendall to bring me here early tomorrow morning. I feel a little weary now.

I prefer to be alone with Father George. My childhood priest will hear my prayer and keep my secret. Kendall and I can pay our respects to Father George in silence. There'll be no strangers here to listen or intrude on our prayers.'

Brianna knew Mieze and her parents would travel home in her cousin's car.

Huddled in the church alcove, Brianna muttered to herself. 'I'll ask Kendall to bring me back here tomorrow. There won't be strangers listening to me, when I tell him my secret as we spend time wandering through these beautiful gardens. He won't be sorry or growl at me, after a good or reasonable night's sleep. I'm eager to see his face. I can't forever keep fobbing him off with little fibs.'

Brianna then recalled Kendall mentioned that he'd approached Peter Bucknell about sponsoring Mieze and her husband to come and live in Australia. Bjorn had also offered and promised to give the couple a job on his property in Yass. That's the least we can do for those two caring people.

The Requiem Mass had taken longer than Brianna expected. Her father had said the main eulogy and spoken of his and Father Brady's long association in County Clare. Kendall and several parishioners read prayers. A Parish priest had spoken in a solemn tone during the Bible readings, followed by their local Archbishop who said the final blessing.

Brianna now recalled the organist playing Father George's favourite tune *The Old Colonial Boy*, as they carried his coffin to the altar. The four pallbearers stood alongside the multi-draped casket. A trio of flags cloaked his coffin. The Irish standard, the French tricolour and a flag, which represented his years working with the Marquands in France and England.

Towards the end of this solemn service Brianna had moved back to the last pew, vacated by a Roman priest. Stress building within her had dimmed her wet eyes. She could neither say anything nor prevent unrestrained tears from blotting her mascara.

A warden, who had detected her sorrow, held Brianna's hand, just as her husband clasped her other one. Neither man spoke a word to her.

Feeling unwell, she excused herself and had passively walked far from the church. Kendall caught up with Brianna after he and the three other pallbearers had shouldered the coffin to the hearse. Supporting her waist, he stood nearby with his wife.

'What's worrying you, darling? You look shocking. I knew it was wrong of me to let you come here this morning. Selfishly wrapped in my own misery, I neglected to understand your anguish and your misery.'

'I'll be okay in a minute. You go back with them, Kendall,' she whispered. 'I needed some fresh air. I found it quite stuffy in the church.'

'Go and leave you here? No. Let's move over to a more sheltered spot. It's peaceful here. Once the cars move on, we can spend some time together strolling through these beautiful church grounds.'

'Has your mother mentioned anything to you, Kendall?'

He didn't answer while watching the remaining pallbearers placing Father Brady's coffin in the hearse. The chief mourners followed, at a respectful distance in an orderly file behind the vehicle. Brianna and Kendall had placed their bunch of native wildflowers alongside her parent's sheath of red roses on top of the coffin.

'Come darling, your father is ready to drive us around to the cemetery.' Guiding her by the arm, Kendall smiled. 'I see my mother has finally parted with her ruby eternity ring. That style suits your finger Bridy.'

'Yes, she gave it to me a while ago. Pity they couldn't fly over for the funeral. Gigi knows my favourite gemstone is a ruby. It came as a surprise, although I recall her saying something about a gift that night in her bed after she'd given Arneka her mother's emerald necklace. I've often wondered if that Nazi had given his first wife Erika's ring to Gigi. He was ultra-mean and didn't want to buy her an engagement ring, I think.'

'Probably! How can we be sure? That miserable criminal would've done anything to have saved a quid.' Kendall frowned; this was one subject he preferred to forget. 'I meant to tell you; I spoke to Kurt about him and Arneka coming to live with us on the Gold Coast. Nothing's conclusive yet. Not until I'd spoken to you. Now you know darling, do you mind? I've been advised to invest in two more blocks of land there. Christian can't come down for some while, so I believe.' Kendall smiled to think how ironic their lives had evolved.

Brianna thought they were Heaven blessed as Kendall's features softened. His sombre moods had grown worse. They frightened her.

'Darling, of course I don't mind. Arneka's a bright spark, she'll enliven all our lives. I'm sure she'll enjoy going on coastal jaunts.' Brianna hesitated, then unexpectedly announced, 'What I wanted to tell you can wait.'

'You're fantasising again. What's worrying you, Bridy? I've never seen you this depressed before. Not the bleakness of my moods, I hope? Now I feel free and able to spend my life loving you, I realise how selfish I've been. Don't turn against me now, because of our silly tiff?'

'How could you think that? I love you deeply and won't let a slight row come between us, darling. I'll explain later when we're our own.'

The officiating priest seemed pleased that this short graveside ceremony was over. Restless, Brianna stood a short distance off. Feeling nauseous, she feared the wretched vomiting would begin again. She been free of the annoying sickness over the last few days and didn't appreciate vomiting in front of strangers. 'Let's stroll around the beautiful gardens, Bridy. Long laborious ceremonies frustrate me. We need to get away from here, because once this crowd disperses, it'll be hectic for drivers, including your father.'

'Yes, let's escape while we can. Please hold my arm over this rough patch of loose soil. I can't and don't need a fall.'

The hour had just turned two when they paused by the churchyard gate. The majority of mourners had gone. A few odd stragglers remained and they too looked like leaving within minutes.

Kendall lovingly declared, 'Come darling, we've this entire cemetery to ourselves, so let's make the best of it. We can take our time. There's something I need to do.'

'What is it, besides kissing your wife … may I ask?'

Kendall stopped and pulled something from his pocket. A small gold cross glistened in the sunlight. 'You might remember Father George gave us this cross on our wedding day. I wish to say a silent prayer by his grave, while dangling this blessed crucifix over the coffin. Do you mind, Sweetie?'

'I'd love you to. I also have a present for him. One you can't see. Our secret you know about already.'

She moved in beside her husband as they approached the artificial grass-covered mound. The gravedigger hadn't arrived to backfill the hole. Passively, they stood with their heads bowed, did their bit and said their silent prayers.

Kendall placed his arm around Brianna's waist in a comforting gesture. 'Bridy, let's come back here tomorrow before we catch our flight home to Australia. It'll be quieter here and peaceful to say our final farewell. Your father's car is pulling into the curb now. Let's hop aboard, before it rains again. I'll hold your heavy coat until you're seated. Move over please. I don't want to miss my footing in this mud.'

She grizzled and moved over. 'What I have to say will mean a lot to you, 'she whispered, sitting on the back seat. 'My only regret is I couldn't tell Father George our wonderful news before he died.'

Passing her the wooden cross Father Brady had made from his mother's pear tree, Kendall sounded confused. 'Tell Father, what? What do you want to confess now? I think you're acting peculiar, Brianna.'

Without saying a word, she plucked the petals from a wild shamrock. Blessing each one she held it to her breast. Clutching her husband's hand, she placed the four-leaf clover against his lips.

'Bless it with a kiss, Kendall. And one for me on this cheek.' After Brianna tucked the shamrock under her pillow. 'Tomorrow we can place that clover on Father's grave.' Astonished over his sigh, she confessed, 'that's if the weather's as sunny as it was after the rain stopped today.' She rubbed the second, four-leafed clover against her cheek as she quietly spoke, 'This wilted shamrock, I'll keep in my missal. It's a shame he can't be with us today.'

She'd written a verse in memory of Father George and Brianna passed the page to Kendall. 'That poem we can say tomorrow, darling.' He watched her wrap the third clover in a handkerchief and then tuck it under the sheet, resting on her breast. This sentimental gesture moved him to tears. 'You haven't read it to me darling. She recited the verse. '*For every flower that dies and goes to heaven, another bud blossoms to take its place.*'

'That's absolutely beautiful, Brianna. I've an idea why you wrote those tender words.' Kendall's head listed on one side and his expression conveyed a mysterious, though pleasant grin. 'The brief ceremony tomorrow will be for our unborn babes. I still can't absorb the idea that I'll be the father of twins. A girl and a boy.'

'Well darling, our precious Father George is in Heaven with his Maker. Someone has to take his place on earth. Two little babes God has blessed his and her parents …'

'Brianna, What a romantic gesture.' Excitement ran rife in Kendall's mind. 'You mean our offspring will be blessed, because of our mourning for him? Oh, you sweet, lovable darling. I'll be proud to be their father.'

Ecstatically, he held Brianna's hand and held it to his lips. 'We'll have two little mischievous imps to adore.'

'Yes, Doctor Gwlynne. Wait until your mother and Pierre hear our news. They'll go berserk with joy. You've made your wife pregnant.' She beamed and her face blossomed with love. 'I'm just over five months, now.'

'Their conception must've occurred while we were travelling through Ireland, not long after our marriage. Funnily enough, we never finished our honeymoon. I suppose now, I will have to make it up to you, once we're home. Our first stop will be the Whitsundays.'

'Kendall, I'd love to travel on a yacht all around those beautiful and majestic islands. And yes, our babes were conceived in the land of my birth.'

'How wonderful. Well, that settles it. Our children shall have Christian names compatible with your exquisitely green haven of Eyre.'

'Yvette Adelle, we'll call our baby daughter. A touch French for Pierre, a dash of Welsh for Gigi. And they'll have twinkles in their eyes from Innes.' She winked. 'Our twins will be born in late December or early January.'

THE FINALE. © Final edit by J.G. T on 19th July 2019.

ACKNOWLEDGEMENTS

I am appreciative of the assistance from Publicious Book Publishing, and my editor, Debbie Watson. Both of whom have saved my sanity, by being helpful in times of crisis.

Mr Patrick Callaghan, my soulmate in Sydney and Mrs Pam Deed my dear friend and scrabble partner 4 many years here on the gold Coast.

To all my friends too numerous to number here. They have mastered my willpower and kept my patience under control even in the times of trials, when I contemplated giving writing away for good. If I have forgotten and anyone here, it was unintentional, and I apologise.